Intermediate Disturbance Hypothesis in Phytoplankton Ecology

Developments in Hydrobiology 81

Series editor
H. J. Dumont

Intermediate Disturbance Hypothesis in Phytoplankton Ecology

Proceedings of the 8th Workshop of the International Association of Phytoplankton Taxonomy and Ecology held in Baja (Hungary), 5–15 July 1991

Edited by
J. Padisák, C. S. Reynolds & U. Sommer

Reprinted from Hydrobiologia, vol. 249 (1993)

Springer-Science+Business Media, B.V.

Library of Congress Cataloging-in-Publication Data

Intermediate disturbance hypothesis in phytoplankton ecology / edited
by J. Padisák, C.S. Reynolds, and U. Sommer.
p. cm. -- (Developments in hydrobiology ; v. 81)
Includes index.
ISBN 978-90-481-4233-0 ISBN 978-94-017-1919-3 (eBook)
DOI 10.1007/978-94-017-1919-3
1. Phytoplankton--Ecology--Congresses. I. Padisák, J.
II. Reynolds, C. S. III. Sommer, Ulrich, 1952- . IV. Series:
Developments in hydrobiology ; 81.
QK933.I58 1993
589.4--dc20 92-41133

ISBN 978-90-481-4233-0

Printed on acid-free paper

Hommage à G. Evelyn Hutchinson

Contents

Hydrobiologia **249**: ix, 1993.
J. Padisák, C. S. Reynolds & U. Sommer (eds), Intermediate Disturbance Hypothesis in Phytoplankton Ecology.

Foreword

The 8th Workshop of the International Association of Phytoplankton Taxonomy and Ecology, held from 5th–15th July 1991 in Baja (Hungary), was intended to discuss current problems in phytoplankton taxonomy and ecology. In the first three days the central ecological issue, the relevance of the Intermediate Disturbance Hypothesis (IDH) to plankton ecology, was discussed in detail. During the following days, invited specialists gave overviews of the taxonomy of green algae and centric diatoms and assisted in the examination and interpretation of fresh material collected during the field trips (Kis-Balaton reservoir, shallow alkaline lakes, Danube). Moreover, 20 lectures and posters concerning the taxonomy and other aspects of phytoplankton ecology were presented and discussed. Data-processing methods of phytoplankton counting and an on-line databank of the published Hungarian algal records were presented with computer demonstrations. On the occasion of the meeting, an exhibition of the original algal drawings by the famous taxonomist, the late Professor Tior Hortobágyi, was opened.

This volume is a collection of papers presented in the IDH session held during the first three days of the meeting. In order to make the discussion as fruitful and as straightforward as possible, an editorial brief (see Appendix 1 in Sommer *et al.*, this volume) was distributed among the participants prior to the meeting.

This volume gives an insight to what a group of contemporary plankton biologists think about the utility, virtues, strengths, and theoretical and practical weaknesses of J. H. Connell's Intermediate Disturbance Hypothesis within the context of phytoplankton ecology. The sequence of papers in this volume moves from particular case studies to more general and, finally, theoretical approaches. Why this topic of plankton ecology was selected for detailed discussion, together with several aspects of the final outcome and conclusions of the meeting, is discussed in two editorial papers.

As is detailed in the introductory editorial paper, Connell's hypothesis is closely related to Hutchinson's famous "paradox of plankton". It is sadly ironic that our workshop should have coincided with G. Evelyn Hutchinson's untimely death and with the thirtieth anniversary of the publication of his paradox. It is no less than with a profound sense of duty that we respectfully dedicate this volume to his memory.

Our special gratitude is due to the Kereskedelmi Bank Ltd., the official sponsor, for the financial support of the workshop, as well as to the cosponsor, the Hungarian SIL group. We wish to extend our sincere thanks to the Balaton Limnological Research Institute of the Hungarian Academy of Sciences (Tihany, Hungary), the Kis-Balaton Laboratory of the West Transdanubian District Water Authority (Keszthely, Hungary), the South Transdanubian District Environmental Protection Authority (Baja, Hungary) and the Hungarian National Science Found for their essential help in many ways.

We are extremely grateful to the referees who worked diligently in improving the scientific quality of the manuscripts.

THE EDITORS

Hydrobiologia **249:** x, 1993.
J. Padisák, C. S. Reynolds & U. Sommer (eds), Intermediate Disturbance Hypothesis in Phytoplankton Ecology.

List of participants

ÁCS Éva, Botanical Department of the Hungarian Natural History Museum, H-1476 Budapest, Pf. 222, Hungary

BORBÉLY György, Biological Research Centre, Szeged, POB 521, Hungary

ALVAREZ COBELAS Miguel, Water Research Centre (CSIC), La Poveda Argenda del Rey, E-28500 Madrid, Spain

CARVAJAL-CHITTY Humberto, Simón Bolívar University, Institute of Natural Resources, Apdo. 89.000, Z.P. 1080-A Caracas, Venezuela

CHORUS Ingrid, Institut für Wasser-, Boden und Lufthygienie des Bundesgesundheitsamtes, B II. 2., Corrensplatz 1, DW-1000 Berlin 33, Germany

DEISINGER Gerhild, Kärtner Institute for Lake Research, A-9020 Klagenfurt, Flatschacher Str. 70, Austria

DESCY Jean-Pierre, Unit of Freshwater Ecology, Department of Biology, FNDP Namur, Rue de Bruxelles 61, B-5000 Namur, Belgium

DOKUIL Martin, Institut für Limnologie, ÖAW, Gaisberg 116, A-5310 Mondsee, Austria

ELORANTA Pertti, Department of Limnology, University of Helsinki, E-house, VIIKKI, SF-00710 Helsinki, Finland

HEGEWALD Eberhard, KFA, IBT 3, P.O.B. 1913, DW-5170 Jülich, Germany

HOEHN Eberhard, Bureau for Limnology Hoehn, Glümerstraße 2a, DW-7800 Freiburg, Germany

HOLFELD Harald, Max-Planck-Institut für Limnologie, August Thienemannstr. 2, P.O.B. 165, DW-2320 Plön, Germany

HOLZMANN Roswitha, Zoologisches Institut der Universität München, Limnologische Station Seeon, Seidlstr. 25, DW-8000 Munich 2, Germany

HORECKÁ Mária, National Institute of Hygiene and Epidemiology, NÚHE, Trnavská 52, 82645 Bratislava, Czecho-Slovakia

JACOBSEN Bodil Aavad, Environmental Biology Laboratory, Baunebjergvej 5, DK-3050 Humlebæk, Denmark

JUHÁSZ-NAGY Pál, Department of Plant Taxonomy and Ecology, Eötvös Loránd University, H-1083 Budapest, Ludovika tér 2, Hungary

KISS Keve Tihamér, Danube Research Station of the Hungarian Academy of Sciences, H-2131 Göd, Jávorka Sándor u. 14, Hungary

KOMÁRKOVÁ Jaroslava, Hydrobiological Institute of the Czecho-Slovak Academy of Sciences, Na Sádkách 7, 370058 Ceské Budejovice, Czecho-Slovakia

KRISTIANSEN Jørgen, Institute for Sporeplanter, University of Copenhagen, Ø. Farinsgade 2D, DK-1353 Copenhagen K, Denmark

LENHART Brigitte, Wasserwirthschafsamt Weilhei, Pütrichstraße 15, DW-8120 Weilheim i. OB, Germany

LEPSTÖ Liisa, National Board of Waters and Environment, Pohj. Rautatiekatu 21.B, P.O. Box 250, SF-00101 Helsinki, Finland

MAKULLA Alexandra, Max-Planck-Insitut für Limnologie, August Thienemannstr. 2, P.O.B. 165, DW-2320 Plön, Germany

MÁTYÁS Kálmán, West Transdanubian Water Authority, Department Kis-Balaton, H-8360 Keszthely, Csik F. sétány 1, Hungary

MIHAILEVIC Melita, Dept. of Ecology, University of Osijek, P.O. Box 144, Ulica J. Jagora 9, 54000 Osijek, Croatia

MOUSTAKA-GOUNI Maria, Institute of Botany, School of Biology, University of Thessaloniki, GR-54006 Thessaloniki, Greece

OLRIK Kirsten, Environmental Biology Laboratory, Baunebjergvej 5, DK-3050 Humblebæk, Denmark

PADISÁK Judit, Botanical Department of the Hungarian Natural History Museum/Balaton Limnological Institute of the Hungarian Academy of Sciences, H-8237 Tihany, Hungary

PRADER Karl, Institute of Botany, University of Innsbruck, Sternwartestr. 15, A-6020 Innsbruck, Austria

RAFN Pia Dorthea, Institute for Sporeplanter, University of Copenhagen, Ø. Farinsgade 2D, DK-1353 Copenhagen K, Denmark

RAJCZY Miklós, Botanical Department of the Hungarian Natural History Museum, H-1476 Budapest, Pf. 222, Hungary

REYNOLDS Colin S., Freshwater Biological Association, NERC Institute of Freshwater Ecology, Windermere Laboratory, Ambleside, GB-LA22 0LP, UK

ROJO Carmen, Department of Microbiology, Univ. Valencia, 46100-E, Burjasot, Valencia, Spain

SCHMIDT Antal, South Transdanubian District Environmental Protection Authority, H-6500 Baja, Széchenyi u. 2, Hungary

SCHWARZ Kurt, Federal Institute of Fishery, A-5310 Mondsee, Scharfling 18, Austria

SELL Anne, Max-Planck-Institut für Limnologie, August Thienemannstr. 2, P.O.B. 165, DW-2320 Plön, Germany

SOMMER Ulrich, Institut für Biologie und Chemie des Meeres, Universität Oldenburg, Postf. 2503, DW-2900 Oldenburg, Germany

SØRENSEN Annie, Environmental Biology Laboratory, Baunebjergvej 5, DK-3050 Humblebæk, Denmark

STOYNEVA Maya, Sofia University "St. Kl. Ohridski", Faculty of Biology, Department of Botany, Sofia-1000, bld. Dr. Zankov 8, Bulgaria

TRIFONOVA Irina, Institute of Lake Research, Russian Academy of Sciences, 196199 Sevastianova 9, St. Petersburg, Russia

Hydrobiologia **249**: 1–7, 1993.
J. Padisák, C.S. Reynolds & U. Sommer (eds), Intermediate Disturbance Hypothesis in Phytoplankton Ecology.

Hutchinson's heritage: the diversity-disturbance relationship in phytoplankton

U. Sommer [1], J. Padisák [2], C. S. Reynolds [3] & P. Juhász-Nagy [4]
[1] *Institut für Biologie und Chemie des Meeres, Universität Oldenburg, Postf. 2503, D-2900 Oldenburg, Germany;* [2] *Botanical Department of the Hungarian Natural History Museum/Balaton Limnological Institute of the Hungarian Academy of Sciences, H-8237 Tihany, Hungary;* [3] *Freshwater Biological Association, NERC Institute of Freshwater Ecology, Windermere Laboratory, Ambleside, LA22 0LP, UK;* [4] *Department of Plant Taxonomy and Ecology, Eötvös Loránd University, H1083 Budapest, Ludovika tér 2, Hungary*

Key words: the paradox of plankton, Intermediate Disturbance Hypothesis, diversity, general ecology, community changes

Abstract

This paper introduces a collection of contributions presented at the 8th Workshop of the International Association of Phytoplankton Taxonomy and Ecology. It compares the substance of with what to limnologists is the more familiar 'paradox of the plankton' posed by G. E. Hutchinson. The utility of Connell's Intermediate Disturbance Hypothesis in plankton ecology is, potentially, more instructive but inherent difficulties in relating response to stimulus have to be overcome. A copy of the brief distributed to contributors before the workshop is appended.

Preamble

All modern students of limnology will have benefited from the influence of the late Professor G. Evelyn Hutchinson. The thoroughness and erudition of his monumental Treatise on Limnology (Hutchinson, 1957a, 1967,1975a) imparted knowledge, understanding and inspiration to all engaged in the study of limnetic and lotic ecosystems. Those of us who have specialized in pelagic ecology hold a special regard for Hutchinson's challenge that the high diversity of species present in limnetic plankton is inconsistent with the assumptions of competitive exclusion in the supposed uniformity of pelagic environments. Indeed, the dilemma has become widely known as the 'paradox of the plankton' (Hutchinson, 1961).

Hutchinson was intrigued by the fact that epilimnetic phytoplankton coexist in a well mixed environment and compete for a very small number of common limiting resources (light and a few nutritional elements). Nevertheless, even in small samples of water, it is not unusual to find more than 30 species of phytoplankton present. Equilibrium concepts predict that competition would eventually select the best fit species to the exclusion of all others (Hardin, 1960) and that only as many species can coexist as there are limiting factors. The search for solution to this apparent paradox has occupied plankton biologists ever since.

Going into a bit deeper into Hutchinson's heritage, it is interesting to observe the thread of reasoning running through his masterly book 'An Introduction to Population Ecology' (Hutchinson, 1978). This begins with a consideration of temporal processes, enriched by a very deep feeling for science history, and goes on to cover such exciting topics as 'Living Together in Theory and Practice' (Chapter 4) or 'How Is Living Nature Put Together?' (Chapter 6). The book is full of very interesting examples and speculations on the possible interconnections between limnic and terrestrial ecology.

In making his perceptive generalizations, it may be noted that, in addition to his own deep intentions, Hutchinson was influenced by a number of friends, students, including Charles Elton, David Lack, R. H. MacArthur, R. H. Whittaker & J. L. Harper. It is interesting to observe, however, that his attitude is always critical, as is shown, for example, in his questioning comments on some statements of Lack (1954) or Slobodkin (1961).

We believe that at least three closely related topics, have been relevant in motivating Hutchinson's appreciation:

(i) the nature of ecological patterns,
(ii) the nature of niche,
(iii) and the nature of coexistence.

Topic (i) can be followed reasonably well from Hutchinson (1953) through MacArthur (1965) to Hutchinson (1975b). Topic (ii), the story of 'niche-ology', is also fairly well accepted, started, in a sense, by Hutchinson (1957b) and continued by the activities of R. H. MacArthur, R. K. Colwell, I. Emlen, I. Futuyama, M. E. Gilpin, I. Roughgarden, who are, directly or indirectly, students of Hutchinson.

The third topic (iii) is the most difficult one: it was also the most pertinent problem of Hutchinson. Following his publications carefully, from the rather early papers (e.g. Hutchinson, 1948, 1951, 1953, 1954) to his most famous syntheses (Hutchinson, 1959, 1961, 1978), we discern a recurrent view that the rebuses of coexistence would be solved through the search for potential optima, the recognision and definition of moving patterns and where the interpretation of which remained within the context of evolutionary ecology.

Needless to say, all the components of such a reasoning have induced several divergent lines of further investigations (e. g. Levins, 1968, 1979; Lewontin & Cohen, 1969, etc.).

In his seminal paper, Hutchinson (1961) suggested several possible explanations for his palanktonic paradox of diversity. Among them was the idea that the boundary conditions of competition change frequently enough to revert competitive hierarchies before exclusion occurs. The proposal was reformulated by Richerson *et al.* (1970) as 'contemporaneous disequilibrium'. Later attempts to solve the paradox include theoretical analyses (Grenney, 1973; Armstrong & McGehee, 1976; Levins, 1979, Ebenhöh, 1988; Grover, 1990) and experimental approaches (Turpin & Harrison, 1980; Robinson & Sandgren, 1983; Sommer, 1984, 1985; Gaedeke & Sommer, 1986; Grover, 1988, 1989, 1991). In spite of these considerable efforts, no universally accepted explanation for the paradox has been forthcoming.

Although freshwater biologists associate the problem of species diversity with Hutchinson, it is fair to say that analogous questions have been prevalent among terrestrial ecologists. For instance, Grime (1973) was among the first to recognize that although competition can be a major causal factor in the maintenance of low diversities in herbaceous vegetation, there is nevertheless a low incidence of competitive species in floristically rich habitats. Wilson (1990) considered twelve possible mechanisms to account for species coexistence in New Zealand plant communities: most of these are equilibrium concepts, including niche diversification and stabilizing coevolution; his non-equilibrium explanations assumed that there had not been enough time for competition to completely exclude relevant species before sudden or gradual alterations intervene to arrest or shift back the development of the association of species.

Still more perplexing is the astonishing diversity observed among tropical forests and in coral reef communities in which an equilibrated climax condition is intuitively anticipated; competitive exclusion appears, again paradoxically, not to have been effective. Connell (1978) argued that, even here, the time taken to recover from damage through change of climatic conditions, landslides, fires or violent storms is such for it to be difficult for competitive exclusion to occur finally. Thus the intervention of factors delaying progress towards, or preventing the attainment of, an equilibrium condition is supposed to be responsible for the maintenance of the diversity. This has become widely known as 'Connell's Intermediate Disturbance Hypothesis'.

The Intermediate Disturbance Hypothesis and its applicability to phytoplankton

Connell's (1978) hypothesis states:

(1) In the absence of disturbance (eternal steady state), competitive exclusion will reduce diversity to minimal levels.

(2) Under very intense disturbance only few populations of pioneer species could establish themselves after each disturbance event. This would also lead to minimal diversity.

(3) If disturbances are of intermediate frequency and/or intensity there will be repeated opportunities for the re-establishment of pioneer populations which would otherwise be outcompeted and the populations of the successful competitors could withstand the disturbance without completely taking over the community. Thus, a peak of diversity should be found at intermediate frequencies and intensities of disturbance.

In his original paper, Connell conceived disturbances as primarily originating from internal processes (e.g. treefall gaps caused by the death of senescent trees). However, thee in no *a priori* reason why disturbances of external origin should have different consequences on species diversity. Obviously, there are instances, when the contribution of internal and external factors cannot be separated easily. For instance, senescence of a tree might not make a tree fall by itself but it may increase its susceptibility to windfall. For our present discussion, we have assumed that the distinction between external or internal origin of disturbance is less important then the occurrence of disturbance *per se*.

It is implicit in IDH that the post-disturbance recovery of communities is strongly directional, in accord with successional theory (Odum, 1969, 1971). Habitats newly- opened by, for instance, storm damage are likely to be recolonized first by invasive species. These will generally be those which can provide the largest number of propagules or which can expand their biomass the most rapidly, assuming that the new conditions are adequate to meet their requirements. In time, however, these colonist species will be replaced by others which are better equipped to live close to the limits of the available resources (McArthur & Wilson, 1967). As the community is assembled, the ratio between the production and the biomass falls simultaneously until, ultimately, the biomass supported is the maximum achievable with the energy available.

The application of successional theory to the seasonally changing structure of phytoplankton assemblages has been attempted by Reynolds (1988). His approach depended upon the recognition of the importance (a) of the temporal scaling, which relates primarily to the life spans of consecutive generations of algae (measured in hours to days rather than months to decades), and (b) that different species of phytoplankton show adaptations to exploit particular parts of the spectrum of environmental variability. Just as on the land, colonist weeds and equilibrial dominants are recognizable in the plankton (see Padisák & G. Tóth, 1991).

Sommer (1991, see also Padisák, 1992) has argued strongly that succession of phytoplankton is analogous to the succession of terrestrial vegetation rather than being a purely seasonal phenomenon. The several months of the plankton growing season, accommodating dozens of generations of phytoplankton, correspond to tens of years in grassland, and to centuries in forest. Variations in growth conditions should necessarily be viewed at a similar scale. Under favourable physical conditions, the intrinsically transient nature of early- and middle phases of planktonic successions, and the self-sustainability of final stages have been demonstrated (Sommer, 1991). It is only the external cycle in climatic and hydrological conditions which resets the plankton succession each year.

As a consequence of short generation times, however, plankton communities are responsive to meteorological variability. The 'weather' experienced by terrestrial vegetation is the 'climate' perceived by phytoplankton. The fluctuations are especially pronounced in shallow, polymictic lakes where apparently random population variations often make the main trend unrecognizable.

Developing the utility of the Intermediate Disturbance Hypothesis

It seemed to us germane to take advantage of the small temporal scales of phytoplankton dynamics to explore the operation of intermediate disturbance and its role in maintaining species diversity. These issues were considered by participants at the 8th Workshop of the International Association of Phytoplankton Taxonomy and Ecology convened at Baja, Hungary, hold between 5–15 July 1991. This volume contains the individual contributions presented at the workshop. All authors were expressly asked to address their presentations within the context of the Intermediate Disturbance Hypothesis (IDH). Because IDH had not been widely accepted as a basis for viewing phytoplankton ecology and, in consequence, compliance with our request may have involved authors in unfamiliar conceptualization, we considered it important to append (Appendix 1) to this introductory paper the brief issue to participants prior to the workshop.

The particular problems upon which resolution was sought included the distinction of disturbance from the undisturbed, the timescales involved and the quantification of their impacts on diversity. However, we were anxious not to impose too rigid a definition of disturbance. In everyday usage, 'disturbance' implies a stimulus whereas the formal attempts to recognize and quantify ecological 'disturbance' (e.g. Pickett *et al.*, 1989) lie exclusively in the response to that stimulus. In order to make the hypothesis testable, the magnitude and frequency of environmental change and of the disturbance-responses have to be quantified. Change can be found in many environmental properties which may be measured on different scales. As long as all factors other than that to be tested are held constant, no problem of commensurability arises. This is usually the case in experimental design. In the analysis of collected field data, however, commensurability becomes a prominent problem. On what basis, for example, does one equate simultaneous changes in incident radiation, temperature, nutrient loading or flushing rate, etc.? Obviously, generally valid conversion factors for empirical measurements cannot be found. By quantifying the biotic response rather than the stimulus, the problem of commensurability is circumvented but it is replaced by a danger of circular reasoning in that the strength of the disturbance is the strength of the biotic response. Does the lack of a measurable response indicate that there was no measurable stimulus? Without appreciation of these serious difficulties of interpretation it is unlikely that meaningful application of the intermediate disturbance concept or its assimilation into ecological theory can be usefully achieved.

Dedication

For many plankton ecologists, Hutchinson's paradox has been a more familiar and a more challenging problem than Connell's hypothesis. Among the significant attempts to address the issue (see above), the tendency for phytoplankton diversity to be increased by environmental variability has been strongly evident. In this respect Connell's Intermediate Disturbance Hypothesis represents a workable framework for superimposing non equilibrium explanations for Hutchinson's paradox.

We acknowledge this close empathy between Hutchinson's and Connell's views. Because of our perception of the prevalence of non-equilibrium dynamics in plankton ecology, we recognize the potentially greater utility of Connell's hypothesis to our understanding of pelagic ecosystems. Nevertheless, the inspiration and challenge to plankton ecologists will long remain that of Hutchinson.

It is sadly ironic that our workshop should have coincided with G. Evelyn Hutchinson's untimely death and with the thirtieth anniversary of the publication of his paradox. It is no less than with a profound sense of duty that we respectfully dedicate this volume to his memory.

Appendix 1

The intermediate disturbance hypothesis

The question of diversity has been one of the most controversial issues in succession theory in general. According to the classical (e.g. Clements, 1916: Plant Succession: an Analysis of the Development of Vegetation. Carnegie Inst. Washington Publ. 242, 1–517) view, diversity of communities increases towards the late phases of succession, while the competitve exclusion theory (if several species compete for the same resource one of them excludes all the others: Hardin, 1960: Science: 131: 1292–1297) predicts that the process should be towards the establishment of a low-diversity equilibrium. The Intermediate Disturbance Hypothesis (IDH) was originally elaborated by Connell (1978: Science 199: 1304–1310) to interpret the high species diversity that can be observed in tropical rain forests and coral reefs. The theory was adapted to phytoplankton succession by Reynolds (1988: Verh. int. Ver. Limnol. 23: 683–691).

Phytoplankton of temperate lakes has been rarely limited simultaneously by more than 1–3 main resources (N, P, Si alone or in any combination, light). Competitive exclusion theory allows the phytoplankton succession to tend towards an equilibrium of 1–3 species, at any phase of its seasonal development. This is obviously inconsistent with the great species richness and high species diversity that can be observed in most cases (e.g. the paradox of the plankton, Hutchinson, 1961: Am. Nat. 95: 137–147). Theoretical approaches (e.g. Tilman, 1982: Resource Competition and Community Structure. Princeton Univ. Press.) and chemostat experiments (Sommer, 1985: Limnol. Oceanogr. 30: 335–346; 1989: Plankton Ecology: Succession in Plankton Communities. Springer; Gaedeke & Sommer, 1986: Oecologia 71: 98–102) indicated that fluctuations in resource availability makes possible the coexistence of more species than is predicted by the competitive exclusion theory. Several field observations indicate the same (e.g. G.-Tóth & Padisák, 1986: J. Plankton Res. 8: 353–363; Padisák *et al.*, 1988: J. Plankton Res. 10: 249–265; 1990: Hydrobiologia 191: 249–254). As follows from the aforementioned theories, experiments and observations, the frequency of disturbances has an essential influence on the diversity of the phytoplankton and on the establisment of the equilibrium state.

According to Reynolds (1988: Verh. int. Ver. Limnol. 23: 683–691):

(1) at frequencies in order of a few hours (<1 generation time) the responses are physiological;

(2) low frequency pulses, separated by intervals of 10 days or more, can initiate a successional sequence;

(3) progressively smaller intermediate scales (200–20 h) interact with the growth rates of phytoplankton species and tend to preserve high species diversity. The attainment of equilibrium is delayed at a sort of 'plagioclimax' (Reynolds, 1984: Freshwat. Biol. 14: 111–142).

Because the above outlined IDH seems to be very useful in understanding many phenomena that we are used to observing in phytoplankton time series, the participants of the last IAP round-table (Munich, August 1989) agreed to hold a detailed discussion on the topic. In order to make this discussion as useful as possible we summarize here the outcome of our several hours of talk about the IDH and about the possibilities of its test.

Long term data series

In order to test the IDH it is necessary to have phytoplankton data (species composition) with weekly, biweekly (but only if the sampling was really very regular over many years) or shorter sampling intervals over several years. In order to identify the frequency of disturbances data on appropriate physical parameters are needed, such as mixing depth, water column stability. Contributions with insufficient data series should not be admitted, except for experimental and theoretical papers relevant to the topic.

Identification of disturbances

In stratifying lakes, increase in the mixing depth would be the ideal indicator of physical disturbance. If meteorological data are used as 'independent variables', it has to be noted that not every wind or cooling event qualifies as a 'disturbance'. In very stably stratified lakes, even a major meteorological event may lead to only a small increase of mixing depth and may be quite ineffective in terms of its impact on succession. The identification of disturbances in nonstratifying lakes is more problematic. Besides physical variables (changes in transparency, turbidity, etc.), biological variables (for example, the contribution of non-planktonic algae to some population measure, the number of diatom frustules, etc.) can register the physical disturbance.

It should be mentioned that 'disturbances' are not necessarily physical in nature. In the context of the IDH and its relation to the competitive exclusion principle, any event that interrupts the approach toward the eventual exclusion is a disturbance. Generally, disturbances would be recognised by a decrease of phytoplankton biomass and probably by an increase in the availability of resources.

Identification of equilibrium

In natural phytoplankton communities, it is often difficult to determine whether a given 'phase' in a seasonal sequence can be considered to be in an equilibrium state or not, due either to a lack of chemical data, or to insufficient sampling frequency, or to any other cause. For practical purposes, provided that (i) 1, 2 or 3 species of algae contribute more than

80% of total biomass, (ii) their existence or coexistence persists for long enough (more than 1–2 weeks) and (iii) during that period the total biomass does not increase significantly, then that phase can be considered to be at equilibrium. Since our knowledge on the nature of equilibrium state in natural phytoplankton is very limited, contributions (for example: how many generations are involved?) to this aspect of the IDH are also invited.

Definition of diversity
In appreciation of the fact that usually data on the dominant species are more reliable than data of the rare species, an index should be used which gives more weight to the dominant species. Therefore, we ask that the Shannon-Weaver Index be used. Some people have strong opinions about whether this index should be based on cell-numbers, 'units' (one colony being one unit) or biomass. We shall not be strict on this point, because we have to use old data and different contributors may well have documented them differently and they will not always be able to reconvert them. If cell number is used, unidentified picoalgae should be excluded, otherwise even a very massive bloom of some large alga (e.g. *Ceratium*) will be 'drowned' by 10^4 to 10^5 picoalgae per ml. Alternatively, partial diversity indices of conventionally defined size- classes could be calculated. Participants with appropriate data sets are encouraged to try numbers-, unit-, biomass- and surface area-based indices in order to test whether they show the same pattern or not. However, use of biomass data is recommended wherever possible. Because the Shannon-Weaver Index is sensitive to the final number of coexisting species, please indicate on your graphs, how many species (in general 1, 2 or 3) provided the final equilibrium.

Time scale of disturbance
This is the most problematic part of our enterprise. We think that most of the contributors (including ourselves) have an insufficient knowledge of the potentials and pitfalls of the different statistical procedures of time-series analyses. We do not think that we should try to solve this problem by sending out 'recipes'. This problem should be discussed at the workshop. This means, that our workshop will not be the end of the discussion procedure but only an intermediate step.

ULRICH SOMMER and JUDIT PADISÁK"

References

Armstrong, R. A. & R. McGehee, 1976. Coexistence of competing species for shared resources. Theor. Pop. Biol. 9: 317–328.

Connell, J., 1978. Diversity in tropical rain forests and coral reefs. Science 199: 1304–1310.

Ebenhöh, W., 1988. Coexistence of an unlimited number of algal species in model systems. Theor. Pop. Biol. 34: 130–144.

Gaedeke, A. & U. Sommer, 1986. The influence of periodic disturbances on the maintainance of phytoplankton diversity. Oecologia 71: 98–102.

Grime, J. P., 1973. Competitve exclusion in herbaceous vegetation. Nature 242: 344–347.

Grenney, W. J.,1973. A theoretical approach to interspecific competition in the phytoplankton community. Am. Nat. 107: 405–425.

Grover, J. P., 1988. Dynamics of competition in a variable environment. Ecology 69: 408–417.

Grover, J. P., 1989. Effects of Si:P supply variability, and selective grazing in the plankton: An experiment with a natural algal and protistan assemblage. Limnol. Oceanogr. 34: 349–367.

Grover, J. P., 1990. Resource competition in a variable environment: Phytoplankton growing according to Monod's model. Am. Nat. 136: 772–789.

Grover, J. P., 1991. Non steady-state dynamics of algal population growth: experiments with two chlorophytes. J. Phycol. 27: 70–79.

Hardin, G., 1960. The competitve exclusion threory. Science: 131: 1292–1297.

Hutchinson, G. E., 1948. Circular cause systems in ecology. Ann. N. Y. Acad. Sci. 50: 211–246.

Hutchinson, G. E., 1951. Copepodology for ornithologists. Ecology 32: 571–577.

Hutchinson, G. E., 1953. The concept of pattern in ecology. Proc. Acad. natn. Sci. Philadelphia 105: 1–12.

Hutchinson, G. E., 1954. Theoretical notes on oscillatory populations. J. Wildlife mngt 18: 107–109.

Hutchinson, G. E., 1957a. A Treatise on Limnology. Vol. 1. Geography, physics, and chemistry. Wiley Intersci. Publ., New York.

Hutchinson, G. E., 1957b. Concluding remarks. Cold Spring Harbour Symp. Quant. Biol. 22: 415–427.

Hutchinson, G. E., 1959. Hommage to Santa Rosalia, or, Why are there so many kinds of animals? Am. Nat. 93: 154–159.

Hutchinson, G. E., 1961. The paradox of plankton. Am. Nat. 95: 137–147.

Hutchinson, G. E., 1967. A Treatise on Limnology. Vol. 2. Introduction to lake biology and the limnoplankton. Wiley Intersci. Publ., New York.

Hutchinson, G. E., 1975a. A Treatise on Limnology. Vol. 3. Limnological Botany. John Wiley & Sons, New York.

Hutchinson, G. E., 1975b. Variations on a theme of Robert MacArthur. In: Cody, M. L. & J. M. Diamond (eds), Ecology and Evolution of Communities. Harvard Univ. Press & Belknap Press, Cambridge, Mass.: 492–521.

Hutchinson, G. E., 1978. An Introduction to Population Ecology. Yale Univ. Press, New Haven, London.

Lack, D., 1954. The Natural Regulation of Animal Numbers. Clarendon Press, Oxford.

Levins, R., 1968. Evolution in Changing Environment. Princeton Univ. Press, Princeton.

Levins, R., 1979., Coexistence in a variable environment. Am. Nat. 114: 765–783.

Lewontin, R. C. & D. Cohen, 1969. On population growth in randomly fluctuating environment. Proc. natn. acad. Sci. U.S. 62: 1056–1060.

MacArthur, R. H., 1965. Patterns of species diversity. Biol. Rev. 40: 510–533.

McArthur, R. H. & E. O. Wilson, 1967. The Theory of Island Biogeography. Princeton Univ. Press, NJ.

Odum, E. P., 1969. The strategy of ecosystem development. Science 164: 262–270.

Odum, E. P., 1971. Fundamentals of Ecology. 3rd. edn. Saunders, Philadelphia.

Padisák, J., 1992. Seasonal succession of phytoplankton in a large shallow lake (Balaton, Hungary) – a dynamic approach to ecological memory, its possible role and mechanisms. J. Ecol.

Padisák, J. & L. G.-Tóth, 1991. Some aspects of the ecology of the subdominant green algae in a large nutrient limited shallow lake (Balaton, Hungary). Arch. Protistenkunde 139: 225–242.

Pickett, S. T. A., I. Kolasa, I. I. Armesto & S. L. Collins, 1989. The ecological concept of disturbance and its expression at various hierarchical levels. Oikos 54: 129–136.

Reynolds, C. S., 1988. The concept of biological succession applied to seasonal periodicity of phytoplankton. Verh. int. Ver. Limnol. 23: 683–691.

Richerson, P., R. Armstrong & C. R. Goldman, 1970. Contemporaneous disequilibrium, a new hypothesis to explain the 'paradox of plankton'. Proc. Natl. acad. Sci. 67: 1710–1714.

Robinson, J. D. & C. D. Sandgren, 1983. The effect of temporal environmental heterogeneity on community structure: a replicated experimental study. Oecologia 57: 98–102.

Slobodkin, L. B., 1961. Growth and Regulation of Animal Populations. Holt, Rinehart and Wilson, New York.

Sommer, U., 1984. The paradox of plankton: fluctuations of phosphorus availability maintain diversity in flow-through cultures. Limnol. Oceanogr. 29: 633–636.

Sommer, U., 1985. Comparisons between steady state and non-steady state competitions: experiments with natural phytoplankton. Limnol. Oceanogr. 30: 335–346.

Sommer, U., 1991. Phytoplankton: directional succession and forced cycles. In H. Remmert (ed.), The Mosaic-Cycle Concept of Ecosystems. Springer Verlag, Berlin: 132–146.

Turpin, D. H. & P. J. Harrison, 1980. Cell size manipulation in natural marine planktonic diatom communities. Can. J. Fish. aquat. Sci. 37: 1193–1195.

Wilson, J. B., 1990. Mechanisms of species coexistence: twelve explanations for Hutchinson's 'paradox of the plankton': evidence from New Zealand plant communities. N. Z. J. Ecol. 13: 17–42.

Hydrobiologia **249**: 9–14, 1993.
J. Padisák, C.S. Reynolds & U. Sommer (eds), Intermediate Disturbance Hypothesis in Phytoplankton Ecology.

Disturbance events affecting phytoplankton biomass, composition and species diversity in a shallow, eutrophic, temperate lake

Bodil Aavad Jacobsen & Peter Simonsen
Laboratory of Environmental Biology ApS, Baunebjergvej 5, DK-3050 Humlebæk, Denmark

Key words: disturbance, species diversity, *Aphanizomenon*, *Daphnia*, grazing pressure

Abstract

The seasonal changes in phytoplankton biomass and species diversity in a shallow, eutrophic Danish lake are described and related to different disturbance events acting on the phytoplankton community.

Both the spring diatom maximum and the summer bloom of the filamentous blue-green alga, *Aphanizomenon flos-aquae* (L.) Ralfs, coincided with low values of phytoplankton species diversity and equitability. Diatom collapse was mainly due to internal modifications as nutrient depletion (Si, P) caused by rapid growth of phytoplankton, and increased grazing activity from zooplankton. A large population of *Daphnia longispina* O.F. Müller in June effectively removed smaller algal competitors, thus favouring the development of a huge summer bloom (140 $mm^3\ l^{-1}$) of *Aphanizomenon flos-aquae*. Heavy rainfall and storms in late July increased the loss of *Apahnizomenon* by out-flow and disturbed the stratification of the lake. These events caused a marked decline in phytoplankton biomass but had no effect on species diversity. A second storm period in late August circulated the lake completely and was followed by a rapid increase in phytoplankton diversity, and a change in the phytoplankton community structure from dominance of large, slow-growing K-selected species (*Aphanizomenon*) to small, fast-growing r-selected species (cryptomonads).

Introduction

Seasonal changes in phytoplankton community structure can be described as an interaction between internal (autogenic) modifications leading to equilibrium conditions, and external (allogenic) disturbances, such as storms and floods, which shift the phytoplankton assemblage to a more primitive, non-equilibrium stage (Reynolds, 1988).

According to Connell (1978), disturbance of intermediate timescale will maintain high species diversity, when introduced to the environment before the community reaches steady-state conditions and competitive exclusion. Gaedeke & Sommer (1986) found in chemostat experiments, that diversity increased only when disturbance frequency exceeded one generation time (3–14 days). In temperate climate meteorological changes mostly operate with a frequency of 5–15 days, which is considered as the most appropriate intermediate disturbance frequency for phytoplankton (Reynolds, 1988), while disturbances from herbivorous zooplankton operate on a frequency of 25–45 days (McCauley & Murdoch, 1987).

The main purpose of the present paper is to describe the effects of different disturbance events on phytoplankton biomass and diversity in Lake Gødstrup, a shallow, eutrophic, Danish lake, especially concerning the development and sudden collapse of the filamentous blue-green alga *Aphanizomenon flos-aquae* (L.) Ralfs.

Study area

Lake Gødstrup is a small, shallow lake situated near the town Herning in the western part of Jutland, Denmark. The surface area is 46 ha and the mean depth 1.8 m. During the greater part of the year, the lake is completely mixed but there is a short stratification period in the summer. The flushing rate is high and varied in 1989 from 4 d in March to 20 d in July (Table 1).

The nutrient load on the lake is considerable, especially of phosphorus (Table 2). The main part of the phosphorus originates from the central sewage treatment plant of Herning, although this contribution has been reduced about eight times since the early 1980's. The phosphorus binding capacity of the lake sediment is large due to a high iron content, resulting in concentrations of phosphorus in the lake water lower than expected considering the loading (Table 2). Concentrations of inorganic, dissolved N, P and Si decreased markedly in connection with the diatom maximum in May. During the *Aphanizomenon* bloom in July, the concentration of inorganic, dissolved N, and to a lesser extent P, dropped to low values while the Si concentration had returned to high level (Fig. 1).

Materials and methods

Sampling and chemical analyses were carried out monthly from February to December 1989

Table 1. Morphometric data of Lake Gødstrup

Surface area, km^2	0.46
Volume, 10^6 m^3	0.83
Mean depth, m	1.8
Maximum depth, m	4.0
Hydraulic retention time, days	4–20
Catchment area, km^2	75

Table 2. External phosphorus and nitrogen loadings, and concentrations of total phosphorus and total nitrogen. Lake Gødstrup 1989.

Loading		
$g\ P\ m^{-2}\ y^{-1}$		25
$g\ N\ m^{-2}\ y^{-1}$		410
Concentration	Range	Mean
$mg\ P\ l^{-1}$	0.12–0.28	0.21
$mg\ N\ l^{-1}$	2.04–6.70	4.09

by the County of Ringkjøbing as part of a general monitoring program. Samples for phytoplankton countings were collected as mixed samples from 3 layers at the deepest site of the lake, and samples for zooplankton countings as mixed samples from 2 layers and 3 stations. All samples were fixed with Lugol's solution with acetic acid.

After sedimentation of the samples, the organisms were enumerated in an inverted microscope. At least 100 individuals of the most common phytoplankton species were counted. Rare and undetermined species were counted as size classes. In zooplankton samples 100 individuals or all animals were counted.

On each sampling date mean cell dimensions of phytoplankton were calculated for all species on the basis of measurements of 10–20 individuals. The species were approximated to simple geometrical shapes (Willén, 1976) in order to estimate cell volumes ($mm^3\ l^{-1}$ = mg fresh weight l^{-1}). Zooplankton biomass (mg fresh weight l^{-1}) was estimated from length-weight equations (Bottrell *et al.*, 1976). When possible 10 individuals of copepodes, ciliates and rotifers and more than 20 individuals of cladocerans were measured.

Diversity of the phytoplankton community (H′) was determined according to the method described by Shannon (1948), using the $\log_2$ equation and species biovolume. Evenness or equitability (E%) was calculated as $100 \cdot H'/H'max$. Species number (n) refers to number of species or groups used in biomass calculations.

Wind velocity data originate from the Danish Meteorological Institute (1989).

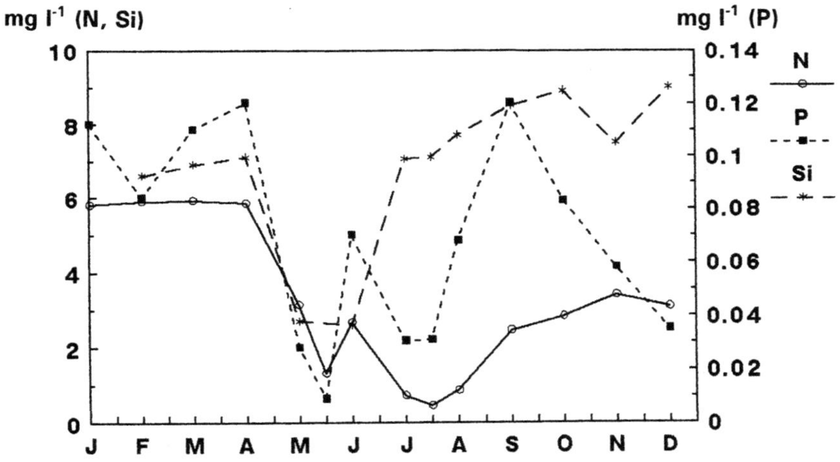

Fig. 1. Seasonal variation in concentrations of dissolved, inorganic nutrients. N = NO2-N + NO3-N + NH4-N. P = PO4-P. Si = silicate-Si.

Results

Phytoplankton (Fig. 2)

Phytoplankton showed two distinct biomass maxima during 1989. A diatom maximum developed in May, at first mainly consisting of small *Cyclotella* species (68%) and later dominated by the pennate diatom *Diatoma elongatum* (Lyngb.) Ag. (95%). In July, the filamentous blue-green alga *Aphanizomenon flos-aquae* formed a huge summer bloom with a maximum biomass of 140 $mm^3\ l^{-1}$. In early August, the population had declined markedly (6.2 $mm^3\ l^{-1}$), but *Aphanizomenon* continued to be the dominant species (99–56%) until it disappeared in November. The number of species being numerous enough for biomass calculations was low, varying from 3 to 9 species/groups.

Zooplankton (Fig. 3)

Zooplankton biomass was low (< 1 mg l^{-1}) until May when a maximum (8.1 mg l^{-1}) mainly consisting of rotifers (*Brachionus calyciflorus* Pallas) appeared. In June, the biomass increased to

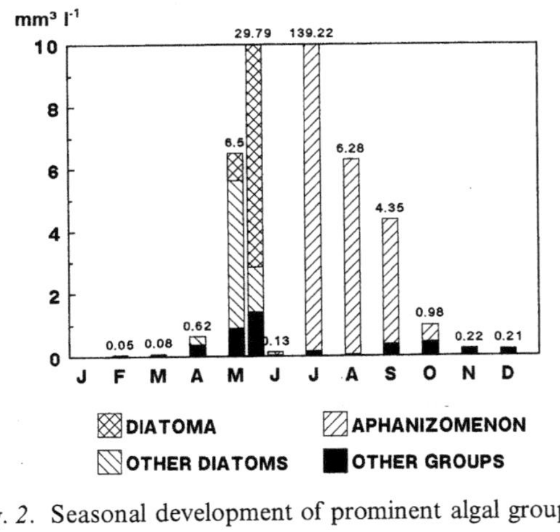

Fig. 2. Seasonal development of prominent algal groups.

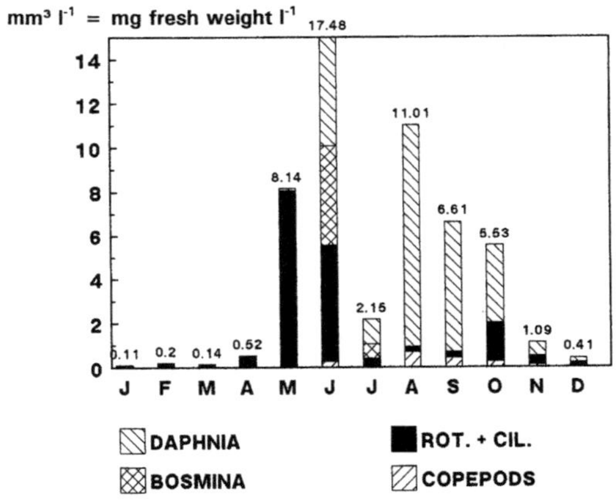

Fig. 3. Seasonal development of prominent zooplankton groups.

17 mg l^{-1} and cladocerans (*Daphnia longispina* and *Bosmina longirostris* O.F. Müller) dominated (68%). Coinciding with the *Aphanizomenon* maximum in July, the zooplankton biomass decreased markedly (2.2 mg l^{-1}), but after collapse of the blue-green algal bloom in August it increased to another maximum (11 mg l^{-1}). This maximum lasted until November and was dominated by *Daphnia longispina* (92–64%).

Diversity (Fig. 4)

Species diversity of the phytoplankton community ranged from 0.02 to 3.9. The lowest values coincided with the phytoplankton maxima in May (*Diatoma*) and July–August (*Aphanizomenon*). For most of the year diversity closely followed the values of evenness. However, low species diversity in the beginning and the end of the year was caused by low species number.

Discussion

Diatom maximum

The rise in phytoplankton biomass to the spring peak in May was driven by rising temperature and light levels combined with high nutrient concentrations. Diversity increased in April due to an increase in species number (Fig. 4). In May, species diversity declined markedly, its minimum coinciding with maximum phytoplankton biomass (Fig. 5) and minimum evenness. In late May *Diatoma elongatum* constituted 90% of the biomass.

Clear-water phase and cladoceran maximum

From late May to early June the phytoplankton biomass decreased from 28 mm^3 l^{-1} to 0.09 mm^3 l^{-1} in 12 days or less, owing to several changes in environmental conditions: continued rise in water temperature, a drop in concentrations of dissolved Si, ortho-P and inorganic N (Fig. 1), and increased grazing pressure from a rapidly growing population of cladocerans (*Daphnia longispina* and *Bosmina longirostris*). In June grazing activity had lead to development of a clear water phase (biomass 0.1 mm^3 l^{-1}). Remineralization caused an increase in concentrations of ortho-P and dissolved inorganic N. Si concentration remained low (Fig. 1).

In this investigation, disturbance by zooplankton seemed to exert a positive effect on diversity by causing a rise in evenness (Fig. 4). Because of the relatively long sampling intervals, it is not possible to conclude whether the following, rapid decline in diversity was caused by continued zooplankton activity or by the development of an immense population of *Aphanizomenon flos-aquae*. Padisák (1993) found a marked decrease in diversity shortly after the clear-water phase and a peak of *Bosmina longirostris*, while the results of

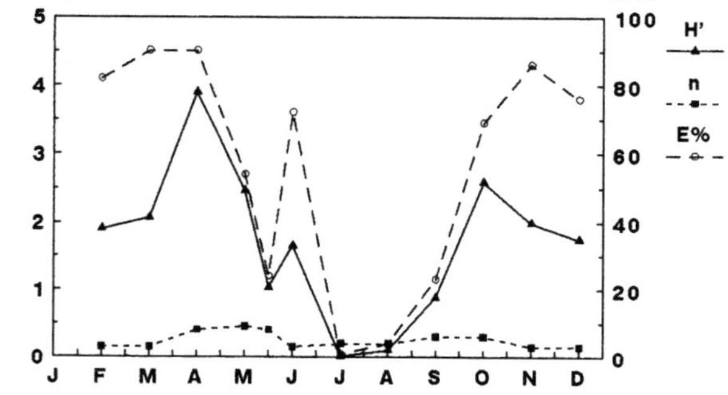

Fig. 4. Seasonal changes in phytoplankton species diversity (H′), species number (n) and evenness (E%).

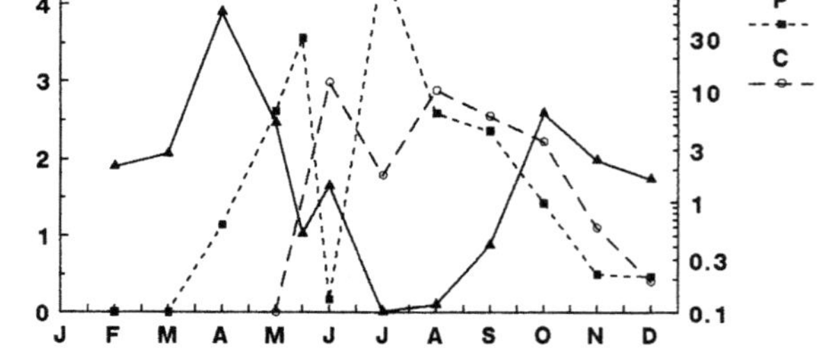

Fig. 5. Seasonal variation of phytoplankton species diversity (H′), phytoplankton biomass (P) and cladoceran biomass (C).

Sommer (1993) were more heterogenous concerning response of species diversity on the clear-water phase.

The Aphanizomenon *bloom*

The large population of *Daphnia* in June certainly had a strong effect on the development of the huge summer bloom of *Aphanizomenon*.

Aphanizomenon often forms distinct maxima in eutrophic lakes when large cladocerans are present (Lynch, 1980; Lynch & Shapiro, 1981; Gliwicz, 1990). *Aphanizomenon* is said to benefit from the feeding activity of cladocerans, which in the spring and early summer remove most potential algal competitors. This is especially the case when *Aphanizomenon* forms 'grass-blade' colonies which make it highly resistant to grazing (Lynch, 1980). In Lake Gødstrup 'grass-blade' colonies were not observed, but as the *Daphnia* species present was medium sized (mean length 610–1400 μm), filaments ranging from 180–330 μm would at least need fragmentation before ingestion by *Daphnia* (Holm *et al.*, 1983; Dawidowicz, 1990).

In addition, the growth of *Aphanizomenon* is often favoured, compared with other algae, by its ability to utilize atmospheric N_2 when the concentration of dissolved inorganic nitrogen is low. Concentrations of dissolved inorganic N lower than 0.45 mg l^{-1} were not measured in Lake Gødstrup during the investigation period (Fig. 1). In addition, the frequency of heterocytes was estimated to be less than 1 heterocyte per 100 *Aphanizomenon* trichomes, indicating minimal resort to N_2-fixation (Neilson *et al.*, 1971).

Environmental conditions in Lake Gødstrup in June were stable (thermal stratification, high nutrient levels) and permitted the phytoplankton community to develop towards an equilibrium stage very close to competitive exclusion. One single species, *Aphanizomenon flos-aquae*, constituted more than 99% of the phytoplankton biomass. This development was followed by a drop in species diversity and evenness to extremely low values (H′ = 0.02, E% = 0.01).

The Aphanizomenon collapse

Factors other than grazing pressure or nutient depletion are needed to explain the rapid decline in *Aphanizomenon biomass* from 140 mm3 l^{-1} July to 6.2 mm^3 l^{-1} in early August.

Considering the negative effect of the high filament concentration on zooplankton filtration (Gliwicz & Siedlar, 1980), it is most unlikely that cladoceran herbivores should have been able to control phytoplankton biomass, even though a second rise in cladoceran biomass was seen from July to August.

Nutrient concentrations were measured biweekly in July and August and did not reach limiting levels (Fig. 1).

Occurrence of wind velocities > 10 m s^{-1} and heavy rain-fall at the end of July caused an increase in flushing rate and disturbed the stratification of the lake after a stagnation period of at least two weeks.

Padisák *et al.* (1988) concluded that sudden changes in phytoplankton community structure can be caused by storms occurring after 5 to 15 days of calm weather. This might be due to storm-induced changes in nutrient availability and/or increased mortality caused by mechanical stress (Padisák *et al.*, 1990). When pre-storm conditions are nutrient limited, and when the circulation event introduces more nutrients to the phytoplankton community, the species composition will shift from large, slow-growing, K-selected, equilibrium species to small, fast-growing, r-selected, pioneer species (Sommer, 1981; Reynolds, 1988). In Lake Gødstrup, nutrient limitation did not occur and *Aphanizomenon* continued to dominate after the storm, causing species diversity to remain low. Loss by out-flow from the epilimnion is therefore a more probable explanation of increased mortality of *Aphanizomenon* leading to the observed decline in biomass. In fact, stratification (although weaker than before) was again or still present in the lake in early August. Either disturbance intensity in late July was too weak to break down steady-state conditions completely, or a new equilibrium stage had started to establish, thus again favouring large, slow-growing K-selected *Aphanizomenon*.

In late August, new storms totally broke down the thermal stratification of the lake. Diversity increased nine-fold (from 0.10 on August 8 to 0.9 on September 5) and continued to increase until October 10, where species diversity was calculated to 2.6. These storms parallel those described by Padisák *et al.* (1988, 1990). Wind-induced mixing provided mechanical stress from resuspended matter on the *Aphanizomenon* population, caused reduction of light accessibility and increased nutrient levels (Fig. 1). The meteorological events were followed by a shift in community structure from a stage close to equilibrium conditions to an earlier successional phase, dominated by small, fast-growing cryptomonads (56%).

Acknowledgements

Thanks are due to Kirsten Olrik and Judit Padisák for useful comments and suggestions, to Kirsten W. Sørensen for correcting the English language and to Jette Poulsen for typing the manuscript.

References

Bottrell, H. H., A. Duncan, Z. M. Gliwicz, E. Grygierek, A. Herzig, A. Hillbricht-Ilkowska, H. Kurasawa, P. Larsson & T. Weglenska, 1976. A review of some problems in zooplankton production studies. Norw. J. Zool. 24: 419–456.

Connell, J., 1978. Diversity in tropical rain forests and coral reefs. Science 199: 1304–1310.

Danish Meteorological Institute, 1989. The climate of Denmark 1989. Dafolo A/S. 131 pp.

Dawidowicz, P., 1990. The effect of *Daphnia* on filament length of blue-green algae. In P. Biró & J. F. Talling (eds), Trophic Relationships in Inland Waters. Developments in Hydrobiology 53. Kluwer Academic Publishers, Dordrecht: 265–268. Reprinted from Hydrobiologia 191.

Gaedeke, A. & U. Sommer, 1986. The influence of the frequency of periodic disturbances on the maintainance of phytoplankton diversity. Oecologia (Berlin) 71: 25–28.

Gliwicz, Z. M., 1990. Why do cladocerans fail to control algal blooms? In R. D. Gulati, E. H. R. R. Lammens, M.-L. Meijer & E. van Donk (eds), Biomanipulation – Tool for Water Management. Developments in Hydrobiology 61. Kluwer Academic Publishers, Dordrecht: 83–97. Reprinted from Hydrobiologia 200/201.

Gliwicz, Z. M. & E. Siedlar, 1980. Food size limitation and algae interfering with food collection in *Daphnia*. Arch. Hydrobiol. 88: 155–177.

Holm, N. P., G. G. Ganf & J. Shapiro, 1983. Feeding and assimilation rates of *Daphnia pulex* fed *Aphanizomenon flos-aquae*. Limnol. Oceanogr. 28: 677–687.

Lynch, M., 1980. *Aphanizomenon* blooms: Alternate control and cultivation by *Daphnia pulex*. Am. Soc. Limnol Oceanogr. Spec. Symp. 3: 299–304.

Lynch, M. & J. Shapiro, 1981. Predation, enrichment and phytoplankton community structure. Limnol. Oceanogr. 26: 86–102.

McCauley, E. & W. W. Murdoch, 1987. Cyclic and stable populations: plankton as a paradigm. Am. Nat. 129: 97–121.

Neilson, A., R. Rippka & R. Kunisawa, 1971. Heterocyst formation and nitrogenase synthesis in *Anabaena* sp. Arch. Microbiol. 76: 139–150.

Padisák, J., L. G.-Tóth & M. Rajczy, 1988. The role of storms in the summer succession of the phytoplankton community in a shallow lake (Lake Balaton, Hungary). J. Plankton Res. 10: 249–265.

Padisák, J., L. G.-Tóth & M. Rajczy, 1990. Stir-up effect of wind on a more-or-less stratified shallow lake phytoplankton community, Lake Balaton, Hungary. In P. Biró & J. F. Talling (eds), Trophic Relationships in Inland Waters. Developments in Hydrobiology 53. Kluwer Academic Publishers, Dordrecht: 249–254. Reprinted from Hydrobiologia 191.

Padisák, J., 1993. The influence of different disturbance frequencies on the species richness, diversity and equitability of phytoplankton in shallow lakes. In J. Padisák, C. S. Reynolds & U. Sommer (eds), Intermediate Disturbance Hypothesis in Phytoplankton Ecology. Developments in Hydrobiology 81. Kluwer Academic Publishers, Dordrecht: 135–156. Reprinted from Hydrobiologia 249.

Reynolds, C .S., 1988. The concept of ecological succession applied to seasonal periodicity of freshwater phytoplankton. Verh. int. Ver. Limnol. 23: 683–691.

Shannon, C. E., 1948. A mathematical theory of communication. Bell Syst. Tech. J. 27: 379–423.

Sommer, U., 1981. The role of r- and K-selection in the succession of phytoplankton in Lake Constance. Acta Oecol. Oecol. Gener. 2: 327–342.

Sommer, U., 1993. Disturbance-diversity relationships in two lakes of similar nutrient chemistry but contrasting disturbance regimes. In J. Padisák, C. S. Reynolds & U. Sommer (eds), Intermediate Disturbance Hypothesis in Phytoplankton Ecology. Developments in Hydrobiology 81. Kluwer Academic Publishers, Dordrecht: 59–65. Reprinted from Hydrobiologia 249.

Willén, E., 1976. A simplified method of phytoplankton counting. Br. phycol. J. 11: 265–278.

Hydrobiologia **249**: 15–24, 1993.
J. Padisák, C.S. Reynolds & U. Sommer (eds), Intermediate Disturbance Hypothesis in Phytoplankton Ecology.

Stress and disturbance in the phytoplankton community of a shallow, hypertrophic lake

Kirsten Olrik & Arnold Nauwerck [1]
Laboratory of Environmental Biology ApS, Baunebjergvej 5, DK-3050 Humlebæk, Denmark; [1] *Institute of Limnology, Gaisberg 116, A-5310 Mondsee, Austria*

Key words: phytoplankton collapse, species diversity, mixotrophy, CO_2-depletion, pH, stress, disturbance

Abstract

The validity of Connell's 'intermediate disturbance hypothesis' in phytoplankton communities was tested on data from a hypertrophic, shallow lake, Hjarbæk Fjord, Denmark.

The present data from Hjarbæk Fjord demonstrate the difficulties in distinguishing stress from disturbance in a phytoplankton community, and show that great changes in the phytoplankton community can take place within few days.

A collapse of blue-green algae in late June 1986 caused remineralization of nutrients and resulted in a rapid increase of fast-growing small chlorococcal green algae and phytoplankton species diversity, without any external disturbances acting on the lake. External disturbances in the form of wind action and brackish water intrusion occurred several days after the onset of these events. Carbon depletion and pH 11.0 were severe stress factors on the phytoplankton community. They were induced by calm, warm weather, but eventually acted as a kind of disturbance to the normally well circulated lake.

Introduction

From the intermediate disturbance hypothesis the following can be stated (Connell, 1978):

1. Species diversity will be reduced to a minimum both in the absence of disturbances (eternal steady state) and during very intensive disturbances.
2. Diversity will be at a maximum when disturbances are at intermediate frequencies and intensities. Under such circumstances, pioneer populations (small, fast growing C-species *sensu* Reynolds, 1984) are able to co-exist with populations of more successful competitors (R- and S-species *sensu* Reynolds, 1984).

In theory and in laboratory experiments, the concepts of 'stress' and 'disturbance' are fairly easily distinguished (Reynolds, 1984, 1988; Sommer *et al.*, this vol.). In natural ecosystems, on the contrary, the impact of a disturbance and its separation from that of a stress event is less clear. Rather, the effects can be both confusing and ambiguous. Often a disturbance in a natural environment is defined by its effects on the biotic community, and quite often this definition results in tautological arguments.

Description of the site studied

Hjarbæk Fjord is a shallow wind-exposed, impounded freshwater lake with a mean depth of 2 m and a surface area of 24 km^2. It receives occasional intrusions of brackish water from the Limfjord in North Jutland, Denmark. In 1986, nutrient loadings were 80 g nitrogen m^{-2} yr^{-1} and 2.6 g phosphorus m^{-2} yr^{-1}, deriving from a large drainage area (1200 km^2) consisting of 75% cultivated land. The mean water retention time in Hjarbæk Fjord in 1986 was approximately 1½ months (Viborg Amtskommune, 1986).

Materials and methods

Samples for phytoplankton, zooplankton and water chemical analyses were taken bi-weekly from April 1986 through June and further 5 times from July 1 through July 9. Samples for chemical analyses were collected bi-weekly until September 10. All sampling was carried out by the Environmental Department of the County of Viborg. Samples for chemical analysis were collected at 0.2 m beneath the surface and 0.5 m above the bottom. Those for analysis of phytoplankton were integrated from samples taken at 0.2 m, at the Secchi depth, and at 2 × Secchi depth. The zooplankton samples were mixtures of water from the whole water column. The samples were collected in two sets, each comprising a 5 l water sample filtered through a 90 μm net and a 0.5 l water sample, sedimented over 24 hrs and concentrated by decantation. All the plankton samples were fixed with Lugol's solution and acetic acid.

The phyto- and zooplankton samples were counted and measured in an Utermöhl inverted microscope, Leitz Labovert with phase contrast (Olrik *et al.*, 1986). Photos were taken in a Leitz Dialux 22 microscope with Nomarski interference contrast. Species diversity was calculated using the Shannon function as $\log^2 x$ based upon volume biomass (Shannon, 1948).

The chemical analyses were performed according to Danish Standards (Dansk Standardiseringsråd, 1975a, b, c; 1977a, b; 1978; 1984; 1985a, b): Oxygen according to DS 277; pH according to DS 287; N compounds according to DS 221 (total N), DS 223 ($NO_2 + NO_3N$), DS 224 (NH_4N); P-compounds by a photometric method according to DS 291 (PO_4P), DS 292 (total P); total alcalinity was measured according to DS 253, and total inorganic carbon was calculated from total alkalinity according to DS 256. Cl was measured potentriometrically according to DS 239. Si was determined according to

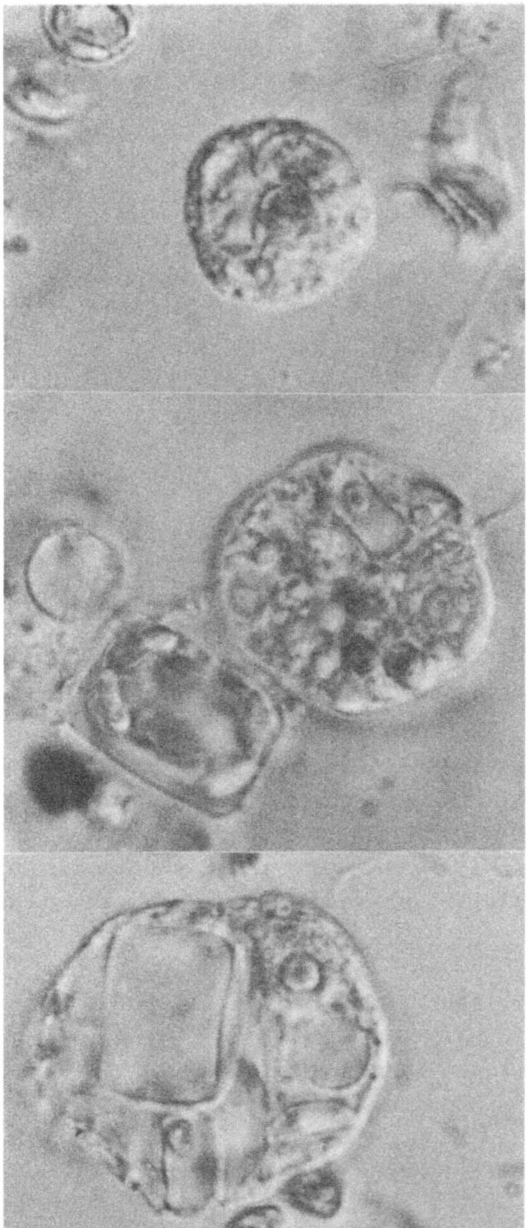

Fig. 1. Individuals of *cf. Ochromonas* with engulfed small centric diatoms, Hjarbæk Fjord 9 April 1986. Magnification 1500 ×. Bar indicates 10 μm.

Table 1. Phytoplankton succession in Hjarbæk Fjord, April–July 1986. Dominant species, total volume biomass ($mm^3 l^{-1}$), dominant species contribution to total biomass, species diversity $\log^2 x$ of volume biomass, and total number of species determined.

HJARBÆK FJORD 1986

Date	Dominant species	Contribution to total biomass	Total biomass $mm^3 l^{-1}$	Species diversity $\log^2 x$	Enumerated number of species	Total number of species determined
April 9			78	1.98	8	29
	Small unicellar centric diatoms	47%				
	cf. Ochromonas	47%				
April 23			8	1.74	6	48
	Gymnodinium sp.	78%				
May 12			49	1.25	7	40
	Diatoma elongatum	83%				
	Small unicellular centric diatoms	13%				
May 28			27	2.52	7	44
	Diatoma elongatum	63%				
	Small unicellular centric diatoms	17%				
June 16			22	2.14	9	47
	Limnothrix planctonica	74%				
	Scenedesmus spp.	9%				
June 25			93	0.67	6	38
	Limnothrix planctonica	93%				
	Scenedesmus spp.	6%				
July 1			15	1.17	6	34
	Scenedesmus spp.	87%				
	Limnothrix planctonica	2%				
July 2			15	2.55	7	50
	Scenedesmus spp.	65%				
	Cryptomonas ad ovata	7%				
	Limnothrix planctonica	3%				
July 3			18	2.75	8	42
	Scenedesmus spp.	60%				
	Cryptomonas ad ovata	9%				
	cf. Chrysamoeba sp.	5%				
July 4			18	2.63	9	28
	Scenedesmus spp.	64%				
	cf. Chrosamoeba sp.	7%				
	Limnothrix planctonica	3%				
	Cryptomonas ad ovata	3%				
July 9			31	2.00	9	37
	Scenedesmus spp.	73%				
	Limnothrix planctonica	6%				
	cf. Chrosamoeba sp.	1%				

Københavns Universitet, Ferskvandsbiologisk Laboratorium (1977).

Daily wind data for the study period were obtained from the Danish Meteorological Institute. They are presented as means of 4 measurements each day at the nearest climatestation, Foulum.

Results

Phytoplankton biomass and succession

Changes in the phytoplankton biomass, the dominant species, the species diversity and species number through the study period are seen from Table 1 and Fig. 2. The phytoplankton biomass was very high, and maxima of different algae rapidly succeeded one another. During the spring maximum (78 $mm^3 l^{-1}$) in early April, small, unicellular centric diatoms and mixotrophic chrysophytes, *cf. Ochromonas* (Fig. 1), dominated, making up 49% and 47% of the biomass, respectively. They were succeeded by a maximum of the dinophyte *Gymnodinium sp.* (6.1 $mm^3 l^{-1}$) which constituted 78% of the total biomass in late April, and a second diatom maximum (49 $mm^3 l^{-1}$) in May, consisting mainly of the pennate species *Diatoma elongatum* AGARDH (63%) and small unicellular centric species (17%). In June a filamentous blue-green alga *Limnothrix planctonica* (WOLOSZ.) MEFFERT became dominant and soon constituted 93% of the biomass of a large maximum (93 $mm^3 l^{-1}$). This excessive blue-green algal biomass collapsed during the last week of June. By July 1, the dominance of the phytoplankton had shifted to chlorococcal green algae, *Scenedesmus spp.*, the biomass of which had increased 2.3 fold during the last 6 days of June, (from 6.3 $mm^3 l^{-1}$ to 15 $mm l^{-1}$). During July 1–9, the biomass increased from 15 $mm^3 l^{-1}$ to 31 $mm^3 l^{-1}$. *Scenedesmus spp.* dominated by 87–60% throughout this period. The subdominating species were cryptophytes (7–9%), a mixotrophic chrysophyte, *cf. Ochromonas* (1–5%), and *Limnothrix planctonica* (2–6%).

Species diversity and species number

Species diversity (Table 1 and Fig. 3) was highest during the second diatom maximum in May (2.5) and during the chlorococcal maximum in July (2.8). Species diversity was at a minimum of 0.67 during the maximum of blue-green algae in late June.

The total number of species was not always consistant with species diversity (Table 1). Total species number was highest during the second diatom maximum in May (47 species) and during the first days of the chlorococcal maximum in July (50 species) and lowest during the first diatom maximum in early April (29 species) and a few days after the chlorococcal green algae had started to build up a maximum in July (28 species). During the blue-green algal maximum when species diversity was at a minimum, the total number of species determined was intermediate (38 species).

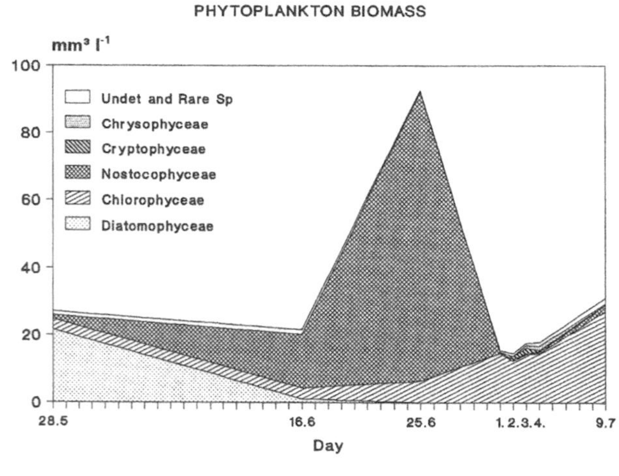

Fig. 2. Biomass of different groups of phytoplankton in Hjarbæk Fjord between 28 May and 9 July 1986.

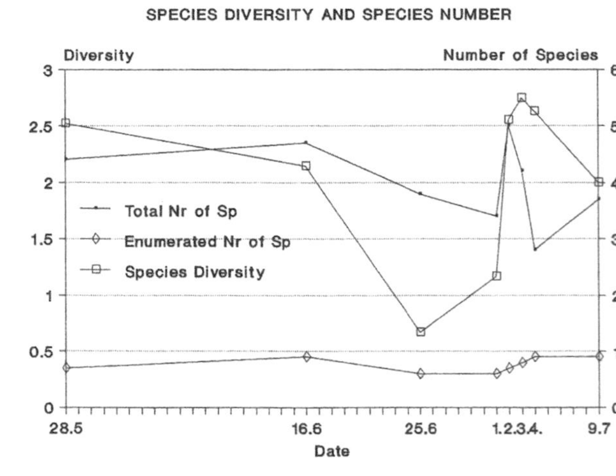

Fig. 3. Phytoplankton species diversity calculated from the biomass; number of species enumerated, and total number of species determined in samples from the period 28 May to 9 July 1991.

Zooplankton biomass and succession

During most of the study period, the zooplankton biomass was remarkably lower than the phytoplankton biomass. Copepods dominated by 79–62% in April, but the biomass was low (0.18–0.20 mg ww l^{-1}). Rotifers (*Brachyonus angularis* GOSSE, *Keratella quadrata* O.F. MÜLLER) dominated by 63–83% in May during the second diatom maximum. Cladocerans (*Bosmina longirostris* O.F. MÜLLER, *Chydorus sphaericus* O.F. MÜLLER) dominated by 61–57% during the blue-green algal maximum in June and during the zooplankton maximum (16 mg ww l^{-1}) and the maximum of chlorococcal green algae July 9. In July, copepods (*Cyclops vicinus* ULIANINE, cyclopoid naupliae) dominated by 76%–49% with a sudden and large biomass of 13 mg ww l^{-1} immediately after the collapse of the blue-green algae, July 14.

Environmental conditions

The close sampling intervals in late June and early July (6 samples in 3 weeks) provides data for an explanation for the collapse of the blue-green algae and their replacement with chlorococcal green algae in late June – early July (Figs 2–3). For this purpose, the environmental factors wind velocity, water temperature, pH, total inorganic C, Cl, inorganic N, total P, and ortho P, are presented in Figs 4–13 (Viborg Amtskommune, 1986).

Wind velocity

Wind velocity (Fig. 4) oscillated between 2–8.5 m s^{-1} during the period May 28–July 9. During this period the wind velocity only exceeded 6 m s^{-1} 4 times before the blue-green algal maximum. Wind velocity was below 6 m s^{-1} during the maximum and collapse of blue-green algae in the last days of June and until July 9 when it increased to 6.7 m s^{-1}; for some hours of the day, it exceeded 10 m s^{-1} (Danish Meteorological Institute).

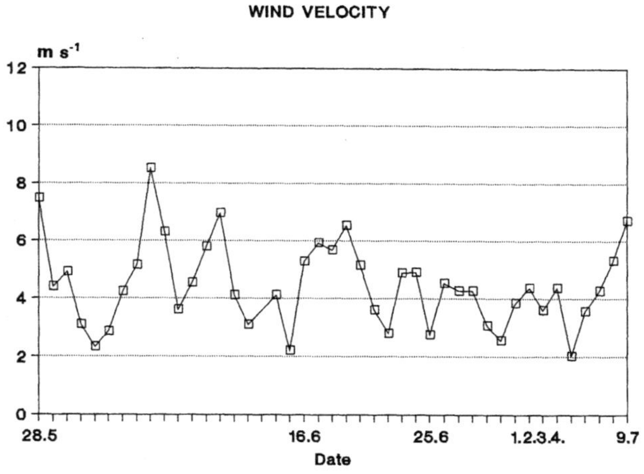

Fig. 4. Wind velocity, m s^{-1}, from climate-station Foulum. Mean of 4 daily measurements between 28 May and 9 July 1991. (Danish Meteorological Institute).

Lake temperature

The lake was unstratified until June 25, when a distinct thermal stratification of the water column was observed. This situation lasted until July 9 when the stratification collapsed (Fig. 5).

pH

Surface water values of pH (Fig. 6) increased to 11.0 during the maximum of blue-green algae but

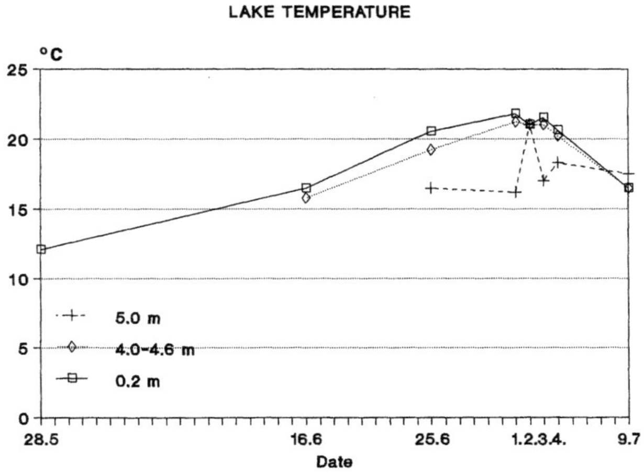

Fig. 5. Lake temperature, 0.2 m beneath surface, 4–4.6 m beneath the surface, and 0.5 m above the bottom (5 m), between 28 May and 9 July 1991 (Viborg Amtskommune, 1986).

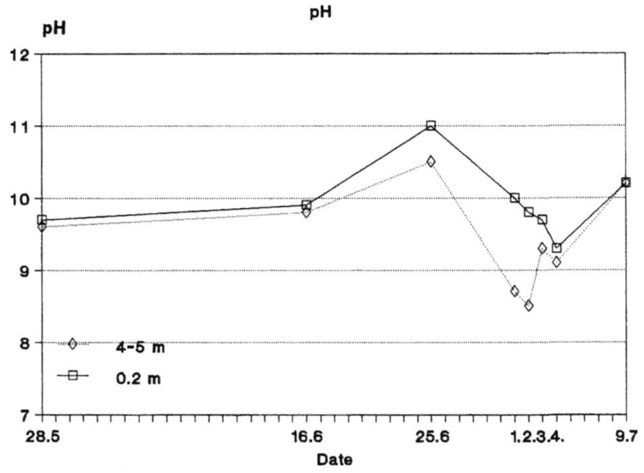

Fig. 6. pH, 0.2 m beneath the surface and 0.5 m above the bottom, between 28 May and 9 July 1991 (Viborg Amtskommune, 1986).

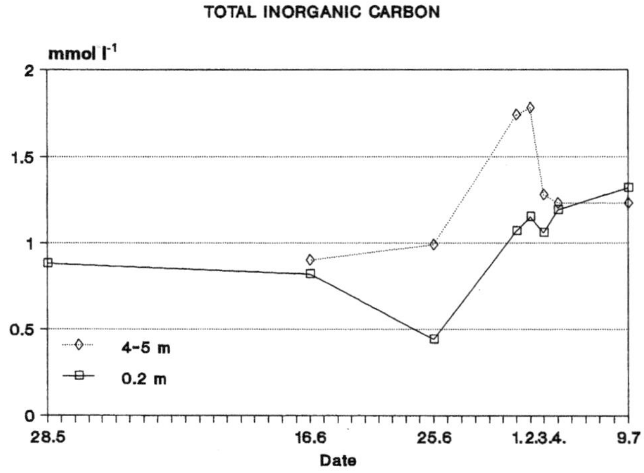

Fig. 7. Total inorganic carbon, mmol l^{-1}, 0.2 m beneath the surface and 0.5 m above the bottom, between 28 May and 9 July 1991 (Viborg Amtskommune, 1986).

decreased to 9.2 after the collapse of the blue-green algae. A further increase to 10.2 was observed on July 9. In the bottom water, pH increased to 10.5 during the blue-green algal maximum, but was considerably lower than the values in the surface water after the collapse of the blue-green algae.

Total inorganic carbon

Total inorganic C (TIC, Fig. 7) was strongly stratified during the stagnation period. TIC was <0.5 mmol l^{-1} during the blue-green algal maximum. Concentration of TIC in the bottom water was markedly higher than at 0.2 m until the water was completely circulated.

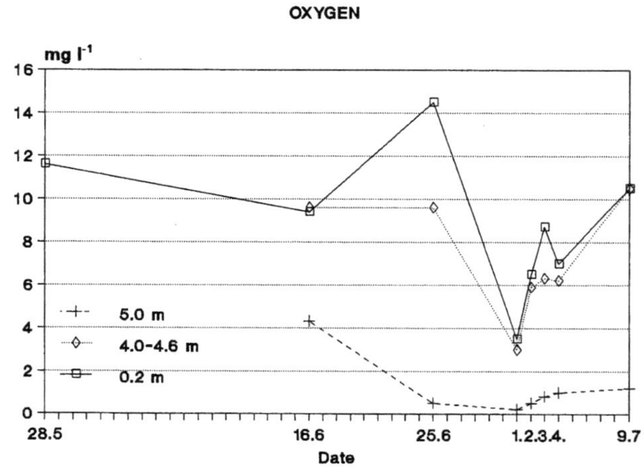

Fig. 8. O_2 mg l^{-1}, 0.2 m beneath surface, 4–4.6 m beneath the surface, and 0.5 m above the bottom (5 m), between 28 May and 9 July 1991 (Viborg Amtskommune, 1986).

Oxygen

Oxygen concentrations (Fig. 8) were strongly stratified during the peak of blue-green algae. Oxygen decreased to less than 4 mg l^{-1} in the whole water column immediately after the collapse of the blue-green algae but quickly increased again during the development of the chlorococcal maximum. In the bottom layer (5.0 m), the oxygen concentration remained <1 mg l^{-1} until July 18.

Chloride

Cl^- (Fig. 9) was below 200 mg l^{-1} until June 25. It remained at that level in the surface layer but by July 2, it had increased to 1200 mg l^{-1} at depths >4 m, indicating a brackish water intrusion through the lock-chamber of the sluice from the Limfjord into Hjarbæk Fjord. After mixing of the water column, the brackish water pocket disappeared, at least at the sampling station.

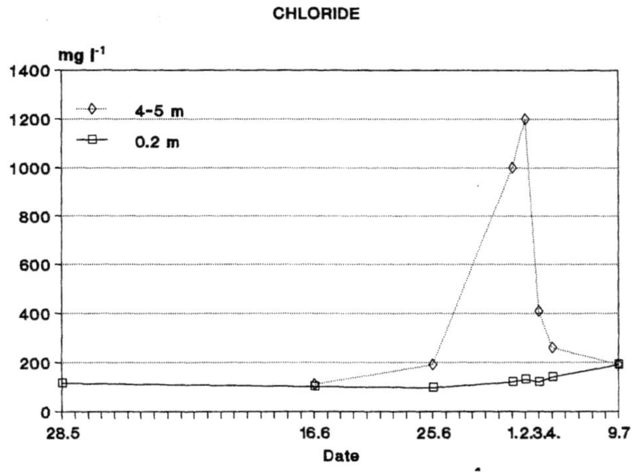

Fig. 9. Cl, mg l^{-1}, 0.2 m beneath surface, 4–4.6 m beneath the surface, and 0.5 m above the bottom (5 m), between 28 May and 9 July 1991 (Viborg Amtskommune, 1986).

Nitrite and nitrate nitrogen

Concentration of $NO_2 + NO_3N$ (Fig. 10) was very high in January (5200 $\mu g\,l^{-1}$) and decreased gradually during spring. By May 28, the concentration was 700 $\mu g\,l^{-1}$, then decreased below 100 $\mu g\,l^{-1}$ during the blue-green algal maximum and below 50 mg l^{-1} during the collapse of the blue-green algae.

Ammonia nitrogen

Concentration of $NH_3 + NH_4N$ (Fig. 11) was 310 $\mu g\,l^{-1}$ in January and 370 $\mu g\,l^{-1}$ during the diatom maximum in early April. From April 23 through June 25, it was $< 10\ \mu g\,l^{-1}$ in the whole water column. Immediately after the collapse of the blue-green algae and the drop in O_2^- concentration, July 1, $NH_3 + NH_4N$ concentration increased to 89 $\mu g\,l^{-1}$ in the upper 0.2 m and to 1790 $\mu g\,l^{-1}$ in the oxygen-depleted bottom meter of the water column (4–5 m).

Total phosphorus

Total P concentration (Fig. 12) was 130-150 $\mu g\,l^{-1}$ in the winter samples and increased to 220–230 $\mu g\,l^{-1}$ during the early April diatom maximum (Viborg Amtskommune, 1986). In the deepest 4–5 m there was a peak of 920 $\mu g\,l^{-1}$ in early July when anoxia in the bottom water had lasted for approximately one week after the collapse of the blue-green algae. From July 1 to July 2, the surface concentration of total P increased from 300 $\mu g\,l^{-1}$ to 510 $\mu g\,l^{-1}$. July 9, the concentration of total P was uniform through the whole water column.

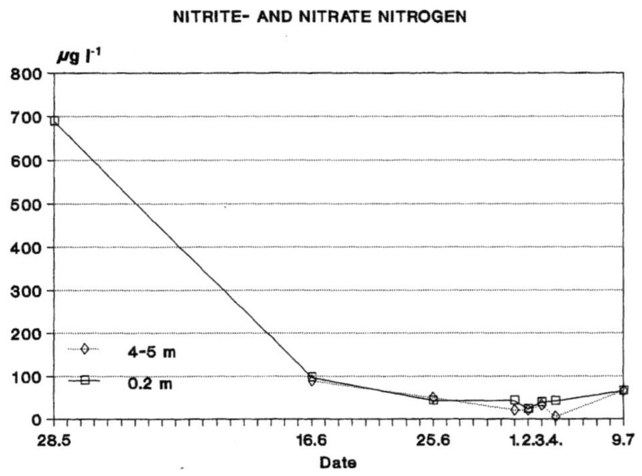

Fig. 10. $NO_2 + NO_3N$, $\mu g\,l^{-1}$, 0.2 m beneath the surface and 0.5 m above the bottom, from the period 28 May to 9 July 1991 (Viborg Amtskommune, 1986).

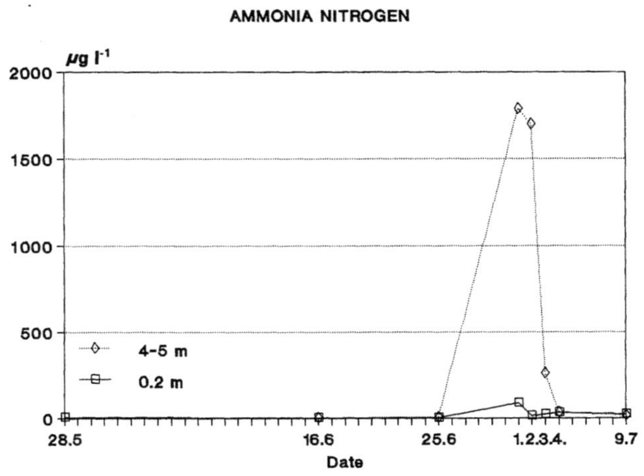

Fig. 11. $NH_3 + NH_4N$, $\mu g\,l^{-1}$, 0.2 m beneath the surface and 0.5 m above the bottom, between 28 May and 9 July 1991 (Viborg Amtskommune, 1986).

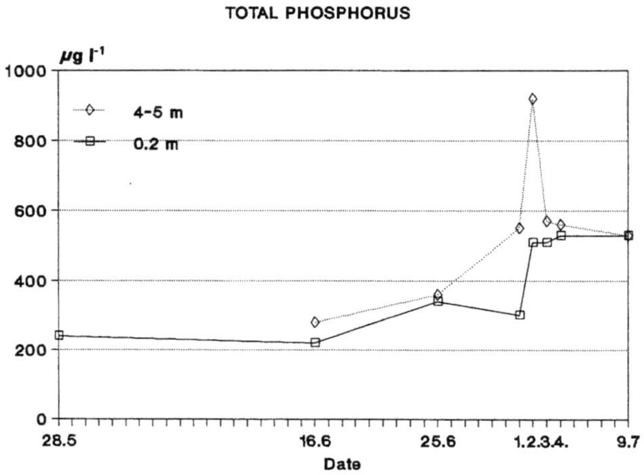

Fig. 12. Total P, $\mu g\,l^{-1}$, 0.2 m beneath the surface and 0.5 m above the bottom, between 28 May and 9 July 1991 (Viborg Amtskommune, 1986).

Orthophosphate phosphorus

PO_4P concentration (Fig. 13) was 79-85 $\mu g\,l^{-1}$ during the winter months and decreased to 16 $\mu g\,l^{-1}$ during the early April diatom maximum. It decreased to $<10\,\mu g\,l^{-1}$ during the second diatom maximum and remained low until after the collapse of the blue-green algae. During one week the concentration in the surface water increased from 26 $\mu g\,l^{-1}$ to 240 $\mu g\,l^{-1}$ and the concentration in the bottom water from 68 $\mu g\,l^{-1}$ to 670 $\mu g\,l^{-1}$.

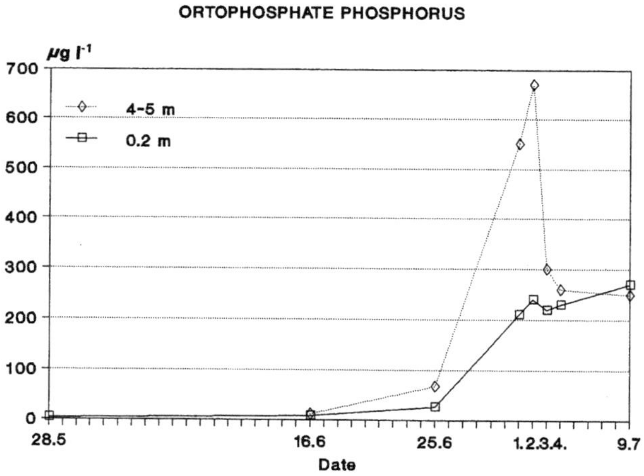

Fig. 13. PO_4P, $\mu g\,l^{-1}$, 0.2 m beneath the surface and 0.5 m above the bottom, between 28 May and 9 July 1991 (Viborg Amtskommune, 1986).

Silicon

Concentration of Si decreased to 700 $\mu g\,l^{-1}$ in early April during the maximum of small unicellular centric diatoms and was 1600 $\mu g\,l^{-1}$ on April 23, when the diatoms had disappeared. Si was completely exhausted during the May diatom maximum (5 $\mu g\,l^{-1}$). Si:TP then fell to 0.03. Si regenerated quickly in the water. The Si-concentration was 1600–13000 $\mu g\,l^{-1}$ and Si:T P was 12–71 during the rest of the year.

Discussion

The disappearance of the maximum of small unicellular diatoms in late April could hardly be explained by depletion of neither N, P, nor Si. The concentration of inorganic N, mainly as $NO_2 + NO_3N$, was 2400 $\mu g\,l^{-1}$; the concentration of PO_4P was 12 $\mu g\,l^{-1}$, and Si-concentration was 1600 $\mu g\,l^{-1}$. Small unicellular centric diatoms provide an excellent food for herbivorous zooplankton. Zooplankton grazing is, nevertheless, a doubtful explanation for the disapperance of the diatom maximum, as zooplankton biomass was still low (0.20 $\mu g\,l^{-1}$) on April 23 when the small unicellular diatoms had already disappeared. However, the mixotrophic chrysophycean flagellate *cf. Ochromonas* is a more probable explanation for the disappearance of the diatoms in late April. By April 9, the biomass of *cf. Ochromonas* was at the same level as the biomass of small unicellular diatoms, and all individuals of *cf. Ochromonas* were observed with at least one diatom within their cells (Fig. 1).

The disappearance of the May diatom maximum of *Diatoma elongatum* and small unicellular centric diatoms was propably due mainly to a collapse due to depletion of Si and PO_4P, the concentrations of which were 0.005 mg Si l^{-1} and 3 μg PO_4P l^{-1}.

It might have been expected that the collapse of blue-green algae, which took place in late June, occurred as a result of external disturbance, eg. strong winds or intrusion of brackish water (Olrik, 1981; Padisák *et al.*, 1988, 1990). Both of these events certainly did occur in the subsequent fortnight of the blue-green algal maximum but the close sampling intervals clearly demonstrated that these events happened several days after the blue-green algal collapse, and that the dramatic increase of fast-growing chlorococcal green algae took place while the water of Hjarbæk Fjord was still stagnant and stratified. The blue-green algal maximum of *Limnothrix planctonica* developed in calm and warm weather simultaneously with a sharp rise in pH and a decline in total alkalinity and TIC. There are references in the literature for selective advantages of blue-green algae over green algae when pH rises above 9.0 (Moss, 1973; Shapiro, 1990). Blue-green algae are favoured by concentrations of low CO_2 availability and/or high pH, as they can actively utilize HCO^{3-} (Espie *et al.*, 1991; Moss, 1973; Olofsson & Woodart, 1977; Shapiro, 1990; Talling, 1976, 1985). N and P were not depleted during the excessive blue-green algal maximum in Hjarbæk Fjord. The collapse of the blue-green algae may have been caused by the rise of pH to 11.0. There is hardly any available inorganic carbon left in the water at this pH-value, where HCO_3^{2-} has shifted to CO_3^{2-}. There was a significant decline in TIC possibly due to precipitation of $CaCO_3$ at the extreme high pH value (Fig. 6–7). At the next cell division of the enormous blue-green algal population (86 $mm^3\ l^{-1}$), carbon would probably have been depleted.

To define a disturbance of an ecosystem, it seems necessary to know the most usual states of that ecosystem. The shallow, eutrophic Hjarbæk Fjord is normally wind-exposed and well circulated. The phytoplankton species diversity, based upon biomass, was high during the second diatom maximum in May and again during the chlorococcal green algal maximum in July. The onset of calm warm weather caused excessive growth of filamentous blue-green algae and depression of phytoplankton species diversity. Carbon limitation was the most probable explanation for the collapse of the blue-green algal population and a consequent release of nutrients from the decaying algae to the water. This event was followed by an increase of phytoplankton species diversity and a renewed growth of fast-growing small C-species having high Michaelis-Menten half-saturation constants for nutrient uptake. This happened without reference to any external disturbance. According to Connell (1978) and Sommer *et al.* (1993), diversity cannot be maintained without disturbances. In the case of Hjarbæk Fjord it could be discussed whether severe stress on the phytoplankton community (high pH and carbon depletion) itself acts as a disturbance on the ecosystem.

Conclusion

Phytoplankton in the highly productive lake Hjarbæk Fjord, Denmark, developed 5 large maxima during the period April-July. A maximum of small, unicellular centric diatoms in early April was presumably grazed away by mixotrophic chrysophycean flagellates and disappeared completely without severe depletion of N, P, and Si. A second diatom maximum in May most likely collapsed due to depletion of Si and PO_4P. An excessive phytoplankton maximum in June (93 $mm^3\ l^{-1}$) mainly consisted of filamentous blue-green algae (93%). In spite of lower growth rate, blue-green algae are superior competitors for chlorococcal green algae at pH values >9.0, as they have an ability to actively utilize HCO_3^-. The blue-green algae in Hjarbæk Fjord probably collapsed in late June as a result of HCO_3^- depletion at pH 11.0. Without any external disturbance acting on the lake, the collapse of blue-green algae caused remineralization of nutrients which resulted in a rapid increase of chlorococcal green algae and increased species diversity of phytoplankton based upon volume biomass by almost 4 times. External disturbances in the form of wind action and brackish water intrusion occured several days after the collapse of blue-green algae and the rapid increase of chlorococcal green algae and species diversity had started. The intrusion of brackish water occurred only once during the study period. Carbon depletion and pH 11.0 were severe stress factors on the phytoplankton community. They were induced by calm, warm weather, but eventually acted as a kind of disturbance to the normally well circulated lake.

Acknowledgements

I am most grateful to The County of Viborg for permission to use its data on water chemistry; to Bodil Aavad Jacobsen, Colin Reynolds, and Judit Padisák for useful comments on the manuscript, and to Jette Poulsen for technical assistance.

References

Connell, J., 1978. Diversity in tropical rain forests and coral reefs. Science 199: 1304–1310.

Dansk Standardiseringsråd, 1975a. Water analysis. Determination of nitrogen content by peroxodisulphate. Dansk Standard DS 221. 1. udg.

Dansk Standardiseringsråd, 1975b. Water analysis. Determination of the sum of nitrite- and nitrate-nitrogen. Dansk Standard DS 223. 1. udg.

Dansk Standardiseringsråd, 1975c. Water analysis. Determination of ammonia-nitrogen. Dansk Standard DS 224. 1. udg.

Dansk Standardiseringsråd, 1977a. Water analysis. Determination of aLkalinity. Dansk Standard DS 253. 1. udg.

Dansk Standardiseringsråd, 1977b. Water analysis. Components of the carbonatesystem (Calculation method). Dansk Standard DS 256. 1. udg.

Dansk Standardiseringsråd, 1978. Water analysis. Determination of pH. Dansk Standard DS 287. 2. udg.

Dansk Standardiseringsråd, 1984. Water analysis. Chloride, potentiometric method. Dansk Standard DS 239. 1. udg.

Dansk Standardiseringsråd, 1985a. Water analysis. Phosphate. Photometric method. Dansk Standard DS 291. 2. udg.

Dansk Standardiseringsråd, 1985b. Water analysis. Total phosphor. Photometric method. Dansk Standard DS 292. 2. udg.

Espie, G. M., A. G. Miller, R. A. Kandasamy & D. T. Canvin, 1991. Active HCO_3^- transport in cyanobacteria. Can. J. Bot. 69: 936–944.

Københavns Universitet, Ferskvandsbiologisk Laboratorium, 1977. Limnologisk metodik. Akademisk Forlag, p. 44.

Moss, B., 1973. The influence of environmental factors on the distribution of freshwater algae: An experimental study. J. Ecol. 61: 157–177.

Olofsson, Jr., J. A. & F. E. Woodard, 1977. Effects of pH and inorganic carbon concentrations upon competition between *Anabaena flos-aquae* and *Selenastrum capricornutum*. Land and Water Resources Institute. University of Maine at Orono, Orono, Maine 04473, 55 pp.

Olrik, K., 1981. Succession of phytoplankton in response to environmental factors in Lake Arresø, North Zealand, Denmark. Schweiz. Z. Hydrol. 43: 6–15.

Olrik, K., A. Nauwerck, A. Sørensen, H. M. Sørensen & S. Lundøer, 1986. Hjarbæk Fjord Undersøgelser 1986–87. Phyto- og zooplankton i Hjarbæk Fjord 1981, 1984, 1985 og april-juli 1986. Viborg Amtskommune Vand- og miljvæsenets rapport 57, 73 pp. + bilag. ISBN 87-87762-26-9.

Padisák, J., L. G.-Tóth & M. Rajczy, 1988. The role of storms in the summer succession of the phytoplankton community in a shallow lake (Lake Balaton, Hungary). J. Plankton Res. 10: 249–265.

Padisák, J., L. G.-Tóth & M. Rajczy, 1990. Stir-up effect of wind on a more-or-less stratified shallow lake phytoplankton community, Lake Balaton, Hungary. In P. Biró & J. F. Talling (eds), Tropic Relationships in Inland Waters. Developments in Hydrobiology 53. Kluwer Academic Publishers, Dordrecht: 249–254. Reprinted from Hydrobiologia 191.

Reynolds, C. S., 1984. Phytoplankton periodicity: the interactions of form, function and environmetal variability. Freshwat. Biology 14: 111–142.

Reynolds, C. S., 1988. Functional morphology and the adaptive strategies of freshwater phytoplankton. In Sandgren, C. D. (ed.), Growth and Reproductive Strategies of Freshwater Phytoplankton. Cambridge University Press, Cambridge: 388–433.

Shannon, C. E., 1948. A mathematical theory of communication. Bell. Syst. tech. J. 27: 379–423.

Shapiro, J., 1990. Current beliefs regarding dominance by blue-greens: The case for the importance of CO_2 and pH. Verh. int. Ver. Limnol. 24: 38–54.

Sommer, U., J. Padisák, C. S. Reynolds & P. Juhász-Nagy, 1993. Hutchinson's heritage: The diversity-disturbance relationship in phytoplankton. In J. Padisák, C. S. Reynolds & U. Sommer (eds), Intermediate Disturbance Hypothesis in Phytoplankton Ecology. Developments in Hydrobiology 81. Kluwer Academic Publishers, Dordrecht: 1–7. Reprinted from Hydrobiologia 249.

Talling, J. F., 1976. The depletion of carbon dioxide from lake water by phytoplankton. J. Ecol. 64: 79–121.

Talling, J. F., 1985. Inorganic carbon reserves of natural waters and eco-physiological consequences of their photosynthetic depletion: Microalgae. In W. J. Lucas & J. A. Berry (eds), Inorganic Carbon Uptake by Aquatic Photosynthetic Organisms. The American Society of Plant Physiologists: 403–420.

Viborg Amtskommune, 1986. Hjarbæk Fjord Undersøgelser 1986–87. Vandkemi-belastning. Viborg Amtskommune Vand- og miljøvæsenets rapport 55, 46 pp. + bilag. ISBN 87-87762-22-6.

Hydrobiologia **249**: 25–32, 1993.
J. Padisák, C.S. Reynolds & U. Sommer (eds), Intermediate Disturbance Hypothesis in Phytoplankton Ecology.

Diversity and succession of the phytoplankton in a small lake over a two-year period

Pertti Eloranta
Department of Limnology, University of Helsinki, E-house, Viikki, SF-00710 Helsinki, Finland

Key words: phytoplankton, diversity, disturbance hypothesis, succession

Abstract

Phytoplankton in the small central Finnish lake, Vasikkalampi, was studied over a two-year period by weekly sampling simultaneously with monitoring of physical and chemical properties of water, solar radiation energy and zooplankton. In the present paper, the fluctuations in phytoplankton diversity were studied in relation to environmental factors. The special aim for the study was to detect a relation between environmental disturbances and phytoplankton diversity.

Introduction

Phytoplankton community structure is regulated by several factors which distinguish among the growth rates of algal species and among the specific rates of loss attributable to grazing, sedimentation and dilution. In nature, there are numerous factors which fluctuate simultaneously, regularly or irregularly, rapidly (in minutes ... in days) or slowly (in seasons ... years).

In cultures of mixed algal species and under steady conditions, succession leads to low diversity and to the dominance of a single species (Tilman *et al.*, 1982). In nature, especially in the water ecosystem of a temperate lake, environmental conditions are rarely so stable that the community would develop such a low-diversity equilibrium. Weather changes fluctuate quickly and drastically alter the underwater light conditions. Occasional rainfalls may cause increasing turbidity and water throughflow in small forest lakes and high wind periods may lead to enhanced nutrient inputs from deeper layers, etc. The disturbances which occur with the intervals of intermediate length may generate a high diversity causing fluctuations in resource availability (Connell, 1978; Sommer, 1985; Gaedeke & Sommer, 1986). The length of intermediate disturbance period is 20...200 h (*ca.* 1...8 d) (Reynolds, 1988).

These disturbances, in the stratified lakes could be indicated by increase in mixing depth, changes in the transparency and turbidity. An increase of nutrient resources could also indicate some new incoming waters or mixing from deeper layers. Rapid decrease in the phytoplankton biomass may also be an indication of disturbance.

The aim of this study was to explore the relationship between phytoplankton diversity and the relevant environmental factors in a small eutrophic lake.

Study lake and methods

The studied lake, Vasikkalampi, is located in the town Jyväskylä, in Central Finland. The lake surface area is *ca.* 3 ha and the drainage basin area *ca.* 25 ha. The lake is surrounded by hills protecting against high winds. The maximum depth is 7.5 m and mean depth 3.8 m. The lake is eutrophic and it has a clear thermal stratification during the summer period, from May to late September. Winter period lasted only *ca.* 2...3 months due to the use of the lake water for cooling of a small thermal power plant. Weak thermal stratification also occurs during the winter months, due to the warm water input to the surface layer of the lake. During the summer period the lake was not used for cooling. The thermal stratification formed in spring months was sharp and stable over the summer period. The physical and chemical properties and the biology of the lake has been described in an earlier series of published papers (e.g. Eloranta, 1980a, b, 1981, 1982a, b,c, 1983; Eloranta & Salminen, 1984).

General fluctuations in nutrient concentrations followed the seasonal changes of phytoplankton. Spring biomass peak was ceased with the very low concentrations of dissolved phosphorus, inorganic nitrogen and silica, but in summer months the productivity was during longer periods limited also by depletion of inorganic carbon at the daytime.

The lake was sampled at roughly weekly intervals for two years, from October 1978 to October 1980. The disturbance hypothesis was tested using the data from the surface water phytoplankton (0–2 m). The diversity (H′) was calculated using the Shannon-Wiener function (Shannon & Weaver, 1949) with the natural logarithms:

$$H' = -\sum p_i \cdot \ln p_i$$

where p_i is the proportional abundance of i th species. Simpson's index (D) was used as the index of dominance (Simpson, 1949).

$$D = \sum p_i^2$$

Both indices were calculated using the biomass (volume) of each species as a measure of abundance. The counting method and counted sample size were kept constant (10 ml): two diagonal stripes with the magnification of 600 × and the whole chamber bottom for larger form with the magnification of 150 × . With the constant counting method the results of species richness (number of taxa per sample) and diversities were kept comparable.

The fluctuation of phytoplankton community structure was compared with several environmental and biological factors. Environmental variables used in the calculations were water temperature, transparency, daily solar radiation energy, mixing depth and mixing depth/transparency ratio. Phytoplankton biomass, zooplankton bio-

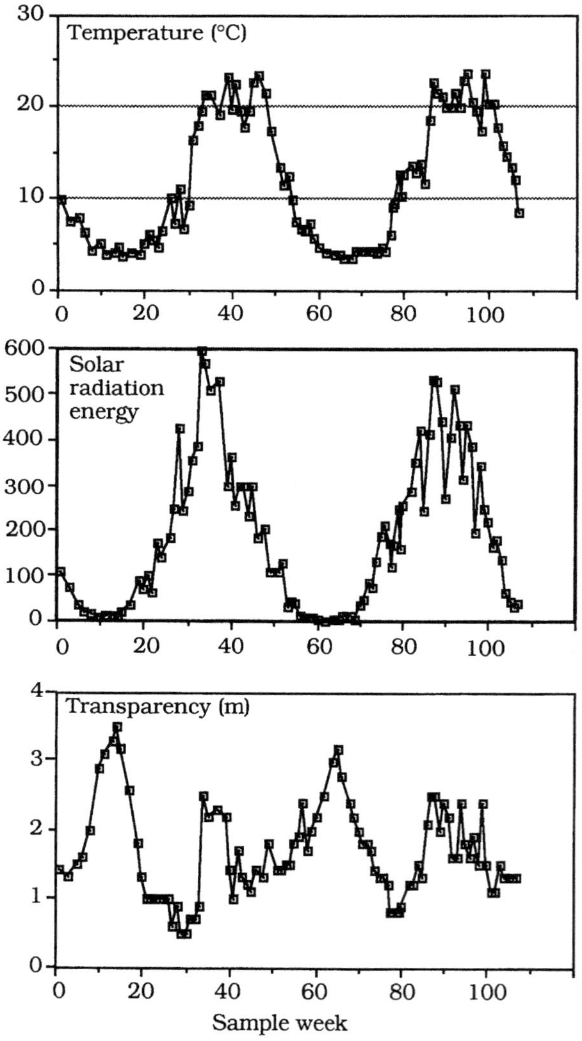

Fig. 1. Variation of water temperature (1 m), daily surface solar radiation energy (cal cm^{-2} d^{-1}) and transparency in the studied lake during the study period from the beginning of October 1978 (week 1) to October 1980 (week 107).

mass, cell density, number of taxa per sample, cell size, number of dominating taxa (= how many taxa were needed to make ≥80% of total biomass) were the biological variables. Because the disturbances may cause fast changes in plankton community, also the change rates of some variables were used in analyses.

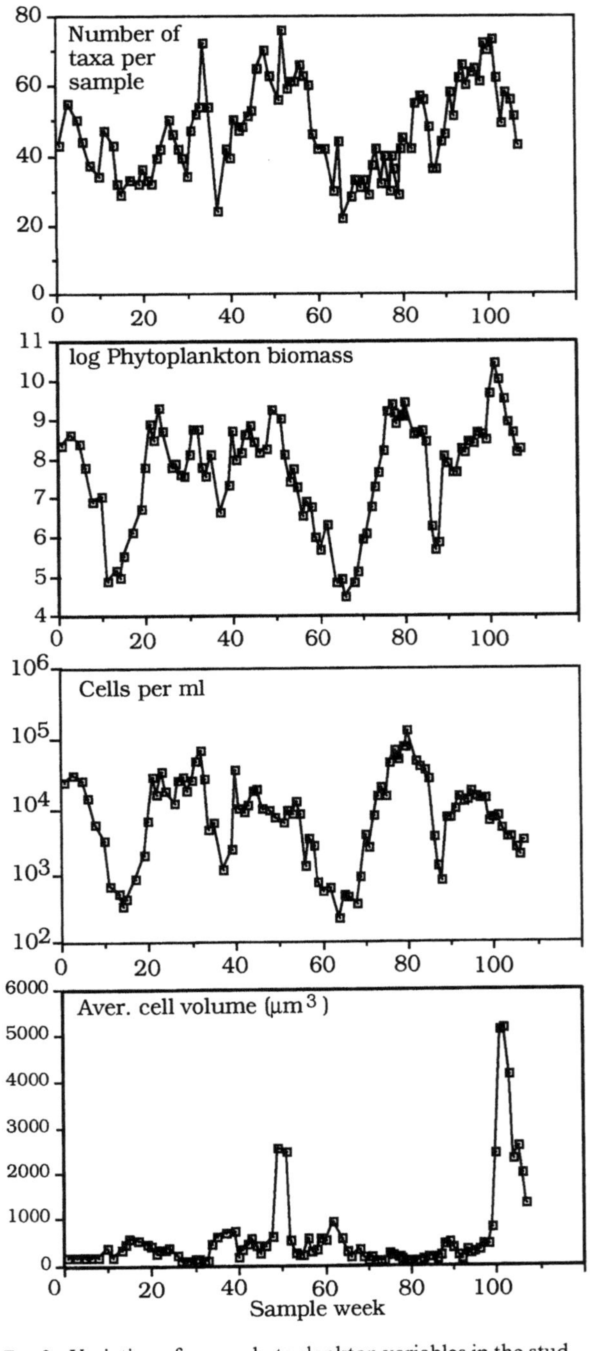

Fig. 2. Variation of some phytoplankton variables in the studied lake during the study period from October 1978 (week 1) to October 1980 (week 107).

Results

Seasonal changes of water temperature and solar radiation were generally rather regular (Fig. 1).

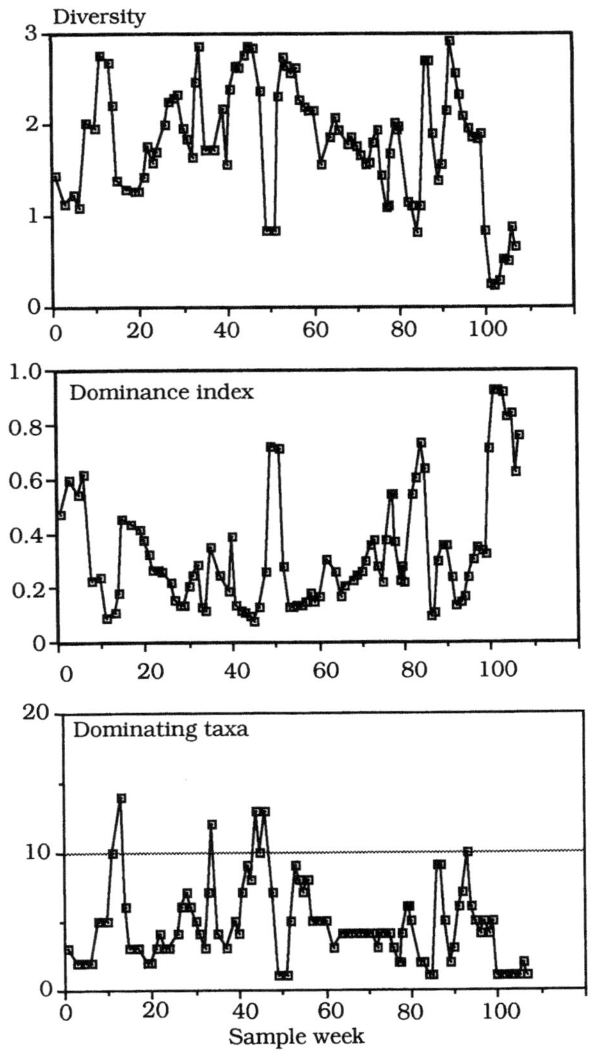

Fig. 3. Variation of phytoplankton diversity (Shannon-Weaver index; biomass basis; natural logarithm basis), dominance index (Simpson's index) and the number of dominating taxa making together ≥80% of total biomass from October 1978 (week 1) to October 1980 (week 107).

Table 1. Significant correlations between the studied variables (Divers = phytoplankton diversity, Domin = dominance index, Temp = water temperature. Trans = transparancy, Energy = daily surface solar radiation, Mixdep = mixing depth, Mix/tra = mixing depth/transparency ratio, Logbm = natural logarithm of phytoplankton biomass, Cellden = phytoplankton cell density per milliliter, Taxa = number of phytoplankton taxa per sample, Dotax = number of dominating taxa making together ≥80% of the total biomass, Logzpl = log of zooplankton biomass, Bmcha = phytoplankton biomass change rate (d^{-1}), Zplcha = zooplankton biomass change rate (d^{-1}), and Temcha = water temperature change rate (d^{-1})).

	Divers	Domin	Temp	Trans	Energy	Mixdep	Mix/tra	Logbm	Cellden	Cellsize	Taxa	Dotax	Logzpl	Bmcha	Zplcha
Divers	1														
Domin	0.892	1													
Temp			1												
Trans	0.202	−0.210		1											
Energy	0.220		0.752	−0.216	1										
Mixdep	−0.215	0.228			−0.299	1									
Mix/tra	−0.266	0.266		−0.628		0.681	1								
Logbm	−0.404	0.484	0.440	−0.756	0.306		0.386	1							
Cellden				−0.566	0.228	−0.214	0.255	0.507	1						
Cellsize									−0.250	1					
Taxa			0.568		0.213			0.431		0.340	1				
Dotax	0.892	−0.765	0.229			−0.262	−0.267	−0.280		−0.381		1			
Logzpl			0.786	−0.249	0.818	−0.374		0.476			0.354		1		
Bmcha	−0.239							0.236				−0.235		1	
Zplcha															1
Temcha	0.255	−0.300			0.251	−0.275			0.294	−0.303		0.225		−0.237	

The winter water temperature was, due to cooling waters from a thermal power plant, a little higher than in the natural Finnish lakes (*ca.* 4–5 °C). The biggest difference between the study lake and the natural lakes was in the light climate due to the very short period of ice cover. Transparency varied between 0.4 and 3.5 m, depending mostly on the density of phytoplankton and zooplankton, especially rotifers. Transparency maxima were measured during winter and during other plankton minima including the summer clearwater period.

Seasonal variation of phytoplankton biomass and cell density followed the classical pattern for an eutrophic lake, with spring and autumn maxima and minima during winter periods and in summer during heavy zooplankton grazing pressure (Fig. 2). The species richness was high during long periods in late autumn from September to October. In early summer the species richness was increasing but decreased markedly with the fast decreasing biomass. The average cell volume had two distinct maxima, which were caused by large diatom *Cyclotella meneghiniana* Kütz. (volume 8500 μm^3). That species dominated during the autumn diatom maxima coincidently with strong water circulation.

Diversity, dominance index (Simpson's index) and the number of dominant taxa fluctuated rather irregularly over the seasons and the whole two-year study period. They were significantly and mutually correlated (Fig. 3, Table 1). These variables had also significant correlations with phytoplankton biomass. During the autumn biomass-maxima, the diversity was at a minimum, owing to the presence of large diatoms. The spring maxima were dominated by the smaller diatoms and also by species from the other algal groups. Thus diversities during spring maxima were somewhat higher. Phytoplankton biomass correlated with most measured variables, mostly due to the seasonal fluctuation of those variables. In spite of the significant negative linear correlation between diversity and phytoplankton mass, high diversities were observed both during high and low biomasses. Without full values from the autumn diatom maxima, the correlation would not been significant (see Fig. 4). Due to the unusual light conditions during the winter period, the number of taxa per sample had not as clear a

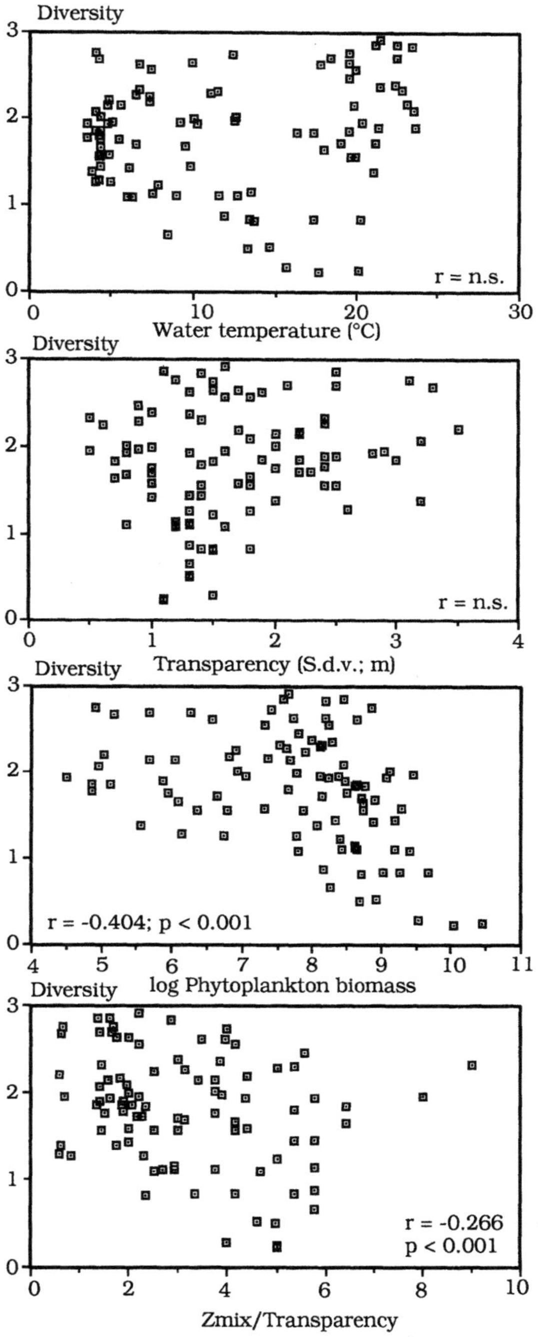

Fig. 4. Relations between phytoplankton diversity and some variables in the studied lake ($n = 93$). Abbreviations: Z_{mix} = depth of the mixing layer, r = n.s.; correlation between the variables not significant).

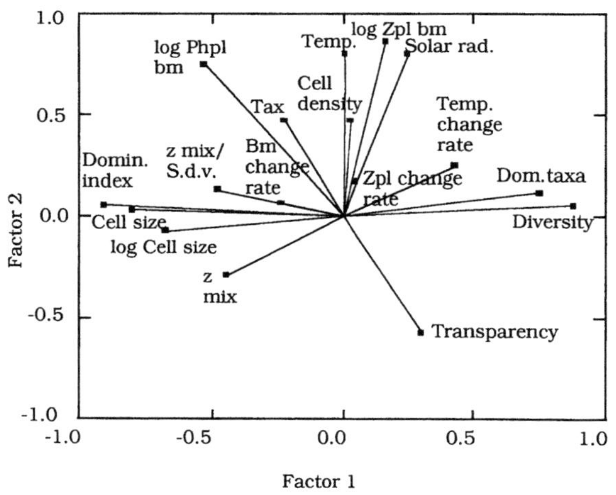

Fig. 5. Results of the principal component analysis showing the relations between diversities and other variables. Abbreviations: Domin. index = dominance index, Temp. = water temperature, Solar rad. = daily surface solar radiation, *Zmix* = mixing depth, z mix/S.d.v. = mixing depth/transparency ratio, log Phpl bm = natural logarithm of phytoplankton biomass, Tax = number of phytoplankton taxa per sample, Dom.taxa = number of dominating taxa making together $\geq 80\%$ of the total biomass, log Zpl bm = log of zooplankton biomass, Bm change rate = phytoplankton biomass change rate (d^{-1}), Zpl change rate = zooplankton biomass change rate (d^{-1}), and Temp. change rate = water temperature change rate (d^{-1}).

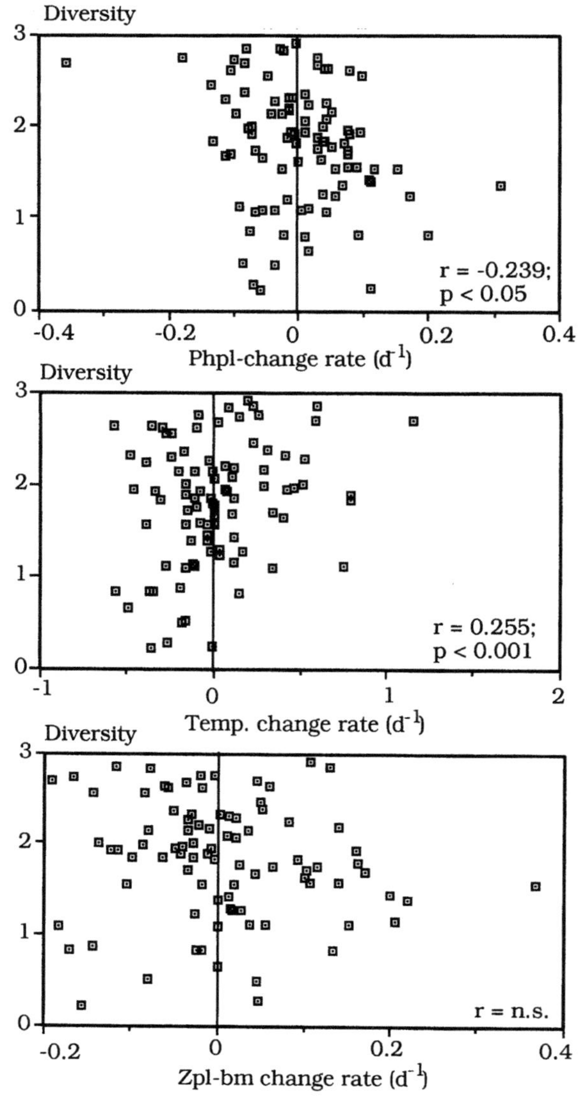

Fig. 6. Relations between phytoplankton diversity and the change rates of phytoplankton biomass, water temperature and zooplankton biomass in the studied lake ($n = 93$).

seasonality as observed in 'normal' ice covered lake in Finland (Fig. 2). However, the midwinter period of low biomass was characterized also by a lower species richness. Because diversity also depends on species richness in the sample, winter minima of species richness were seen as lower diversities at the same time. In the score plot from results of the principal component analysis the variables connecting to the diversity made the first factor and the other factor was made by the variable connecting with the seasonality (Fig. 5).

Diversity could be low or high when phytoplankton biomass decreased or increased fast, but there was a weak negative correlation between diversity and the biomass change rate (Fig. 6). The corresponding correlation with zooplankton change rate was not significant. Increasing temperature was often connected with higher diversity.

The variation of phytoplankton diversity could be explained by several factors, which varied from time to time but it was not possible to show any disturbance-diversity relations (Fig. 7).

Discussion

The intermediate disturbance hypothesis (IDH) is based on the idea that disturbances with intermediate frequency and intensity in the environment prevents achievement of the outcome of

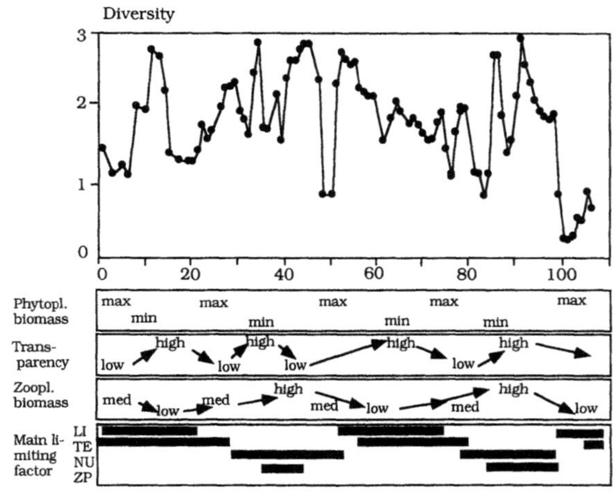

Fig. 7. Variations of phytoplankton diversity in relation to fluctuations of biomasses, water transparency and limiting and regulating factors (LI = light, TE = temperature, NU = nutrients, ZP = zooplankton).

competitive succession and its equilibrium stage characterized by low diversity (Connell, 1978). According to this theory, also very frequent disturbances would lead to low diversities. The richness of phytoplankton community was paradoxical for Hutchinson (1961). He already wondered how, in a well-mixed planktonic environment, so many taxa could live simultaneously when the number of limiting factors (light, nutrients) might not be very high.

The competition theory works well in experimental conditions and leads to the development of a low diversity community, with one strongly dominating species. However, the factors affecting growth rates in natural temperate lake are numerous. Some of them have large scale regular fluctuation causing general seasonality (light, temperature), but also them have also short term fluctuations depending on weather changes (wind, cloudiness, rainfalls). All these fluctuations may cause secondary variations in nutrient concentrations (all inorganic major and minor nutrients as well as organic nutrients like amino acids, vitamines, carbohydrates). Zooplankton grazing is often selective and it depends on actual species structure and life cycle of each species. Hutchinson assumed the epilimnetic environment to be a well-mixed, homogenous environment. That may be true in flat, shallow and open lakes, but not in lakes with variable depth, form and hydrology in temperate zone with varying weather conditions.

In the Finnish oligotrophic and mesotrophic lakes, the phytoplankton community structure reaches typically its highest stability and diversity during the second half of the summer in late July and August. During this period, most algal groups are represented in the phytoplankton and the species richness per sample has its maximum (e.g. Eloranta, 1974, 1976). This means that general environmental stability in late summer period with stabilized water temperature and relatively similar solar radiation leads in these lakes towards higher diversity, not towards a more simple community. The common idea that diversity in oligotrophic lakes is higher than in eutrophic lakes is not necessarily true. When compared lakes with different levels of trophy, diversity is increasing from oligotrophy to mesotrophy and upto lower levels of eutrophy. Simple communities are generated in hypertrophic waters. Low diversity in eutrophy may be found for methodological reasons. The sample size used for phytoplankton counting is often only 5–10 ml from eutrophic lakes because the high cell density, whereas settled samples from oligotrophic lakes are 50–100 ml. Due to the smaller sample size counted from eutrophic lakes, the number of taxa found in sample is lower, although the actual number of taxa in the community would be much higher (Eloranta, 1978). During the winter period, light and temperature are the main limiting factors for the primary producers. Under the ice, however, disturbances are missing. Community structure is going to lower species richness decreasing diversity by that way, but the evenness of the community is often rather high.

In the studied small lake, diversity fluctuated depending on various different factors and the role of disturbances was impossible to demonstrate. We can assume that the general level of taxa found in samples were caused just by many forms of frequently occurring disturbances. However, it was not possible to demonstrate decreasing trend of diversity during stable periods.

References

Connell, J., 1978. Diversity in tropical rain forests and coral reefs. Science 199: 1304–1310.

Eloranta, P., 1974. Studies on the phytoplankton in Lake Keurusselkä. Ann. bot. fenn. 11: 13–24.

Eloranta, P., 1976. Species diversity in the phytoplankton of some Finnish lakes. Ann. bot. fenn. 13: 42–48.

Eloranta, P., 1978. Effects of the size of sample counted in phytoplankton analyses. Ann. bot. fenn. 15: 169–176.

Eloranta, P., 1980a. Winter phytoplankton in a pond warmed by a thermal power plant. Ann. bot. fenn. 17: 264–275.

Eloranta, P., 1980b. Annual succession of phytoplankton in one heated pond in central Finland. Acta Hydrobiol. 22: 421–438.

Eloranta, P., 1981. Yearly succession of the phytoplankton in an ice-free pond in central Finland. Schweiz. Z. Hydrol. 43: 20–33.

Eloranta, P., 1982a. Periphyton growth and diatom community structure in a cooling water pond. Hydrobiologia 96: 253–265.

Eloranta, P., 1982b. Seasonal succession of phytoplankton in an ice-free pond warmed by a thermal power plant. In V. Ilmavirta, R. I. Jones & P.-E. Persson (eds), Lakes and Water Management. Developments in Hydrobiology 7. Dr W. Junk Publishers, The Hague: 87–91. Reprinted from Hydrobiologia 86.

Eloranta, P., 1982c. Zooplankton in the Vasikkalampi pond, a warm water effluent recipient in Central Finland. J. Plankton Res. 4: 813–837.

Eloranta, P., 1983. Physical and chemical properties of pond waters receiving warm-water effluent from a thermal power plant. Wat. Res. 17: 133–140.

Eloranta, P. & R. Salminen, 1984. Phytoplankton primary production in a eutrophic cooling water pond. In C. Heip (ed.), Biology of Meiofauna. Developments in Hydrobiology 26. Dr W. Junk Publishers, Dordrecht: 267–274. Reprinted from Hydrobiologia 118.

Gaedeke, A. & U. Sommer, 1986. The influence of the frequency of periodic disturbances on the maintenance of phytoplankton diversity. Oecologia 71: 98–102.

Hutchinson, G. E., 1961. The paradox of the plankton. Am. Nat. 95: 137–145.

Reynolds, C. S., 1988. The concept of biological succession applied to seasonal periodicity of phytoplankton. Verh. int. Ver. Limnol. 23: 683–691.

Shannon, C. E. & W. Weaver, 1949. The mathematical theory of communication. Univ. of Illinois Press, Urbana, 117 p.

Simpson, E. H., 1949. Measurement of diversity. Nature 163: 688.

Sommer, U., 1985. Comparisons between steady state and non-steady state competition: experiments with natural phytoplankton. Limnol. Oceanogr. 30: 335–346.

Tilman, D., S. S. Kilham & P. Kilham, 1982. Phytoplankton community ecology: the role of limiting nutrient. Ann. Rev. Ecol. Syst. 13: 349–372.

Hydrobiologia **249**: 33–42, 1993.
J. Padisák, C.S. Reynolds & U. Sommer (eds), Intermediate Disturbance Hypothesis in Phytoplankton Ecology.

Phytoplankton succession and diversity in a warm monomictic, relatively shallow lake: Lake Volvi, Macedonia, Greece

Maria Moustaka-Gouni
Institute of Botany, School of Biology, Aristotle University of Thessaloniki, GR-540 06 Thessaloniki, Greece

Key words: phytoplankton succession, nanoplankton, diversity, vertical distribution, seasonality, Intermediate Disturbance Hypothesis

Abstract

In Lake Volvi, phytoplankton dominance was examined in relation to the main phases of the fluctuating physical state of the water column and nutrient levels. Four stages of algal succession were identified. The successional pattern was, in general, similar from year to year. External disturbances such as increased wind mixing and high floods had the effect of setting the succession back to an earlier stage.

Nanoplanktic diatoms and flagellates dominated during the first stages of succession. The succession was running from r-selected species in early stages to K-strategists in summer and towards a mixed community in the terminal stage.

Diversity was expressed by the indices of Gleason and Shannon (based on physical units and biomass). Interpretation of species diversity took into account the differential variations of its two components, the number of species and the evenness of their distribution. Diversity was more affected by evenness than by number of species.

Diversity increased in late stages of succession corresponding to the complexity of the community (high number of species, high evenness). Physical disturbances influenced positively the diversity. The maximal diversity appeared in the transition periods between the compositional changes generated by disturbances and the true successional development. Disturbances may generate a more uniform distribution of diversity throughout the whole water column.

The role of physical disturbances in increasing the phytoplankton diversity may be demonstrated from comparisons of diversity – evenness – species number of the same stages under different meteorological conditions.

Introduction

Several recent papers have summarized information on the seasonal succession of phytoplankton and have proposed general patterns for lakes of different trophic status (e.g. Reynolds, 1982, 1984; Sommer *et al.*, 1986). However, less information is available about the phytoplankton periodicity in warm, relatively large and shallow lakes, such a lake is Lake Volvi.

Successions can be set back by external disturbances, such as floods, storms or episodes of increased wind mixing occurring at low frequency intervals, >10 days (Reynolds, 1988). Disturbances which occur with the interval of intermediate length (200–20 h) may generate a high diversity (Gaedeke & Sommer, 1986; Reynolds, *loc. cit.*) according to the 'Intermediate Disturbance Hypothesis' of Connell (1978) concerning the maintenance of high community diversity in coral reefs and tropical rain forests.

Physical disturbances, such as episodes of increased wind mixing and high floods, were identified in Lake Volvi. An increase in the availability of limiting resources, a significant contribution of non-planktic algae in the euphotic zone and a sharp decrease in phytoplankton biomass accompanied the changes in physical variables.

The purpose of this paper is to describe the algal succession in relation to changes in the environment. The relative dominance of nanoplankton and the distribution of diversity are presented and linked to succession. The effect of physical disturbances on the seasonal progressions in species dominance and diversity is discussed.

Methods

Samples were collected from a central station where the lake is 15 m deep, using a 2 l, 1.3 m long sampler at depths every 1.3 m from surface to the bottom. In addition samples were collected from the water layer 0–0.1 m. Sampling started in March 1984 and was carried out at approximately monthly or biweekly intervals (during the warm period) until August 1986. Algae were counted under an inverted microscope using Utermöhl's (1958) sedimentation method. Determination of the phytoplankton biomass was based on the calculation of the volume of each species from appropriate geometric formulae assuming a specific gravity of one. Algae with a diameter of $> 20\ \mu m$ were considered as netplankton forms.

Diversity was expressed by the following indices:

1) Gleason's (Margalef, 1960)

i) based on physical units (individuals)

$$d_u = \frac{S\text{-}1}{\ln N},$$

ii) based on biomass

$$d_b = \frac{S\text{-}1}{\ln B}$$

where S is the number of species in the sample; N is the number of the individuals; and B is the total biomass

2) Shannon's (Odum, 1980)

i) based on physical units (individuals)

$$H'_u = -\sum_{i=1}^{s} \frac{n_i}{N} \log_2 \frac{n_i}{N},$$

ii) based on biomass

$$H'_u = -\sum_{i=1}^{s} \frac{b_i}{B} \log_2 \frac{b_i}{B}$$

where n_i is the number of the individuals and b_i the biomass of the i-th species; s is the number of species in the sample; and

$$N = \sum_{i=1}^{s} n_i \text{ and } B = \sum_{i=1}^{s} b_i$$

The theoretical maximum of diversity in any sample can be expressed as $H'_{max} = \log_2 s$

and eveness (or equitability) can be expressed as

$$E = \frac{H'}{H'_{max}}$$

Simpson's index, used as the index of dominance (Odum, 1980), was based on:

i) physical units (individuals)

$$D_u = \sum_{i=1}^{s} \left(\frac{n_i}{N}\right)^2$$

and ii) biomass

$$D_b = \sum_{i=1}^{s} \left(\frac{b_i}{B}\right)^2$$

where N_i is the number of the individulas and b_i the biomass of the i-th species; s is the number of species in the sample; and

$$N = \sum_{i=1}^{s} n_i \text{ and } B = \sum_{i=1}^{s} b_i$$

Both diversity and dominance indices were calculated for each sample: the value of each successional stage represents the average of the relative samples.

Background

Lake Volvi, located in Northern Greece, has a surface area of 68.6 km^2, a maximum depth of 19 m and a mean depth of 13.5 m. The site has been described previously by Moustaka-Gouni (1988b).

Meteorological conditions, especially rainfall, varied widely during the study period (March 1984–August 1986). Unusually heavy rains fell in late spring and summer of 1986 and high floods occurred in contrast to the drought of 1985. In 1984, rainfall was moderate and evenly distributed.

Lake Volvi is a warm monomictic lake. During the stratification period (June–August) the epilimnion is 11 m thick and microstratification often develops within it. By August, the deepening of the thermocline is initiated and isothermal conditions prevail in winter and early spring. During stratification, the hypolimnion is devoid of oxygen.

The annual mean depth of the euphotic zone (4 m) has been found to be always smaller than the epilimnion. In the euphotic zone, pH remained above 8 throughout the year (max. 9.3 in summer).

Concentrations of dissolved phosphorus in the euphotic zone never exceeded 10 $\mu g\ l^{-1}$. The highest concentrations were recorded in winter while in April, prior to the onset of thermal stratification, phosphorus dropped to very low concentrations (especially in 1986 to $< 1 \mu g\ PO_4P\ l^{-1}$). The highest concentrations of nitrate nitrogen (about 200 $\mu g\ NO_3N\ l^{-1}$) occurred in late winter–early spring, while in April they started decreasing. At the end of April and throughout the warm period (May–October), the whole water column became almost entirely depleted of nitrate but in summer of 1986 the nitrate concentration increased after heavy rainfall. Very low concentrations of silica were recorded in spring (465 $\mu g\ SiO_2Si\ l^{-1}$) and autumn (146 $\mu g\ SiO_2Si\ l^{-1}$).

Critical changes in the stratification regime, the warming and cooling of the water, and in the nutrient conditions occurred in early April, late May, early September and early November. Thus, four main phases, each characterized by different physical and chemical conditions, can be detected (Fig. 1).

A total of 140 taxa have been identified in the phytoplankton of Lake Volvi; 21 of these contributed more than 10% of total biomass at some time. Total algal biomass attained peaks ranging from 1.5 to 13.0 g m^{-3} while the mean annual values varied from 2.2 to 3.6 g m^{-3}. Cyanophytes, diatoms and cryptophytes were the most important constituents of the lake's phytoplankton (Moustaka-Gouni, 1988a).

Results and discussion

Succession stages

Relating the seasonal progressions in species dominance (only those species which contributed more than 10% of total biomass at some time) to the different phases of physical and chemical conditions (Fig. 1), specific associations result and four stages of succession can be recognized (Fig. 2).

Within a year, each stage of succession was characterized by the development of the following organisms:

Stage I. *Stephanodiscus – Rhodomonas lens* Pascher & Ruttner stage (November–March)
During the period of high turbulence (November–March), with low temperature and nutrients at their maximum the small, relatively fast growing species *Rhodomonas lens*, *Stephanodiscus* cf. *astraea* (Ehr.) Grun. and *Stephanodiscus hantzschii* Grun. were dominant (e.g. Sommer, 1981). In 1986, additional species was the large desmid *Closterium aciculare* T. West.

This successional stage, regular in occurrence

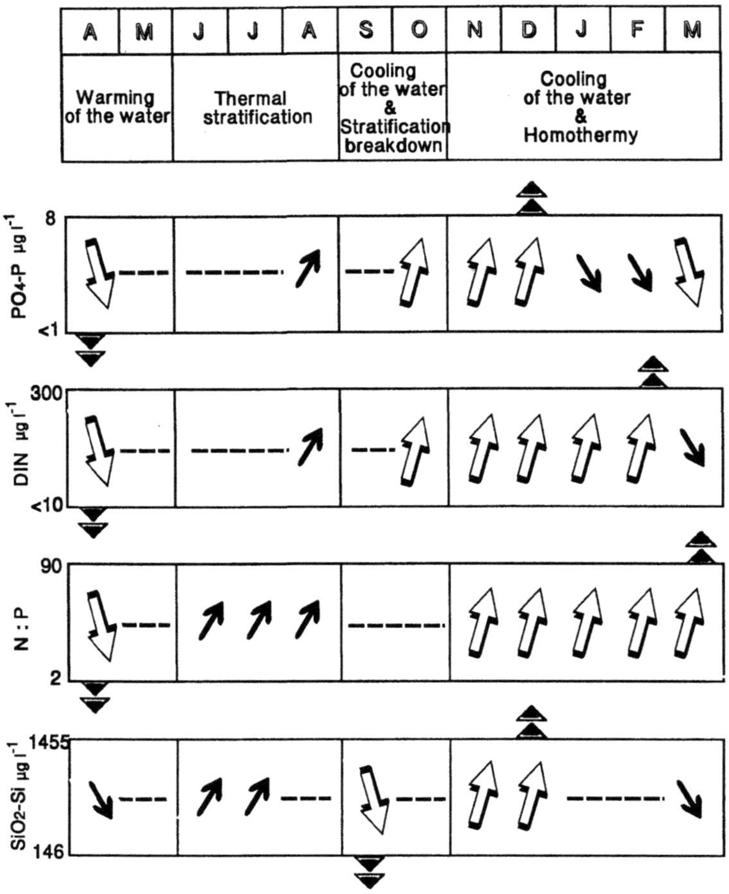

Fig. 1. Main phases of the fluctuating physical state of the water column and nutrient levels; nutrient ranges are shown to the left and major, minor and minimal shifts in nutrient concentrations and the N:P ratio are indicated by large arrows, small arrows and dashed lines respectively; times of maxima and minima are also marked.

and species composition, was associated with the winter phytoplankton biomass maximum. In February 1986, the drop in biomass, due probably to the high turbidity (Zeu <2 m), was not accompanied by changes in species dominance.

Stage II. Small flagellates stage (April–May)

In each period of water warming, when nutrient concentrations were very low, phytoplankton dominance was characterized by *Rhodomonas minuta* var. *nannoplanktica* Skuja.

In 1984 with N:P >5 (atomic ratio), the dominant algae were *R. lens* and *R. minuta*. *R. lens* belonging to the first stage continued its dominance and preceded the *R. minuta* maximum. In 1985 with N:P <5, *R. minuta* was accompanied by *Chrysochromulina parva* Lackey (in April) and the nitrogen-fixing cyanophyte *Anabaena flos-aquae* Bréb. ex Born. & Flah. (in May). It seems that *A. flos-aquae* started the blue-greens stage earlier. In 1986 with N:P >5, *R. lens* built the second maximum simultaneously with an increase of *R. minuta*.

Rhodomonas decline in early April 1986 coincided with large populations of the rotifers *Polyarthra* and *Keratella* (1000 ind. l^{-1}). Instead of *Rhodomonas, Nitzschia acicularis* W. Smith and *S. hantzschii* dominated. Wind induced stirring-up of the sediment probably favoured the strong population increase of *N. acicularis* (e.g.

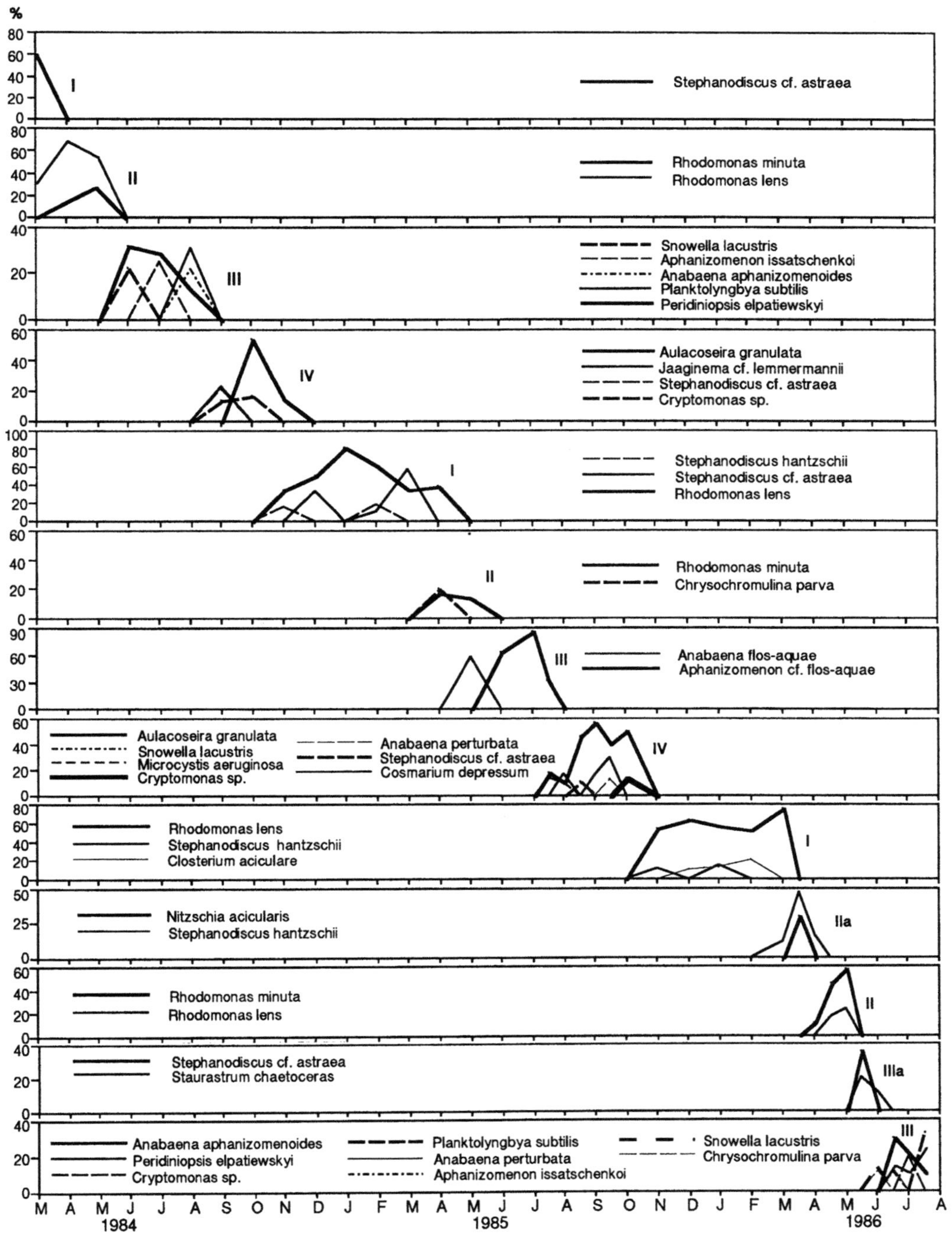

Fig. 2. Succession of phytoplankton species in Lake Volvi, as percentage contribution to biomass in the euphotic zone.

see Padisák *et al.*, 1990). An unusual significant contribution of non-planktic species in the euphotic zone was noticed (Moustaka-Gouni, 1988a). This particular *Nitzschia -Stephanodiscus* phase (IIa) might be separated in 1986.

Stage III. Blue-greens stage (June–August)
During the stratification period of 1984, with N:P > 5 and phosphorus at very low levels, phytoplankton was characterized by the dominance of the cyanophytes: *Snowella lacustris* Kom. &

Hind., *Anabaena aphanizomenoides* Forti, *Aphanizomenon issatschenkoi* (Usák.) Prosk.-Lavr., *Planktolyngbya subtilis* (West) Anagn. & Kom. and the dinophyte *Peridiniopsis elpatiewskyi* (Ostenf.) Bourr. The same species dominated during the period late June–mid August 1986, when similar nutrient conditions (N:P > 5) prevailed. Associated species were the cyanophyte *Anabaena perturbata* Hill and the flagellates *Chrysochromulina parva* and *Cryptomonas sp.*. This stage in 1986 appeared later since heavy rains accompanied by turbulence prevented the blue-greens dominance (e.g. see Padisák *et al.*, 1990).

Storms caused more turbid water, richer in nutrients (especially NO_3N), conditions which probably stimulated the increase of *S.* cf. *astraea* and *Staurastrum chaetoceras* (Schröd.) G. M. Smith, species characteristic of well mixed water columns (Reynolds, 1982). This particular *Staurastrum-Stephanodiscus* phase (IIIa) might be separated in 1986.

In the very dry year, 1985, the blue-greens stage appeared earlier (in late May), since severe nitrogen deficiency and availability of phosphorus led to the dominance of the nitrogen-fixing cyanophyte *Anabaena flos-aquae*. In June, *Aphanizomenon* cf. *flos-aquae* Ralfs ex Born. & Flah. became dominant and held its dominant position in the phytoplankton until mid-August.

Stage IV. *Aulacoseira* stage (September–October) The physical conditions altered between August and September with cooling of the water and stratification breakdown. In September, nutrients were still in low levels and only in October phosphorus and ammonia nitrogen increased significantly. Algae which are particularly tolerant of high frequency disturbances in optically-deep mixed layers such as *Oscillatoria*, *Aulacoseira* and *Cryptomonas* dominated (see Reynolds, 1987).

In 1984, *Jaaginema* cf. *lemmermannii* (Wolosz.) Anagn. & Kom. built the first maximum followed by *Aulacoseira granulata* (Ehr.) Sim. To the same stage belonged the small-celled algae *Cryptomonas* sp. and *Stephanodiscus* cf. *astraea*.

In 1985, no clear separation between the blue-green and *Aulacoseira* stages was observed. The deepening of the epilimnion and the increase of N:P ratio (due to the rise of ammonia nitrogen) favoured the dominance of *A. granulata* and *Cosmarium depressum* Näg. in August. *A. granulata* held its dominance until silica became depleted (end of October). The cyanophytes *Microcystis aeruginosa* (Kütz.) Kütz. and *Anabaena perturbata* formed blooms in calm days of October. The small celled species *Cosmarium depressum*, *Stephanodiscus* cf. *astraea*, *Snowella lacustris* and *Cryptomonas* sp. were co-dominating during the *Aulacoseira* stage.

This successional stage was associated with the autumn phytoplankton biomass maximum.

Nanoplankton contribution

Nanoplankton showed maximal development in winter and early spring (Fig. 3a), when it constituted nearly the total phytoplankton biomass (2.71 from 2.72 g m^{-3} in 1984; 3.71 from 3.72 g m^{-3} in 1985 and 2.5 from 2.6 g m^{-3} in 1986).

The seasonal contribution of nanoplankton to the total phytoplankton followed similar trends in all three years (Fig. 3b). Especially, a tendency to lower values in stages III and IV is shown. The minimum value (15.7%) in 1985 differing from the higher percentages in 1984 and 1986 (40.7 and 38.2% respectively) is correlated with the *Aphanizomenon* cf. *flos-aquae* water bloom.

In Fig. 3b it is also shown that in stage II the nanoplankton percentages differ from year to year. The lower contribution in 1985 (53%) coincided with the minimum values of N:P ratio (< 5) which led to the development of the N_2-fixing cyanophytes. Nanoplankton contribution to the phytoplankton was greater in 1984 and 1986 (95 and 68.3%, respectively) when phosphorus may have been the limiting nutrient.

In summary, the nanoplankton dominance in Volvi was observed: i) during the cold period of the year with nutrients at their maximum (stage I) and ii) during the warming of the water with nutrients at their minimum, especially with phosphorus in nearly undetectable concentrations (stage II).

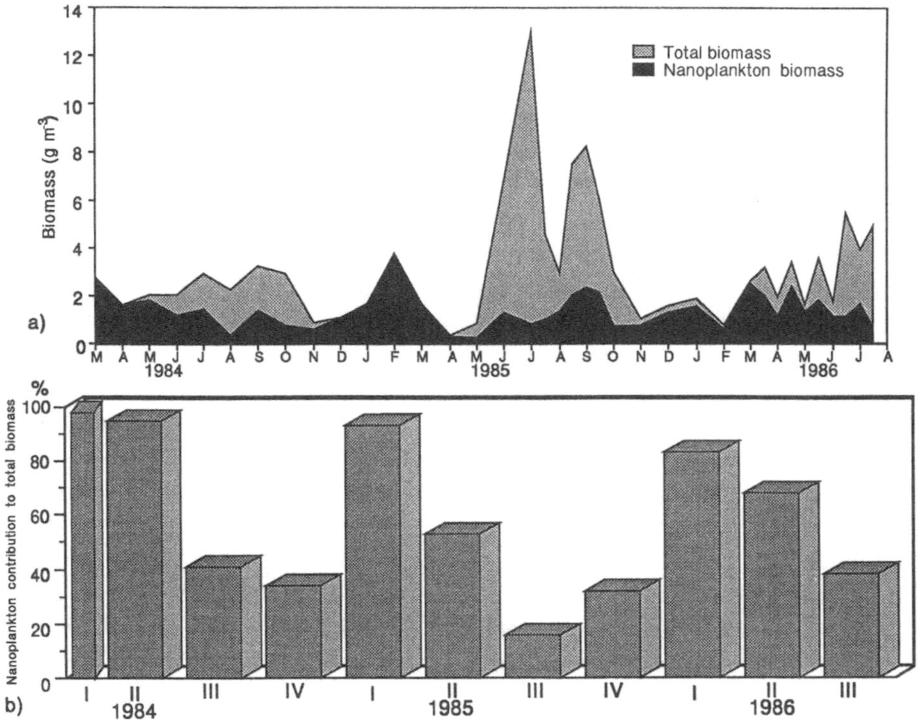

Fig. 3. a) Monthly fluctuations in nanoplankton and total phytoplankton biomass and b) successional cycles of nanoplankton contribution to the total phytoplankton biomass.

Diversity

Diversity indices d and H′ (based on unit and biomass), as well, the dominance index D, varied rather irregularly throughout the year and the whole study period but they were significantly correlated (Fig. 4). The highest diversity ($d_u = 2.92$, $d_b = 3.25$; $H'_u = 3.65$, $H'_b = 3.85$) was observed in late August 1985 and coincided with the lowest values of dominance ($D_u = 0.11$, $D_b = 0.1$). This may be related to the complexity of the community (species number at maximum, high evenness, 4 co-dominant species). The lowest values of index d ($d_u = 0.93$, $d_b = 1.19$) recorded in May 1984 may have resulted from the lowest number of species. Function H′, displayed minimum ($H'_u = 0.61$ in May 1986, $H'_b = 0.94$ in July 1985) when phytoplankton had strong dominant species (low evenness). Linear relationships between indices H'_u, H'_b and number of species, as well as, evenness showed that diversity in Lake Volvi is more affected by evenness ($r = 0.94$ and $r = 0.96$ respectively, $n = 37$) than by species number ($r = 0.59$ and $r = 0.52$ respectively, $n = 37$).

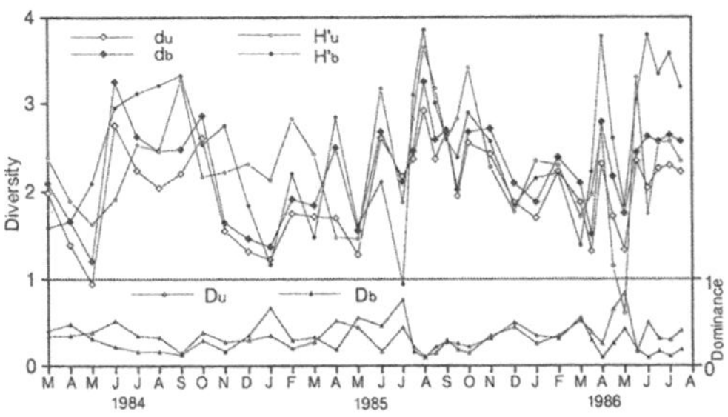

Fig. 4. Variations of phytoplankton diversity (indices d_u, d_b and H'_u, H'_b) and the dominance index (D_u, D_b) during the study period.

Both components of species diversity increased in all cases of physical disturbances (August 1985, April and June 1986). These data for Lake Volvi fit well to the 'Intermediate Disturbance Hypothesis' (Connel, 1978), which claims that disturbances have a positive influence on the diversity of community, as long as they are intermediate in frequency and intensity (e.g. see Gaedeke & Sommer, 1986). However, due to biweekly sampling frequency it was not possible to determine the time scale of these disturbances. The highest increase rates of diversity (H_b') resulting in peak values (> 3.5) were observed (Fig. 4) in the periods with external disturbances. The peaks appeared in the transition periods between the compositional changes generated by disturbances and the true successional development.

Function H_b' had a significant negative correlation ($r = -0.42$; $p < 0.001$; $n = 227$) with phytoplankton biomass. A tendency towards increased diversity at lower biomass levels was observed. The pattern of vertical distribution of diversity (Fig. 5) generally reflects the inverse relationship between these variables. Diversity was distributed fairly uniformly with depth throughout the water column, during periods with low (< 2 g m^{-3} f.w.), evenly distributed biomass. Vertical differences in diversity were very marked during periods of water blooms (October 1985, May 1986). Despite the uneven distribution of biomass, diversity was homogenously distributed throughout the water column, as evenness, when the physical disturbances were identified. From these findings, it appears that physical disturbances may generate a more uniform distribution of total biomass among the coexisting species throughout the whole water column.

Since time scales involved in species diversity analysis, are very significant (Legendre, 1973) variations of indices were analysed as a whole over each stage of succession, that means on a time scale compatible with that of fluctuation in the environmental diversity.

Both, unit (d_u, H_u') and biomass (d_b, H_b') based indices showed similar patterns and fairly stable trends in 1984, 1985 and 1986 (Fig. 6a). Diversity increased in late stages of succession and displayed minimum in the first (for H_b', d_b) or in the second (for H_u', d_u). Each year, the decrease of function H_b' in the first stage coincided with the late autumn cooling of the water and reflected the climatic impact on phytoplankton succession (e.g. see Sommer, 1991; Padisák, 1993). The early succession stages (I, II), which showed the lowest diversity generally due to low number of species, were associated with lower environmental stability and higher productivity (e.g. see Margalef, 1968; Legendre, 1973).

The number of niches increased with succession (stages III, IV) thus the diversity was expected to increase corresponding to the complexity of the community (e.g. Margalef, 1960). However, in stage III of 1985, diversity was low despite the high number of species. In this stage, severe nitrogen deficiency led to the dominance of

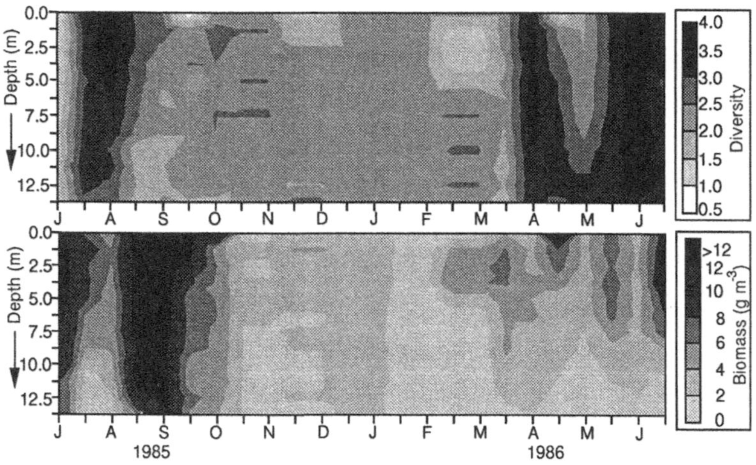

Fig. 5. Depth-time distribution of phytoplankton diversity (H_b') and total biomass throughout the year when the disturbances were recognized.

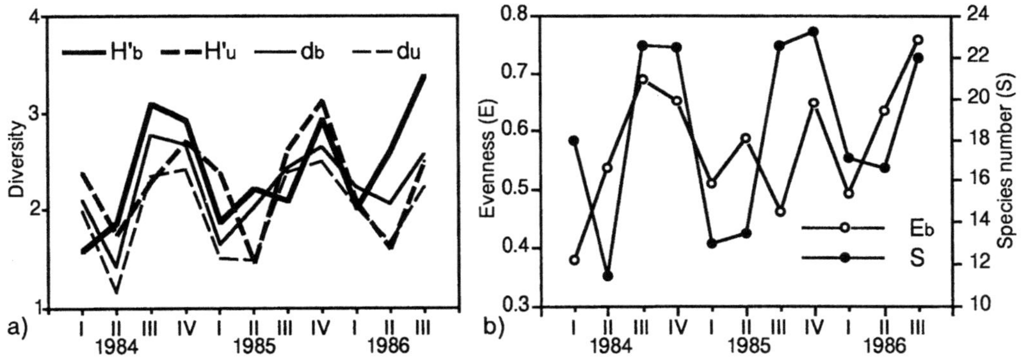

Fig. 6. Successional cycles of a) diversity indices du, H_u' (based on units) and d1Zb, H_b' (based on biomass) and b) species number and evenness.

Aphanizomenon cf. *flos-aquae* for several weeks. By the time the highest phytoplankton biomass (13 g m^{-3}) had developed, the contribution of this alga exceeded 85% (see Fig. 2), so only this species was equilibrating. If the number of co-equlibrating species can be used as a complexity measure (see Padisák, *loc. cit.*) it is clear that diversity and complexity are closely related in Volvi.

As diversity is shaped by the number of species and their evenness, these components were analysed along the time axis as diversity (Fig. 6b). Both components showed fairly stable trends in the studied years, increasing late in succession. The decrease of evenness in stage III of 1985 reflects the strong dominance of *Aphanizomenon*.

Comparing the diversity-evenness-species number of the same stages under different meteorological conditions (rainfall, wind), which prevailed in the study years, the climatic impact on phytoplankton diversity can be seen. Higher diversity, evenness and number of species were found in 1986 (stages II and III), when heavy rains fell and strong winds blew. These results may demonstrate the role of physical disturbances in increasing the phytoplankton diversity.

Conclusions

In Volvi, the phytoplankton successional cycles displayed many features of the sequences observed in other lakes (Reynolds, 1984; Sommer *et al.*, 1986). The succession may be summarized as running from r-selected species in early stages (I, II) to K-strategists in summer (stage III) and towards a mixed community in the terminal stage (IV).

The successional pattern was generally similar from year to year. However, stages were not always dominated by the same species due to widely varied meteorological conditions which largely affected variables important to the phytoplankton. In addition, external disturbances such as enhanced wind mixing and high floods had the effect of setting the succession back to an earlier stage (see Reynolds, 1987,1988).

The early stages (I, II) were characterized by nanoplankton dominance. Its lower contribution in stage II of 1985 coincided with very low values of N:P ratio which led to the N_2-fixing cyanophytes.

Diversity indices showed, in general, similar patterns and fairly stable trends in 1984, 1985 and 1986. Diversity increased in late stages of succession (higher number of niches) corresponding to the community complexity.

The highest increase rates of diversity resulting in peak values were observed in the periods with external disturbances. Diversity was distributed uniformly throughout the water column during periods of low, evenly distributed biomass and on occasions when the physical disturbances were recognized.

Higher diversity, evenness and number of species were found in 1986 (stages II and III), when heavy rains fell and strong winds blew.

Acknowledgements

I am very grateful to Dr. U. Pollingher, Israel Oceanographic and Limnological Research Ltd., for her valuable comments on the manuscript and her constant interest in this work. I'm thankful, as well, to Mrs A. Kiratzidou-Dimopoulou for assisting in the preparation of the text in the English language.

References

Connell, J., 1978. Diversity in tropical rain forests and coral reefs. Science 199: 1304–1310.

Gaedeke, A. & U. Sommer, 1986. The influence of the frequency of periodic disturbances on the maintenance of phytoplankton diversity. Oecologia 71: 98–102.

Legendre, L., 1973. Phytoplankton organization in Baie des Chaleurs (Gulf of St Lawrence). J. Ecol. 61: 135–149.

Margalef, R., 1960. Temporal succession and spatial heterogeneity in phytoplankton. In A. A. Buzzati-Traverso (ed.), Perspectives in marine biology. Univ. Calif. Press, Berkeley, Los Angeles: 323–349.

Margalef, R., 1968. Perspectives in ecological theory. Univ. of Chicago Press, Chicago.

Moustaka-Gouni, M., 1988a. Seasonal variations, annual periodicity and spatial distribution of phytoplankton in Lake Volvi. Doctoral dissertation. Univ. of Thessaloniki.

Moustaka-Gouni, M., 1988b. The structure and dynamics of the phytoplankton assemblages in lake Volvi, Greece. I. Phytoplankton composition and abundance during the period March 1984 - March 1985. Arch. Hydrobiol. 112: 251–264.

Odum, E., 1980. Ecology. Hold Rinehart & Winston, London.

Padisák, J. 1993. The influence of different disturbance frequencies on the species richness, diversity and equitability of phytoplankton in shallow lakes. In J. Padisák, C. S. Reynolds & U. Sommer (eds), Intermediate Disturbance Hypothesis in Phytoplankton Ecology. Developments in Hydrobiology 81. Kluwer Academic Publishers, Dordrecht: 135–156. Reprinted from Hydrobiologia 249.

Padisák, J., L. G. Tóth & M. Rajczy, 1990. Stir-up effect of wind on a more-or-less stratified shallow lake phytoplankton community, Lake Balaton, Hungary. In P. Biró & J. F. Talling (eds), Tropic Relationships in Inland Waters. Developments in Hydrobiology 53. Kluwer Academic Publishers, Dordrecht: 249–254. Reprinted from Hydrobiologia 191.

Reynolds, C. S., 1982. Phytoplankton periodicity: its motivation, mechanisms and manipulation. Rep. Freshwat. Biol. Ass. 50: 60–75.

Reynolds, C. S, 1984. Phytoplankton periodicity: the interactions of form, function and environmental variability. Freshwat. Biol. 14: 111–142.

Reynolds, C. S., 1987. The response of phytoplankton communities to changing lake environments. Schweiz. Z. Hydrol. 49: 220–236.

Reynolds, C. S., 1988. The concept of biological succession applied to seasonal periodicity of phytoplankton. Verh. int. Ver. Limnol. 23: 683–691.

Sommer, U., 1981. The role of r- and K-selection in the succession of phytoplankton in lake Constance. Acta Oecol., Oecol. gen. 2: 327–342.

Sommer, U., 1991. Phytoplankton: directional succession and forced cycles. In H. Remmert (ed), The Mosaic-Cycle Concept of Ecosystems. Springer Verlag, Berlin: 132–146.

Sommer, U., M. Z. Gliwicz, W. Lampert & A. Duncan, 1986. The PEG – model of seasonal succession of planktonic events in fresh waters. Arch. Hydrobiol. 106: 433–471.

Utermöhl, H., 1958. Zur Vervollkommnung der quantitativen Phytoplankton Methodik. Mitt. int. Ver. Limnol. 9: 1–38.

Hydrobiologia **249**: 43–57, 1993.
J. Padisák, C.S. Reynolds & U. Sommer (eds), Intermediate Disturbance Hypothesis in Phytoplankton Ecology.

Hypertrophic phytoplankton and the Intermediate Disturbance Hypothesis

C. Rojo[1] & M. Alvarez Cobelas[2]
[1]*Dept. Microbiología, Fac. Biología, Univ. Valencia, E-46100, Burjasot, Valencia, Spain;* [2]*Centro de Investigaciones del Agua (CSIC), La Poveda, Arganda del Rey, E-28500 Madrid, Spain*

Key words: disturbance, diversity, equilibrium, phytoplankton, succession rate

Abstract

The provisions of Connell's Intermediate Disturbance Hypothesis (IDH) were investigated in relation to the behaviour of freshwater phytoplankton in a hypertrophic lagoon, paying special attention to the link between species-diversity and environmental disturbances. Phytoplankton diversity yielded different indices depending on the basic unit of calculations (cells, particles, phytoplankton units, biomass). Although their ranges were approximately equal, they did not covary so could not be considered mutually substitutable. For the purpose of IDH testing, biomass diversity was chosen.

Equilibrium states were considered to obtain in those periods with a very high fraction of total phytoplankton biomass, shared by no more than three phytoplankton species. Disturbances were considered as counterparts of equilibrium states. Disturbance factors were mostly abiotic, environmental features of the lake operating on the phytoplankton community at different time scales (co-occurring and with 1- and 2-weeks' lags). These scales may relate to the time required to establish phytoplankton community structure.

IDH could be suspected not to hold for the phytoplankton of this hypertrophic lake, which experienced seven near-equilibrium phases and six disturbance periods throughout the study. As a rule, equilibrium states lasted longer than disturbance periods. The expected relationships between both disturbance intensity or frequency were not shown. Furthermore, no relationship was demonstrated between diversity (and hence IDH) and the phytoplankton community change rate. Wind stress probably played a minor role in triggering disturbance events. Disturbances were shown partly to promote small-sized phytoplankton communities.

Finally, a plea for studying hypertrophic phytoplankton in greater detail is stressed if its responses to disturbances are to be fully understood.

Introduction

In this study, our aim was to test the Intermediate Disturbance Hypothesis (IDH hereafter), using a data set gathered in a study on the phytoplankton of a hypertrophic, freshwater lagoon (El Porcal, Madrid). The IDH explicitly states that 'Diversity is higher when disturbances are intermediate on the scales of frequency and intensity' (Connell, 1978: 1303). For the sake of brevity and since an excellent review on Hutchinson's plankton paradox (Hutchinson, 1960) and its reduction, the IDH, is included in this volume (Sommer, *et al.* 1993), we shall not deal with the details of the theory here.

Besides the IDH we investigate also some related ideas, including whether diversity is increased if community change occurs rapidly and whether disturbances result in smaller cell sizes than occur at equilibrium states.

Study site

The study was carried out in El Porcal lake, which is located in the Jarama river watershed, close to Madrid (Spain). The lake is a gravel-pit, small area (45 ha) and shallow (maximal depth: 9 m; average depth: 3 m) and lies very close to the river course (20 m distant). No surface inflow or outflow exists, water level being maintained by groundwater seepage. Other features of the lake are reported in Alvarez Cobelas (1991) and Alvarez Cobelas *et al.* (1986, 1987, 1990): the lake is highly hypertrophic, since the nearby river is very heavily polluted. This results in high nutrient and chlorophyll *a* content with shallow euphotic depths (Table 1).

Materials and methods

Field sampling

Sampling was undertaken weekly with a peristaltic pump from May 1984 to June 1985 at a 6 m deep station of the lake. Samples for phytoplankton identification and counting were collected from several depths throughout the euphotic column and were fixed shortly after collection with Lugol's Iodine. Also some live samples were collected for '*in vivo*' identification.

Secchi-disk measurements, vertical PAR attenuation (LI-COR instrument and scalar sensor) and water-temperature profiles (YSI instrument) were made on each sampling date. Water samples for chemical analyses were collected from the same depths of phytoplankton sampling, such analyses commencing within half an hour after field sampling. Dissolved-nutrient analyses were performed following APHA (1985) procedures.

Average daily wind speeds were measured at a meteorological station, 3 kms distant.

Table 1. El Porcal lake limnological features

Site:	40 °N 3 °W
Water inflow:	Groundwater seepage
PAR attenuation coefficient:	1.4–7.2 m^{-1}
Secchi depth:	0.4–1.0 m
Dissolved inorganic carbon:	6.0–45.0 mg l^{-1}
Dissolved inorganic nitrogen:	0.1–6.0 mg l^{-1}
Soluble reactive phosphorus:	undetect. – 3.30 mg l^{-1}
Soluble reactive silicon:	0.5–4.2 mg l^{-1}
Chlorophyll *a*:	20–860 mg m^{-3}
Areal gross primary productivity:	0.06–0.60 g C m^{-2} h^{-1}
Dominant phytoplankton groups: (biomass)	Crypto-, Chlorophytes Cyanobacteria

Preliminary calculations

The Brunt-Väisäla frequency (to be used as a measure of water column stability) was calculated from the temperature profiles following the method of Mortimer (1974). Concentration of dissolved nutrient and their atomic ratios were calculated. Euphotic depths were calculated on the assumptions that the Lambert-Beer extinction law held true and that it comprised the layers down to 1% underwater PAR irradiance (Kirk, 1983). Mixing depths were calculated after temperature data (uncorrected for solids since these were around 1 g l^{-1} throughout), assuming a surface mixed layer to be well mixed when its density gradient was lower than 6 10^{-5} kg l^{-1} m^{-1}. Based upon both euphotic and mixing depths, a Z_{eu}:Z_m ratio was also calculated. Finally, average values of nutrient contents for surface mixed layers were calculated.

Phytoplankton identification and counting

Identifications were performed on live material using LM techniques. Some controversial species (i.e., *Nitzschia* spp.) were also examined with SEM. Both microscope procedures and taxonomy are reported in Rojo (1990).

Counting, biovolume and biomass estimations were carried out with a Zeiss inverted microscope following the recommendations of Rott (1981). At least 100 individuals of most abundant phytoplankton species were counted in any sample whereas at least 20 specimens of every species were measured for biovolume calculations on any sampling date. Picoplankton (i.e., rod- or sphere-shaped cells without any differential inner feature and sizes above 1 μm) was also counted and measured; recent comparisons between this method and a fluorescence-microscopy method (Weisse, 1988) on samples of the same lake supported the hypothesis that these cells are autotrophic and their numbers are not statistically different from those counted with the inverted-microscope method (unpublished data).

Also counting of compounded aliquot samples was carried out with a COULTER counter. Details of this method are fully reported in Rojo (1990).

Diversity calculations

As diversity may be considered to be a function based on discrete values, according to Information Theory (Shannon & Weaver, 1963), the data set for calculating it should comprise discrete variables (individual units, cells, etc.). For the purpose of Competitive Exclusion Theory (Hardin, 1960), diversity must be calculated on phytoplankton biomass since the Competitive Exclusion Theory and the IDH are closely related (the latter prevents the former); therefore, it is recommended to use biomass data. However, a common criticism of this use of biomass as a basis for diversity calculations is that biomass is a continuous variable and this should preclude such use. In fact, this might be true if total biomass was estimated as, for example, chlorophyll *a*. Since the usual calculation of phytoplankton biomass is based upon volume-weighted cell numbers, we are truly dealing with a discrete variable.

Here we have used the well known Shannon-Wiener formula but the data sets to which it has been applied are quite distinct. The reason for this is to find out whether diversity indices based upon different features of a given phytoplankton community are substitutable and hence, for example, biomass-based diversity, which is seldom measured in phytoplankton studies, can be compared with phytoplankton unit-based diversity, which is more common in phytoplankton studies (Moss, 1973; Elber & Schanz, 1989). Since picoplankton is not usually included in diversity calculations, we have also considered this approach for comparing purposes with more common measures of diversity of the same phytoplankton community. Furthermore, we have tried to find out whether water volume-based calculations (i.e., averages for the euphotic column) yield the same results as areal-based calculations (integrated values). Lastly, our aim has also been to find out whether a less frequent counting method (Coulter Counter) could be related to diversity values obtained by means of the inverted microscope.

Thus, in the volume-based calculations the p_i values in Shannon & Wiener formula are comprised by the ratios of averaged numbers (cells, units, particles, biomass) of a given taxon to total averaged numbers for the euphotic column. In the integrated diversity indices the p_i values come from the ratios of areal numbers of a given taxon to total areal numbers for the euphotic column. The data sets have been the following:

1) particles (averaged),
2) particle biomass (averaged),
3) particle size classes (averaged),
4) microseston particles (particles above 20 μm, Dussart, 1965) (averaged),
5) microseston biomass (averaged),
6) cell counts with picoplankton (integrated for the euphotic column),
7) cell counts without picoplankton (integrated),
8) cell counts with picoplankton (averaged for the euphotic column),
9) phytoplankton units (either cells, trichomes, colonies, etc.) with picoplankton (averaged),
10) biomass with picoplankton (integrated),
11) biomass without picoplankton (integrated),
12) biomass with picoplankton (averaged),
13) phytoplankton size classes (averaged), these classes being the same as in 3).

The rationale for the size classes is given in Rojo (1990).

Identifying disturbances

We assume that a disturbance occurs external to the phytoplankton community and so it must be sought in either the abiotic and biotic environment of phytoplankton. Of course, this opens a broad field to be searched but, as we show below, redundancy is often found. A likely disturbance could be a sudden change in any of the environmental variables relevant to phytoplankton dynamics. We assume a likely disturbance to occur when a given environmental factor suddenly increases or decreases by two Standard Deviations from its average ambient value hitherto on the preceding dates following the previous likely disturbance. This poses the problem of how to ascertain the boundary conditions of 'stability' of such a factor on the earlier dates. We have assumed that early values of a given environmental factor are stable if they do not change more than 1 SD about the mean value for the whole study period. An example of this method can be seen in Fig. 1. We have discarded nutrient depletions as disturbances since they are most likely to occur as a result of nutrient competition by phytoplankton and therefore on the way to equilibrium states.

These methods are not entirely satisfactory since they may exaggerate the number of significant disturbances. So disturbance periods are to be defined as intervals between equilibrium periods; therefore, disturbance periods are considered here as non-equilibrium periods.

Disturbances do have at least two important measures, i.e. Intensity and Frequency. Disturbance Intensity is measured as

$$\left| I = \frac{\Delta v}{\Delta t} \right|,$$

where Δv is the change in a given environmental variable acting as a likely disturbance factor during a given time period (Δt). In order to compare

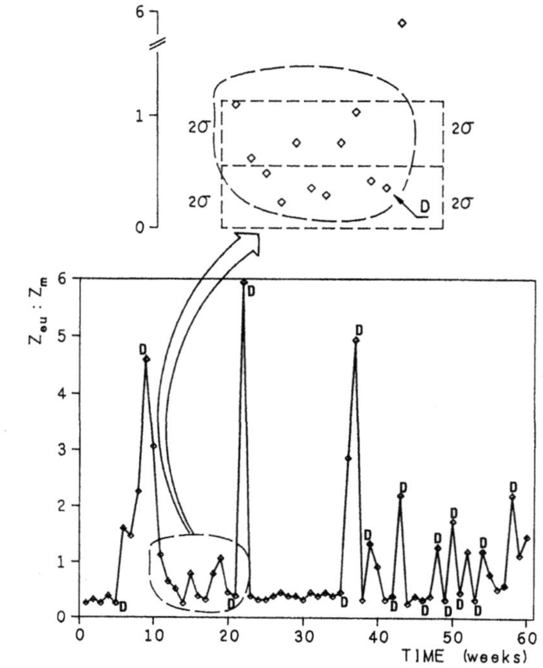

Fig. 1. How to detect a likely disturbance event?. 'D' letters indicate likely disturbance events after the criterion of 2 SD outside has been met (see Materials and methods).

intensities of different environmental factors, relative measurements (%) of intensity are chosen.

Disturbance Frequency is calculated as the number of weeks existing between two likely disturbance events in a given environmental factor, this figure being compared to the number of weeks in the year (52), to render the frequency in the year (unit: cycles y^{-1}) for any likely disturbance event.

Equilibrium states

An equilibrium state is approached as competitive exclusion operates in favour of the dominant species. Disturbances usually prevent competitive exclusion from occurring (Sommer, 1989). So equilibrium states can be suggested when biomass is partitioned among but a few species that are co-dominant. As Competitive Exclusion Theory predicts (Hardin, 1960), at the equilibrium only one species per resource could dominate. So, providing that only 3 or 4 resources really matter to phytoplankton, the outcome of competitive exclusion should be three or four biomass-dominating species. We have identified equilibrium states when at least 80% phytoplankton biomass is shared by no more than three phytoplankton taxa.

Other phytoplankton community features

Diversity does not entirely reflect community changes since a given diversity value could be caused by the same number of species but different species sets. Also changes in biomass fractions of given species could be counterbalanced by opposite changes in other species, thereby rendering the same biomass diversity. This is the main pitfall of Lewis' (1978) approach to a succession rate index and Reynolds' (1980) derivation. Another succession rate index, such as that by Jassby & Goldman (1974), has been already criticized on the grounds of lacking directionality of the succession process and unequal weighting of species (Lewis, 1978). Lastly, the succession rate index by Armstrong (in Williams & Goldman, 1975) relies heavily on diversity concepts, then suffering the same drawbacks already discussed above. Using any of the preceding indices, a true change could have occurred but they cannot take account of it. Instead, a true biomass community change could be calculated by comparing biomass similarity in every two consecutive phytoplankton communities. The more similar they are, the less they have changed. The approach is close to that of Elber & Schanz (1990) but they have used phytoplankton individuals instead of biomass as the basic unit. Thus, a community dissimilarity could be an inverse function of such similarity and if such a new function is divided by the time period occurred between both contiguous sampling dates, a true community change function can be obtained. There are many methods to get knowledge on community similarity (Legendre & Legendre, 1979). Here we have chosen that of correlation between consecutive communities, such correlation being on biomass estimates instead of cell numbers or individuals. Therefore, the independent variable would be community biomass partitioned into species in the date *i*th whereas the dependent variable would be community biomass partitioned into species in the date *i*th + 1. The inverse value of the correlation coefficient thus obtained, divided by the time elapsed between samplings is then considered to be a succession rate index (d^{-1}).

An autocorrelation function on this community change rate (undetrended) will suggest an average periodicity of community change in the time domains of statistically significantly autocorrelations.

On the other hand, we have calculated a community size-weighted index for any sampling date as follows:

$$SI = \frac{\sum_{i=1}^{n} (\bar{s}_i \cdot N_i)}{\sum_{i=1}^{n} N_i}$$

where $\bar{s}_i$ is the average Equivalent Diameter (Mayzaud *et al.*, 1984) for the *i*th species and N_i is the number of phytoplankton units (cells, filaments, colonies) counted for the *i*th species. The reason why we have used this measure of size is that it is the more frequently used measure of size in phytoplankton studies (Peters, 1983).

Autocorrelations and cross-correlations have been tried either on phytoplankton diversity or environmental variables *vs* phytoplankton biomass diversity. Since no such time series is stationary, they have usually been detrended by differencing or using ARIMA models (Chatfield, 1984).

Results

The descriptive limnology of abiotic, environmental features throughout the study is represented in Fig. 2. El Porcal lake is classified as warm monomictic, with a stratification period typically from late May to early September. The autumn overturn is usually completed within a few days. The water column stability clearly reflects the

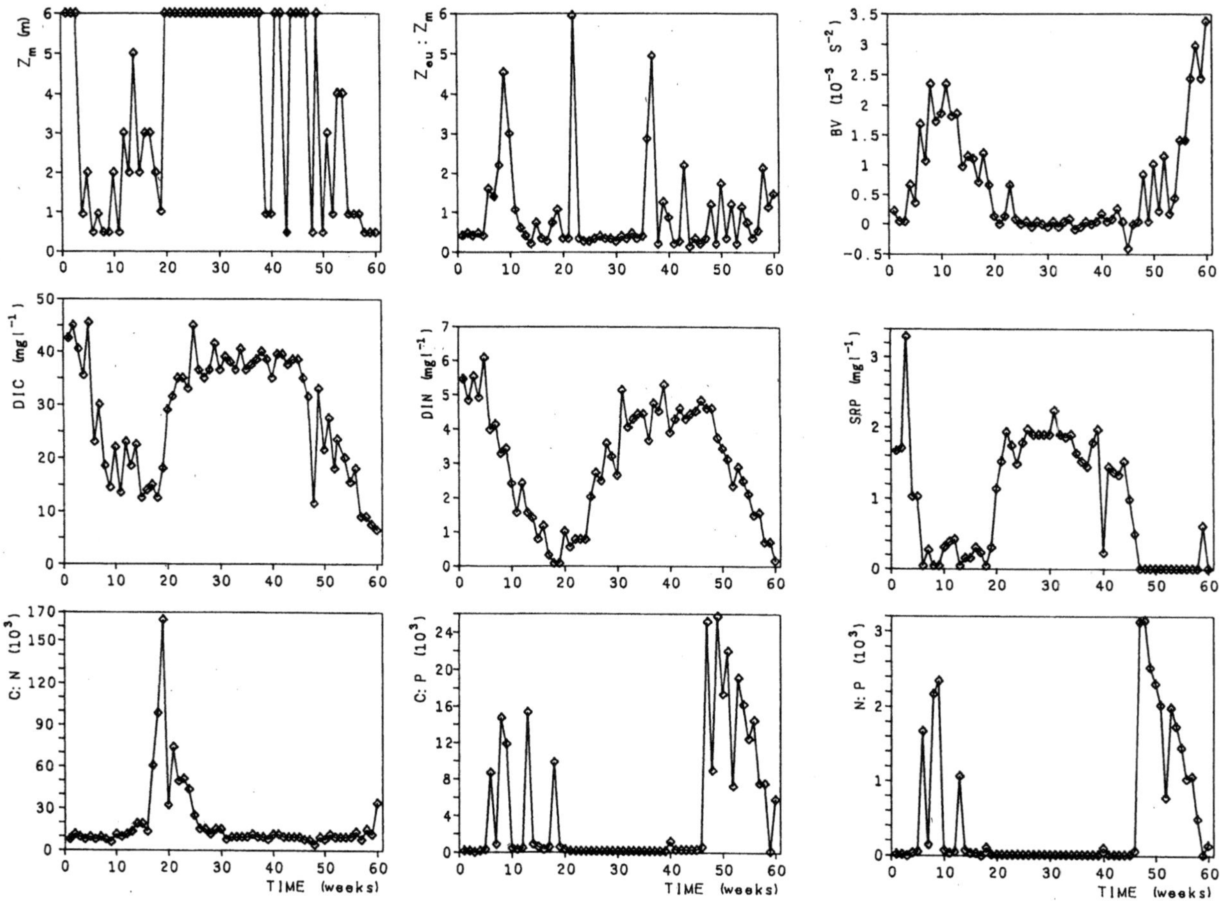

Fig. 2. Time course of environmental, abiotic factors in El Porcal lake throughout the study period. Z_m = mixing depth; Z_{eu}:Z_m = euphotic depth-to-mixing depth ratio; BV = Brunt-Väisäla frequency (a measure of vertical column stability); DIC = Dissolved Inorganic Carbon; DIN = Dissolved Inorganic Nitrogen; SRP = Soluble Reactive Phosphorus.

thermal periods of the lake but the time courses of mixing depth and euphotic depth-to-mixing depth ratio appear to display more complex patterns, with changes occurring very often. All nutrients seem to follow the same basic time course: depletion as stratification develops, replenishment after overturn and steady states for most of the mixing period; Soluble Reactive Phosphorus is depleted more rapidly than any other nutrient. As a result, phosphorus-dependent ratios exhibit the similar time course, with likely transitions from nitrogen-limited phytoplankton growth to phosphorus-limited growth. The carbon-to-nitrogen atomic ratio appears to be very constant, except at the autumn overturn, when a ten-fold increase is observed.

Concerning the time courses of phytoplankton indices (Fig. 3), i.e. the average for the euphotic column, cell numbers closely track the number of phytoplankton units. Nevertheless, pelagic particles do show a distinct time pattern of distribution with more pulses than either phytoplankton cell numbers or units. The time course of phytoplankton biomass (as biovolume) also appears to differ from those of cells or phytoplankton units, with peaks at the end of Winter, 1985, due to green algae and early stratification due to Cyanobacteria.

Phytoplankton diversity calculations show strikingly different results (Table 2) since the highest diversities are attained by microseston whereas the lowest ones are those of total seston.

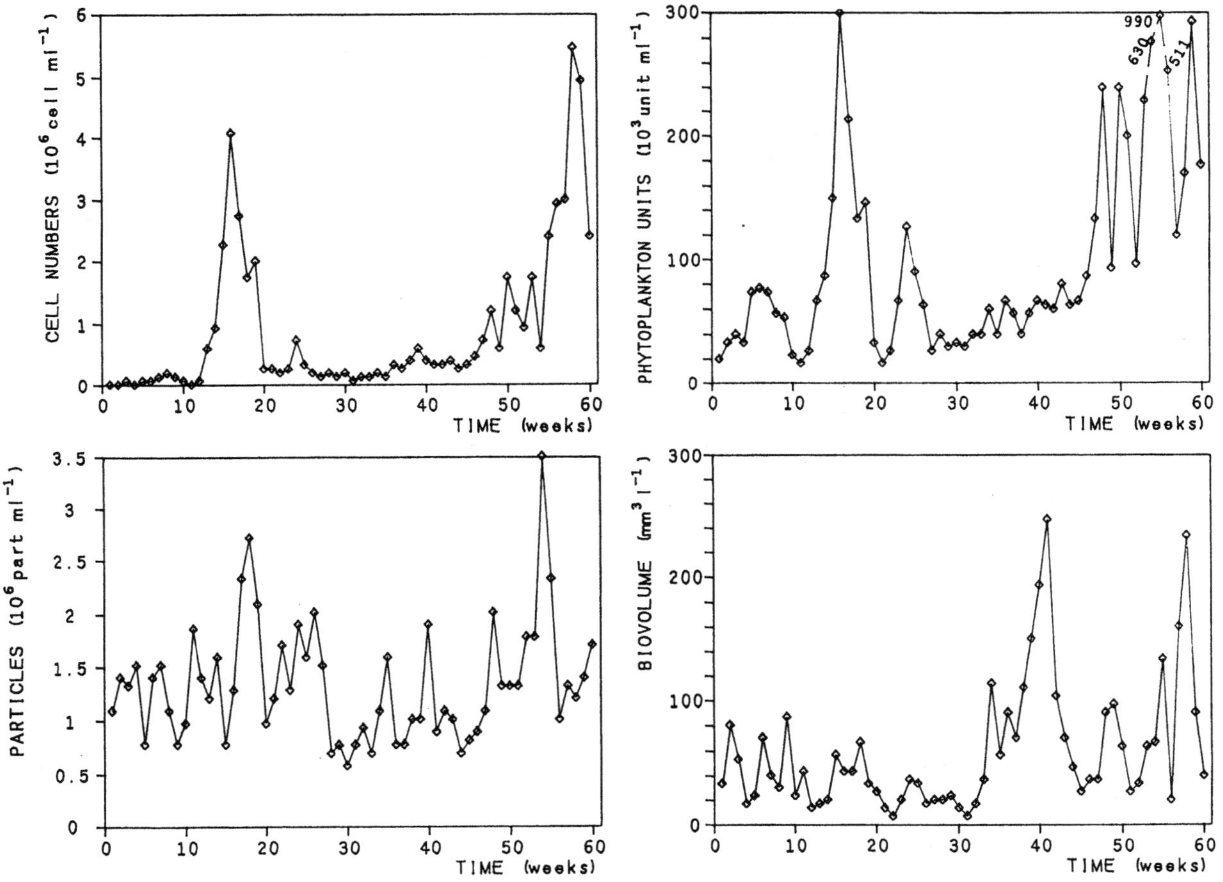

Fig. 3. Time course of El Porcal phytoplankton as judged by either cells, units, particles or biomass.

Diversities on cell numbers and biomass lie in the same range, be picoplankton included or not. Diversity on phytoplankton units appear to be higher than the latter (biomass and cell number). However, most coefficients of variation range from 36 to 50% which certainly suggest low diversity variabilities throughout the study.

When statistical correlations are carried out among all diversity types, three distinctly correlated blocks of diversity appear (Table 3), namely, diversities within the seston type, diversities within the cell type and diversities within the biomass type. Very interestingly, no statistically significantly correlations at the probability level chosen occur between these type diversities. Only diversities on phytoplankton units appear to be weakly, albeit significantly, correlated with diversities on phytoplankton cells. When picoplankton is included in the calculations, diversity is significantly correlated within type diversities, thereby suggesting that it covaries with the remainder of the phytoplankton community. Volume-based diversity appears to follow the same time courses as areal-based diversity within type diversities. Lastly, no relationship has been found between particle- and phytoplankton-based diversity, indicating that the tripton fraction in this lake occurs with time course other than that of phytoplankton.

So we have chosen one of each basis for diversity to be represented, i.e., averaged cell, unit, seston and biomass diversity (Fig. 4). It is easy to see that the two former resemble each other, with almost the same peaks and valleys, whereas seston diversity does vary less and does not show a clear-cut time course. Biomass diversity shows

Table 2. Descriptive features of Diversities. Units clearly differ, i.e., particles (bits part^{-1}), biomass (bits mm^{-3}), cells (bits cell^{-3}), size class (bits μm^{-1}), phytoplankton unit (bits unit^{-1}). SD: Standard Deviation. CV: Coefficient of variation.

	Mean	SD	CV
Average seston particle	0.69	0.25	0.36
Average seston biomass	0.79	0.28	0.36
Average seston size class	0.78	0.28	0.37
Average microseston particle	2.64	0.73	0.28
Average microseston biomass	3.57	2.59	0.72
Areal cell (with picoplankton)	1.47	0.69	0.47
Areal cell (without picoplankton)	1.38	0.67	0.49
Average cell	1.49	0.69	0.46
Average phytoplantkon unit	2.16	0.80	0.37
Areal phytoplankton biomass (with picoplankton)	1.56	0.69	0.44
Areal phytoplankton biomass (without picoplankton)	1.57	0.66	0.41
Average phytoplankton biomass	1.63	0.65	0.40
Average phytoplankton size class	1.45	0.57	0.39

very dramatic 'ups' and 'downs' in the time domains of a week and scarcely tracks those of phytoplankton cells or units. Out of the four diversities involved only cell- and unit-diversity show statistically significantly autocorrelations, occurring in the 1–2 weeks time domain.

Six disturbance events appear to materialize during the study (Table 4), with varying periods of persistence (from 1 to 7 weeks). Most of them appear to trigger biomass diversity increases. The third disturbance would correspond to the thermal overturn and could not be considered strictly as intermediate but it nevertheless induced an increase in biomass diversity. Concerning disturbance factors, we envisage them operating on El Porcal phytoplankton communities at very differ-

Table 3. Correlation matrix of diversities. * indicates correlations statistically signficant at 95% significance level. Abbreviations in brachlets are the rows' abbreviations.

	ASP	ASB	ASSC	AMP	AMB	AC (wp)	AC (np)	AC	AU	AB (wp)	AB (np)	AB	APSC
Average seston particle	1												
Average seston biomass	0.99	1											
Average seston size class	0.99	1	1										
Average microseston particle	−0.20	0.13	0.16	1									
Average microseston biomass	−0.18	0.04	0.15	0.28	1								
Areal cell (with picoplankton)	0.11	0.03	−0.15	0.08	0.13	1							
Areal cell (no picoplankton)	0.07	0.07	0.12	0.04	0.16	0.93	1						
Average cell	0.09	0.12	−0.08	0.12	0.13	0.93	0.99	1					
Average unit	0.15	0.07	0.05	0.03	0.08	0.64	0.61	0.65	1				
Areal biomass (with picoplankton)	−0.15	0.10	0.11	0.11	0.14	0.05	0.16	0.13	0.18	1			
Areal biomass (without picoplankton)	0.06	0.12	0.21	0.09	0.02	0.07	−0.09	0.14	0.16	0.96	1		
Average biomass	0.05	0.16	0.07	0.08	0.14	0.12	−0.03	0.21	0.08	0.85	0.88	1	
Average phytoplankton size class	−0.07	0.18	−0.15	0.04	0.12	0.13	0.06	0.20	0.09	0.93	0.98	0.85	1

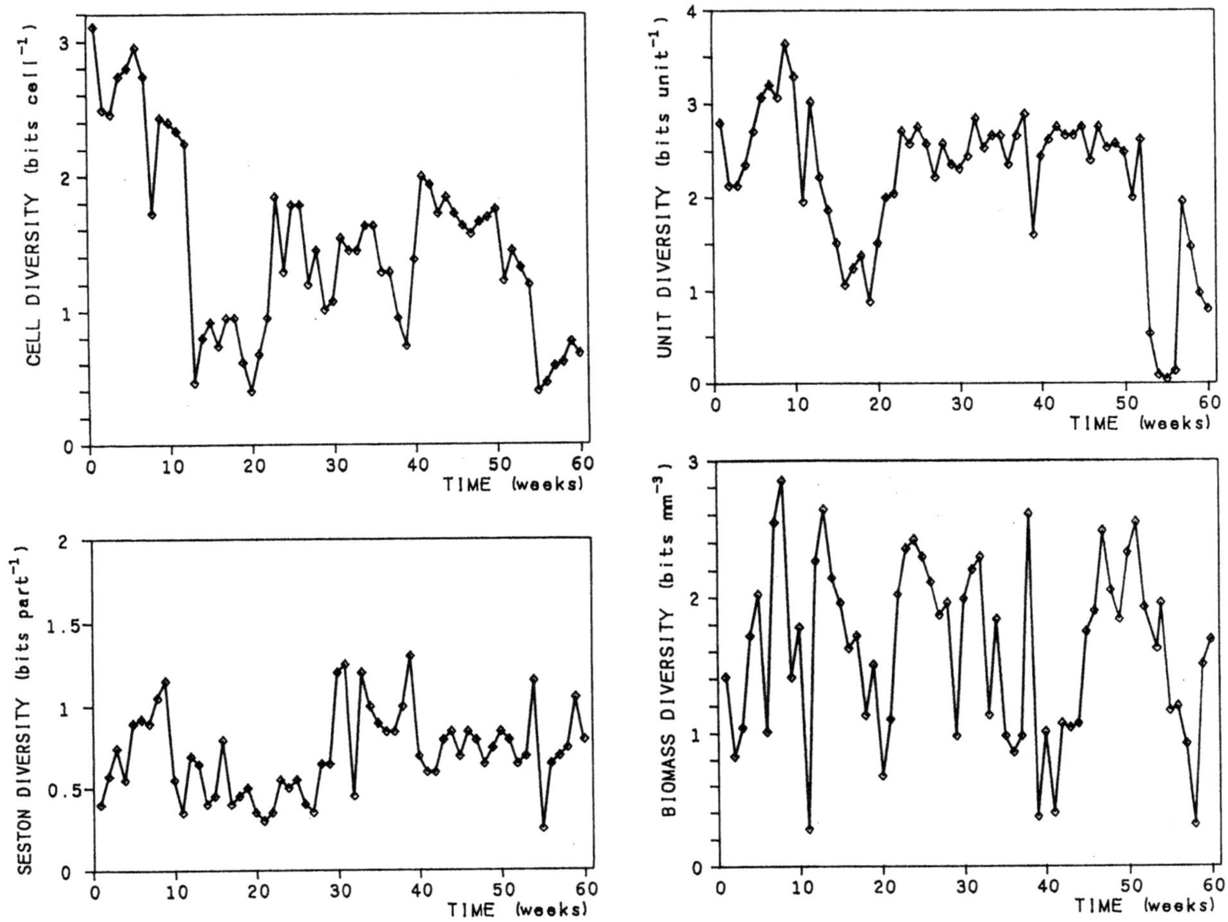

Fig. 4. Temporal plots of main averaged, phytoplankton diversities in El Porcal lake.

ent time scales since some effects are delayed up to two weeks (such as the Soluble Reactive Phosphorus increase triggering the second disturbance period or the water column stability decrease two weeks before the third disturbance period) while others can occur in parallel with the changing

Table 4. Features of disturbance periods. $\Delta H'$: biomass diversity change throughout a given disturbance period. W = week; C:P = Inorganic Carbon-to-Phosphorus ratio; N:P = Inorganic Nitrogen-to-Phosphorus ratio. The remaining symbols as in Fig. 2. Numbers in parentheses are weekly time lags. Arrows indicate whether the disturbance factor is increasing or decreasing.

Equilibrium displacement ($\Delta H'$)	Moment	Persistence (weeks)	Disturbance factors
+ 0.70	W3–W4	7	BV (0)↑
+ 2.02	W11–W12	3	SRP (2)↑, C:P (2)↓, N:P (2)↓
+ 1.92	W21–W22	2	Z_{eu}:Z_m (0)↑, z_m (2)↑, (BV (2)↓, DIC (2)↑, SRP (2)↑, C:N (2)↓
+ 0.12	W31–W32	1	DIN (1)↑
− 0.05	W32–W43	1	Z_m (0)↑
+ 0.48	W49–W50	3	DIC (1)↑, N:P (1)↓

environmental factor (such as the increase in water column stability in the first disturbance period or the mixing depth increase in the fifth period). Another problem posed by this analysis is that a given disturbance factor can operate on phytoplankton diversity with distinct time lags. An example of this could be the mixing depth increase occurring two weeks before the third disturbance and co-occurring in the fifth disturbance period.

When either disturbance intensity or frequency are plotted *vs* biomass diversity (Fig. 5), no clear patterns can be envisaged that resemble those of Connell's hypothesis (see his Figs. 1, 2B, 1978). If either intensity or frequency *vs* diversity plots are lagged one or two weeks, the same picture holds true (not shown).

The criterion used to define equilibrium states allow us to suggest seven equilibrium states, of which at least one (the sixth) shows a significant diversity increase resulting from the equal biomass codominance of two phytoplankters (*Cryptomonas* and *Pteromonas*). The persistence of equilibrium states is highly variable, from one to ten weeks, but most lie in the range of 6 to 10 weeks. The number of generations of taxa involved at the end of an equilibrium ranges from less than one to four but the mode lies in 1 generation. The species number involved in equilibria throughout the study has been six, namely, the greens *Dictyosphaerium tetrachotomum* Printz and *Pteromonas angulosa* Lemmermann, the Cryptophyte *Cryptomonas erosa* Ehrenberg, the Diatom *Cyclotella meneghiniana* Kützing and the Cyanobacteria *Oscillatoria* (*Planktothrix*) *agardhii* Gomont and *Oscillatoria* (*Limnothrix*) *redekei* van Goor.

As a summary of the foregoing results, Fig. 6 shows the seven equilibrium phases and the six disturbance periods. Disturbance periods are very variable in length, lasting from 1 to 7 weeks, and a given disturbance is not always followed by a significant diversity increase (such as in the fourth and the fifth disturbance periods). The environmental factors most likely triggering disturbances

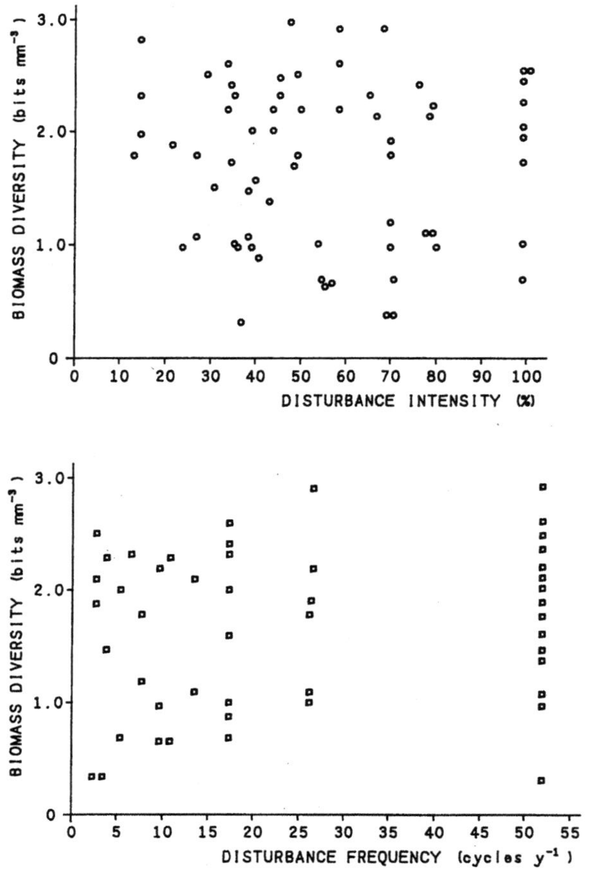

Fig. 5. Disturbance frequency (upper pannel) and disturbance intensity (lower pannel) *vs* phytoplankton biomass diversity in El Porcal lake. Both data sets (disturbance features and diversity) co-occur.

Table 5. Features of equilibrium states. Symbols as in Table 4. Diversity changes cover each entire equilibrium period. The algae shown are the biomass-dominant ones at the end of each equilibrium state.

Equilibrium ($\Delta H'$)	Start	Persistence (weeks)	Number of generations at the end of equilibrium
− 0.38	W1	2	*Cryptomonas*: 1 *Dictyosphaerium* < 1
− 1.52	W10	1	*Cryptomonas*: 4
− 1.04	W14	7	*Cryptomonas* < 1
+ 0.11	W23	8	*Cryptomonas* < *1* *Cyclotella* < 1
− 1.20	W32	10	*Dictyosphaerium*: 3
+ 0.82	W43	6	*Cryptomonas* < 1 *Pteromonas* < 1
− 0.25	W52	8	*Cryptomonas*: 1 *Oscillatoria agardhii*: 1 *O. redekei*: 3

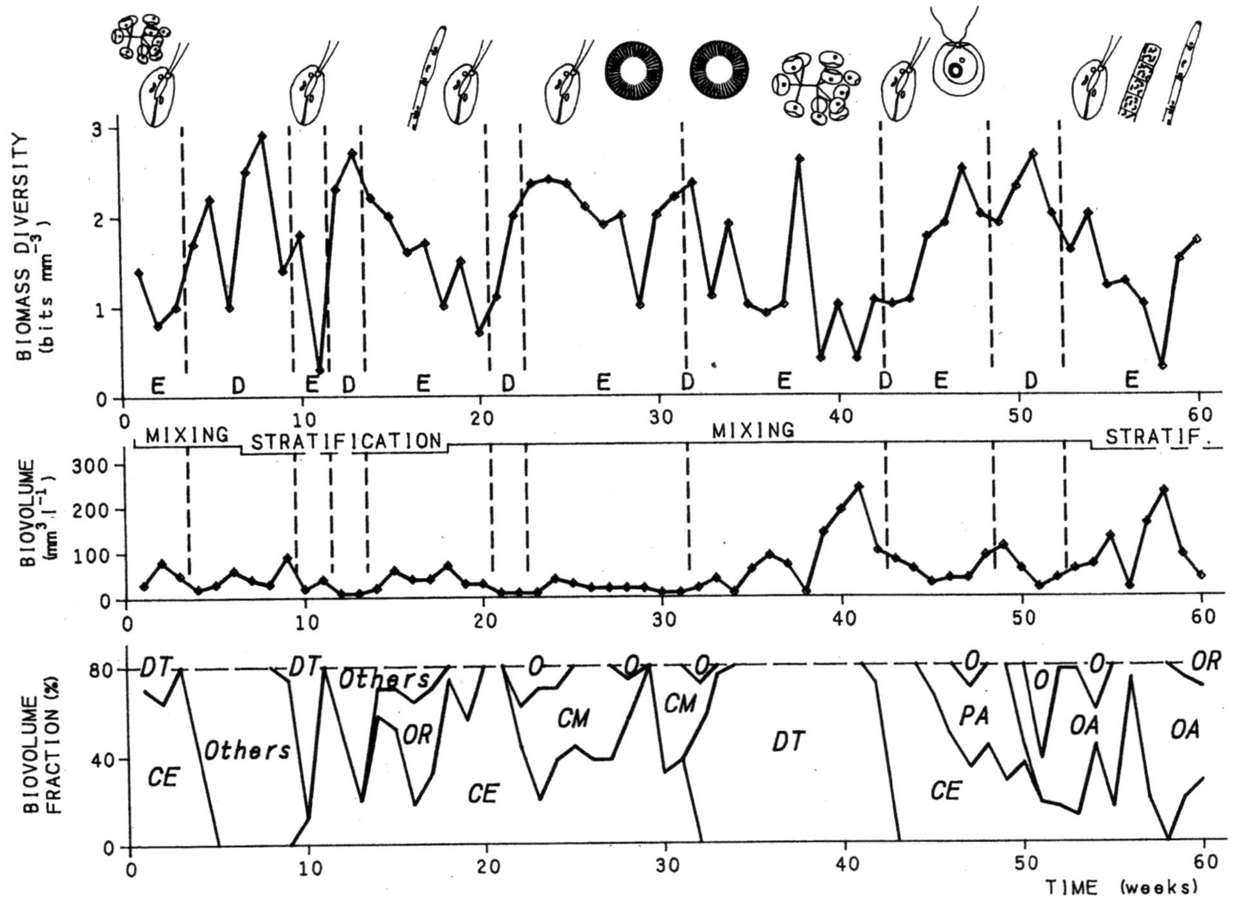

Fig. 6. Upper pannel: disturbance periods (D) and Equilibrium states (E) in El Porcal phytoplankton community throughout the study; medium pannel: time course of total phytoplankton biomass; lower pannel: phytoplankton species, fractional biomass in El Porcal lake. Notice the overwhelming dominance of a few species for most of the time. CE: *Cryptomonas erosa*; CM: *Cyclotella meneghiniana*; DT: *Dictyosphaerium tetrachotomum*, O: Other algae; OA: *Oscillatoria agardhii*; OR: *Oscillatoria redekei*; PA: *Pteromonas angulosa*.

in the phytoplankton community have been an increase of water column stability in the first period, an increase of Soluble Reactive Phosphorus content in the second, the Autumn overturn in the third, an increase of Dissolved Inorganic Nitrogen in the fourth, an increase of the mixing depth in the fifth and an increase of Dissolved Inorganic Carbon content preceding the sixth disturbance period. Most of the times the alga involved in equilibrium states has been *Cryptomonas erosa*, with the other phytoplankton species (such as *Cyclotella* and *Dictyosphaerium*) dominanting in no more than two equilibrium phases. On average, equilibrium states do last more than disturbance periods but phytoplankton biomass is not either constant or decreasing during equilibria. This is shown in the fifth and sixth periods when biomass is indeed increasing. Another conflicting feature of this analysis is a diversity augmentation in the sixth equilibrium phase resulting from the equal biomass codominance of *Cryptomonas* and *Pteromonas*.

On the whole, biomass diversities occurring during disturbance periods are higher than those in equilibrium states (t-test, $t_s = 2.59$, $p > 0.95$).

When trying cross-correlations between environmental features and biomass diversity, only Dissolved Inorganic Nitrogen is statistically and significantly cross-correlated to diversity, with 3 weeks time lag, this correlation being positive.

No relationship ($p<0.95$) has been found between biomass diversity and average wind speed on the days preceding sampling. This has been also true within any disturbance- or equilibrium period.

Community change rate, based upon biomass data, has ranged 0.14–0.21 d^{-1} and showed a decreasing trend throughout (Fig. 7). However, abrupt changes (i.e., higher than 0.05 d^{-1} $week^{-1}$) have not been recorded. Its autocorrelation function suggests 4-weeks' periodicities

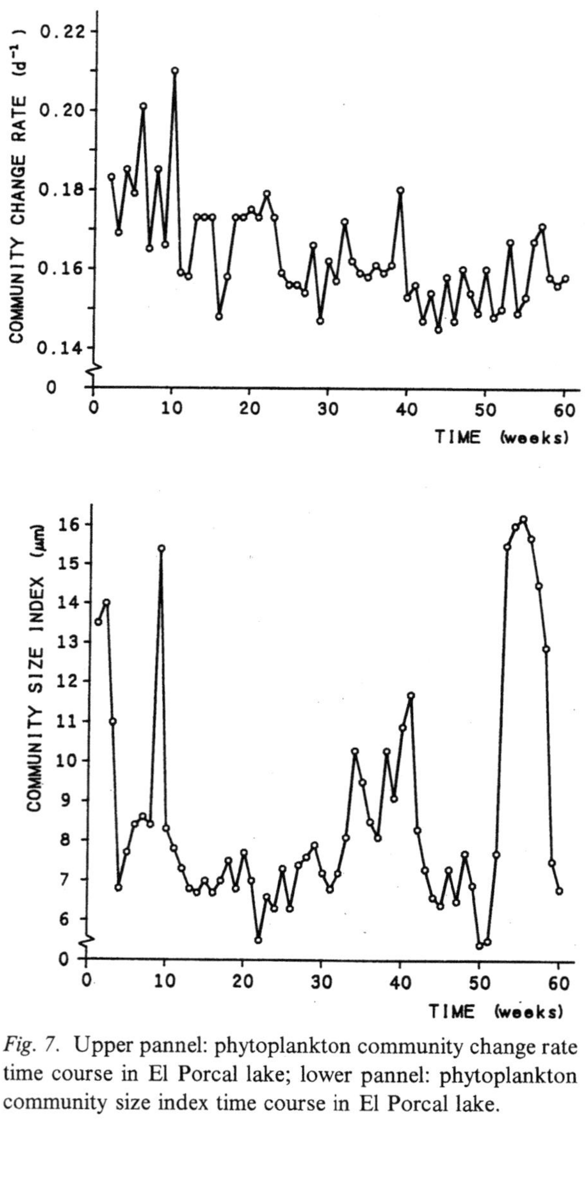

Fig. 7. Upper pannel: phytoplankton community change rate time course in El Porcal lake; lower pannel: phytoplankton community size index time course in El Porcal lake.

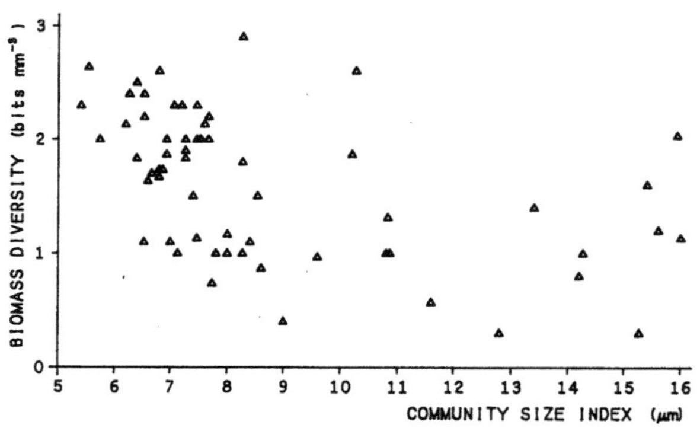

Fig. 8. Phytoplankton community size index *vs* biomass diversity plot.

which implies 15 distinct community changes during the study. Nevertheless, no statistically significantly correlation has been found between community change rate and biomass diversity ($p<0.95$).

Community size shows three distinct peaks, reflecting the dominance of some 'canopy' species, namely, *Dictyosphaerium tetrachotomum* (two former peaks) and *Oscillatoria agardhii* (the third peak, Fig. 7).

Finally, there appears to be a negative relationship between community size index and biomass diversity ($r=-0.47$, $p>0.95$, Fig. 8).

Discussion

Perhaps the main problem with Connell's IDH is that he refers species richness instead of true informational diversity (see his Fig. 2: 1305). Species richness in phytoplankton is very difficult to ascertain (i.e., Kalff & Knoechel, 1978); Bartha & Hajdu, 1979). Admittedly, diversity in freshwater phytoplankton relies heavily upon evenness, species richness having only a minor weight in the diversity value (Sager & Hasler, 1969). However, other studies intending to test IDH also use species richness as the dependent variable (Lubchenco, 1978; Sousa, 1979) instead of informational diversity. This misleading approach is also shared by Tilman (1982) in his otherwise influential book. Anyway, it should be mentioned that those studies have been carried out on sessile communities.

Perhaps other non-sessile communities, such as phytoplankton, could behave in a different manner. This is another idea which should be tested since the data to hand are conflicting indeed: Gaedeke & Sommer (1986) have found species richness to co-vary with biomass diversity.

The IDH, in the way it has been formulated here, does not hold true for El Porcal lake phytoplankton. Despite this, several periods are envisaged (the so-called equilibrium states) with only a few species sharing most of the phytoplankton biomass. It is likely that the different time lags and environmental factors triggering disturbance events could operate depending upon phytoplankton community structure. Throughout this study period, seven distinct phytoplankton community structures have been suggested (Rojo, 1990). Perhaps both the resilience and the susceptibility to environmental changes (likely disturbance factors) of such structures could be differential, these features explaining the patterns observed in time lags and disturbance factors.

Certainly, it is a surprising fact that intermediate disturbance intensities (but not frequencies) have been proved to enhance diversity in field studies (Lubchenco, 1978; Sousa, 1979; Sommer, 1991) whereas intermediate disturbance frequencies do the same in laboratory studies (Geadeke & Sommer, 1986; Grover, 1989). Of course, the strength of disturbance should be more thoroughly investigated in laboratory studies but this does not dismiss the fact that disturbance frequency in the field could be easily ascertained. In fact, the less frequent disturbance events in temperate lakes (i.e., in particular the autumn overturn) have also been shown to trigger increases in diversity (Sommer, 1993: Plüss See 89, Behler See 90).

On the other hand, diversity data sets of phytoplankton in other lakes are common. Comparisons could be both feasible and desirable but there are two drawbacks that prevent going further. Firstly, very few of them have been carried out with weekly (or shorter) resolution. Secondly, they all use phytoplankton units as the basis for diversity calculations (Sager & Hasler, 1969; Moss, 1973; Elber & Schanz, 1989). This procedure could hardly be used to test IDH since numbers do not take into account true dominance as measured by biomass. Our study has clearly shown that no covariation exists for most types of diversity index. This precludes the use of indices other than biomass diversity and so makes it difficult to compare other phytoplankton diversity studies with our own. With all these pitfalls in mind, Moss' (1973, Fig. 2) data on hypertrophic Abbot's Pond (U.K.) suggest sharp peaks and deep valleys in diversity that could be considered to represent disturbance events and equilibrium states. Also, Fott *et al.* (1974, Fig. 4, lower pannel) show a data set on fractional phytoplankton biomass from the Jezárko fish pond (Czechoslowakia) that could be interpreted within this framework, with disturbance events being much shorter than equilibrium phases. However, we do not know whether the relationship disturbance-diversity in hypertrophic ecosystems could conform to IDH since there are still many gaps that have to be filled before a thorough understanding of the process emerges. A deeper knowledge on hypertrophic phytoplankton is therefore needed and this is what we advocate here.

In this study, there appears to be no relationship between community change and biomass diversity. Diversity changes occur irrespective of rates of community change. This is quite different from what has been found by Reynolds (1984) in some lakes of the English Lake District, since he suggests a slight, albeit significant, relationship between community change rates and diversity. So community change rates could up to a point promote diversity changes, and the faster community changes, the higher diversity increases. This is not the case for El Porcal lake phytoplankton. In fact, community change rates in this lake appear to be higher than others recorded in tropical (Lewis, 1978; Richerson & Carney, 1988) and temperate lakes (Reynolds, 1980) for most of the time, but the variability of El Porcal phytoplankton change rates is much lower. These very high rates of community change are not new in limnology, they are common in hypertrophic ecosystems (Uhlmann, 1980) but result in transient states of community organization. So this could explain why no relationship between diversity and community change has appeared for El Porcal lake, which is a hypertrophic ecosystem.

Lastly, providing that high diversity is more or less associated with disturbance events and low diversity is related with equilibrium states in El Porcal lake, our results suggest that higher diversities (and hence disturbances) are related with lower cell sizes. So disturbances promote small-celled species in the community, a result also supported by Padisák *et al.* (1988) in their study on the effects of storms on Lake Balaton (Hungary) phytoplankton community.

Acknowledgements

Sampling and preliminary analyses could have not been carried out without Angel Rubio and José Luis Velasco; we are very grateful to them. Many ideas of Ulrich Sommer, Judit Padisák, Pál Juhász-Nagy and Manuel Serra have inspired the outcome. Colin Reynolds, Ulrich Sommer, Judit Padisák and an anonymous referee have suggested many useful ideas on the manuscript. Marcelo Izquierdo has kindly provided wind speed data. Figures have been drawn by Fernando Delgado. We also acknowledge Colin Reynolds and Julian Michel for linguistic help with English. And last but not least we are indebted to Angel Baltanás for word proccessing.

References

Alvarez Cobelas, M., 1991. Optical limnology of a hypertrophic, gravel-pit lake. Int. Revue ges. Hydrobiol. 76: 213–223.

Alvarez Cobelas, M., A. Rubio & F. J. Acosta, 1986. Difusión vertical y estabilidad térmica en una laguna hipereutrófica. Limnetica 2: 1–9.

Alvarez Cobelas, M., A. Rubio, M. Araúzo, P. Alarcón & V. Alario, 1987. Morfometria y composición química de una laguna de gravera. Limnetica 3: 91–95.

Alvarez Cobelas, M., A. Rubio & J. L. Velasco, 1990. Chemical limnology of a hypertrophic, gravel-pit lake. Ann. Limnol. 26: 97–108.

APHA, 1985. Standard methods for the examination of waters and wastewaters. 16th ed. Washington DC. 1268 pp.

Bartha, Zs. & L. Hajdu, 1979. Phytoplankton community structure studies on Lake Velence, Hungary.-I. Diversity. Acta. bot. Acad. Sci. Hung. 25: 187–222.

Connell, J. H., 1978. Diversity in tropical forests and coral reefs. Science 199: 1302–1310.

Chatfield, C., 1984. The analysis of time series. An introduction. Chapman and Hall, London. 286 pp.

Dussart, B. H., 1965. Les différents catégories de plancton. Hydrobiologia 26: 72–74.

Elber, F. & F. Schanz, 1989. The causes of change in the diversity and stability of phytoplankton communities in small lakes. Freshwat. Biol. 21: 237–251.

Elber, F. & F. Schanz, 1990. The influence of a flood event on phytoplankton succession. Schweiz. Z. Hydrol. 52: 330–344.

Fott, J., V. Korínek, M. Prazáková, B. Vondrus & K. Forejt, 1974. Seasonal development of phytoplankton in fish ponds. Int. Revue ges. Hydrobiol. 59: 629–641.

Gaedeke, A & V. Sommer, 1986. The influence of the frequency of periodic disturbances on the maintenance of phytoplankton diversity. Oecologia 71: 25–28.

Grover, J. P., 1989. Effects of Si:P supply ratio, supply variability, and selective grazing in the plankton: an experiment with a natural algal and protistan assemblage. Limnol. Oceanogr. 34: 349–367.

Hardin, G., 1960. The competitive exclusion principle. Science 131: 1292–1297.

Hutchinson, G. E., 1960. The paradox of the plankton. Am. Nat. 95: 137–145.

Jassby, A. D. & C. R. Goldman, 1974. A quantitative measure of succession rate and its application to the phytoplankton of lakes. Am. Nat. 108: 688–693.

Kalff, J. & R. Knoechel, 1978. Phytoplankton and their dynamics in oligotrophic and eutrophic lakes. Ann. Rev. Ecol. Syst. 9: 475–495.

Kirk, J. T. O., 1983. Light and photosynthesis in aquatic ecosystems. Cambridge University Press, Cambridge. 401 pp.

Legendre, L. & P. Legendre, 1979. Ecologie numérique, I: le traitement multiple des données écologiques. Masson, Paris. 197 pp.

Lewis jr., W. M., 1978. Analysis of succession in a tropical phytoplankton community and a new measure of succession rate. Am. Nat. 112: 401–414.

Lubchenco, J., 1978. Plant species diversity in a marine intertidal community: importance of herbivore food preference and algal competitive abilities. Am. Nat. 112: 23–39.

Mayzaud, P., S. Taguchi & P. Laval, 1984. Seasonal patterns of seston characteristics in Bedford Basin, Nova Scotia, relative to zooplankton feeding: a multivariate approach. Limnol. Oceanogr. 29: 745–762.

Mortimer, C. H., 1974. Lake hydrodynamics. Mitt. int. Ver. Limnol. 20: 124–197.

Moss, B., 1973. Diversity in freshwater phytoplankton. Am. Midl. Nat. 90: 341–355.

Padisák, J., L. G.-Tóth & M. Rajczy, 1988. The role of storms in the summer succession of the phytoplankton community in a shallow lake (Lake Balaton, Hungary). J. Plankton Res. 10: 249–265.

Peters, R. H., 1983. The ecological implications of body size. Cambridge University Press, Cambridge. 329 pp.

Reynolds, C. S., 1980. Phytoplankton assemblages and their periodicity in stratifying lake systems. Holarct. Ecol. 3: 141–159.

Reynolds, C. S., 1984. The ecology of freshwater phytoplankton. Cambridge University Press, Cambridge, 384 pp.

Richerson, P. J. & H. J. Carney, 1988. Patterns of temporal variation in Lake Titicaca, a high altitude tropical lake. II. Succession rate and diversity of the phytoplankton. Verh. int. Ver. Limnol. 23: 734–738.

Rojo, C., 1990. Estructura de la comunidad fitoplanctónica de la laguna hipertrófica 'El Porcal' (Madrid). Ph. D. thesis. Univ. Valencia, Valencia (Spain). 389 pp.

Rott, E., 1981. Some results from phytoplankton counting intercalibrations. Schweiz. Z. Hydrol. 43: 34–62.

Sager, P. E. & A. D. Hasler, 1969. Species diversity in lacustrine phytoplankton. I. The components of the index of diversity from Shannon's formula. Am. Nat. 103: 51–59.

Shannon, C. E. & W. Weaver, 1963. The mathematical theory of communication. Univ. Illinois Press, Urbana (USA). 117 pp.

Sommer, U., 1989. The role of competition for resources in phytoplankton succession. In U. Sommer (ed.), Plankton ecology, succession in plankton communities. Springer Verlag, Berlin. pp. 57–106.

Sommer, U., J. Padisák, C. S. Reynolds & P. Juhász-Nagy, 1993. Hutchinson's heritage: the diversity-disturbance relationship in phytoplankton. In J. Padisák, C. S. Reynolds & U. Sommer (eds), Intermediate Disturbance Hypothesis in Phytoplankton Ecology. Developments in Hydrobiology 81. Kluwer Academic Publishers, Dordrecht: 1–7. Reprinted from Hydrobiologia 249.

Sousa, W. P., 1979. Disturbance in marine intertidal boulder fields: the nonequilibrium maintenance of species diversity. Ecology 60: 1225–1239.

Tilman, D., 1982. Resource competition and community structure. Princeton University Press, Princeton. 296 pp.

Uhlmann, D., 1980. Stability and multiple steady states of hypereutrophic ecosystems. In J. Barica & L. Mur (eds), Hypertrophic Ecosystems. Dr W. Junk Publishers, The Hague: 235–247.

Weisse, T., 1988. Dynamics of autotrophic picoplankton in Lake Constance. J. Plankton Res. 10: 1179–1188.

Williams, N. J. & C. R. Goldman, 1975. Succession rates in lake phytoplankton communities. Verh. int. Ver. Limnol. 19: 808–811.

Hydrobiologia **249**: 59–65, 1993.
J. Padisák, C.S. Reynolds & U. Sommer (eds), Intermediate Disturbance Hypothesis in Phytoplankton Ecology.

Disturbance-diversity relationships in two lakes of similar nutrient chemistry but contrasting disturbance regimes

Ulrich Sommer
Institut für Biologie und Chemie des Meeres, Universität Oldenburg, Postf. 2503, D-2900 Oldenburg, Germany

Key words: diversity, disturbance, phytoplankton, carrying capacity, competitive exclusion

Abstract

Phytoplankton diversity was studied in two North German lakes of comparable nutrient chemistry but different exposure to winds. In both lakes, phytoplankton was primarily N-limited but diatoms were Si-limited. Plu*ß*see had a very constant mixing depth during summer, while week-to-week changes of several meters were quite common in the more exposed Behler See. In Plu*ß*see, phytoplankton biomass during summer came closer to the carrying capacity as defined by the available total N. In Plu*ß*see there was a marked decline of diversity during the summer maximum of biomass, while this decline was less pronounced in Behler See. It is concluded that disturbances which prevented phytoplankton from reaching the carrying capacity also maintained a high level of diversity. A negative response of diversity to undisturbed conditions became apparent, after phytoplankton biomass had exceeded about 5% of the carrying capacity.

Introduction

Attempts to explain the coexistence of phytoplankton species and the avoidance of 'competitive exclusion' (i.e. attempts to solve Hutchinson's 'paradox of the plankton', 1961) fall into two categories: The equilibrium theory of competition (Petersen, 1975; Tilman, 1982) predicts that different species can coexist by being limited by different resources. Disequilibrium approaches (Richerson *et al.*, 1970; Grenney, 1973; Armstrong & McGehee, 1976; Levins, 1979; Ebenhöh, 1988; Grover, 1990) predict that environmental variability permits the coexistence of species competing for the same resource. Both hypotheses are by no means mutually exclusive and might be additive in their potential contribution to 'solve Hutchinson's paradox'. Culture experiments have supported both: Experiments with undisturbed continuous cultures (e.g. Tilman, 1977; Sommer, 1983) have supported the first prediction; experiments with discontinuous nutrient supply (e.g. Turpin & Harrison, 1980; Robinson & Sandgren, 1983; Sommer, 1984, 1985; Gaedeke & Sommer, 1986; Grover, 1988, 1989, 1991) have supported the second. The experiments by Gaedeke & Sommer (1986) have moreover supported Connell's (1978) 'intermediate disturbance hypothesis' by finding a diversity peak at disturbance intervals of *ca.* 3 generation times. Both frequent (intervals <1 generation time) disturbances of low intensity and rare but strong disturbances lead to low diversity levels.

Obviously, both resource partitioning and disturbance at the appropriate scale can increase the number of coexisting species and diversity. Therefore, comparative field studies with the purpose of testing the intermediate disturbance hypothesis have to be carefully designed to avoid the influence of different numbers of limiting resources. In this study, I attempted to solve this problem by selecting two lakes which are similar in their nutrient chemistry (same number of limiting resources) but dissimilar in their exposure to wind (external disturbance).

Methods

The two lakes (Plußsee and Behler See) were sampled during the vegetation period of 1990. For Plußsee also data from another study in 1989 were also available (Sommer, 1991b).

Mixed phytoplankton samples were taken at weekly intervals from the epilimnion of both lakes and fixed with Lugol's iodine. Phytoplankton was counted according to the Utermöhl standard technique. At least 800 individuals were counted per magnification. Dissolved nutrients (soluble reactive phosphorus, nitrate, nitrite, ammonium, dissolved silicate) were measured according to standard techniques. The particulate fractions of biogenic elements were measured on filters: Carbon and nitrogen were measured in a CN-analyser (Carlo-Erba) after filtration onto glass-fiber filters, particulate Si after filtration onto Nuclepore filters, and particulate P after filtration onto cellulose-nitrate membrane filters. For more details on the chemical methodology see Sommer (1991a, 1991b).

The diversity of phytoplankton was calculated using a biomass-related version of Shannon's H' (for a comparison of diversity indices see Washington, 1984):

$$H' = -\sum p_i \times \log_2 p_i$$

where pi is the relative importance of species i, here expressed by biomass (B_i/B_{tot}). Because of the low efficiency of the Utermöhl technique in counting picoplankton ($<2\ \mu m$) and because of the difficulties in their species identification, picoplankton were neglected. H' is thus a partial diversity covering phytoplankton $>2\ \mu m$.

Measurement of disturbance was more problematic, because natural phytoplankton is subject to different kinds of disturbances which are scarcely comensurable. There is no generally acceptable common scale for changes in physical conditions, in nutrient supply, or in grazing pressure. Therefore, a biotic response variable was sought by which the impact of disturbance could be measured. This variable should be mathematically unrelated to diversity, otherwise the analysis of disturbance-diversity relationships would have become tautological. Here I chose the ratio between phytoplankton biomass (B) and the potential maximal biomass ('carrying capacity' in the classic sense, K). A similar approach, using a slightly different terminology, has been suggested by Reynolds (1991). By implication, disturbance is defined as any factor which prevents phytoplankton biomass from reaching K. This is a broad definition of 'disturbance' which includes physical (e.g. mixing events) and biotic factors (e.g. grazing). A maximum estimate of K was obtained from the total (dissolved plus particulate) fraction of the limiting nutrient. It was assumed that K is reached, when all of the limiting nutrient is incorporated into phytoplankton biomass and as much as possible biomass is made of it. This happens if the cell quota (Droop, 1973) is minimal (q_0). Since nitrogen is the primary limiting nutrient in both lakes, the estimate of K was based on N. A series of field studies in North german lakes (Sommer, 1991a, 1991b) had yielded the following estimate for natural populations:

$$q_0 = 0.037(\pm 0.14;\text{S.D.})\text{molN/molC}.$$

The upper boundary of K (in mol C) was therefore estimated as 27 N_{tot}. To calculate the quotient B/K, biovolume was converted into carbon according to Rocha & Duncan (1985).

Results and discussion

Physical and chemical conditions

Plußsee and Behler See were selected for this study because of their similarity in nutrient chemistry and their dissimilarity in exposure to wind and thus external disturbances. Both lakes are situated in the vicinity of Plön and are only 3 km apart. Plu*ß*see is much smaller (0.14 km^2) and strongly sheltered by surrounding moraine hills. Behler See (3.3 km^2) has a bigger wind fetch and is less sheltered. Due to their close vicinity they are subject to the same general weather conditions but the impact of wind events should be stronger on Behler See. Thus they are subject to the same frequency, but different intensities of external disturbances. This difference is clearly reflected in the stratification regime (Fig. 1).

Behler See (sampled weekly from March to September 1990) stratified later in the spring and started to circulate much earlier in autumn. The mixing depth (z_m) during the spring and summer stratified period was highly variable with minima of *ca.* 2.5 m and maxima of *ca.* 7 m. Week-to-week differences of *ca.* 4 m were quite common. Plu*ß*see (sampled weekly from February to early December 1989 and from March 1990 to December 1990) stratified nearly 2 months earlier and started to circulate 2 months later. The mixing depth during spring and summer never exceeded 5 m; week-to-week differences in spring and summer were usually small and never exceeded 1.5 m. The differences between 1989 and 1990 were minor with the notable exception that summerly mixing conditions (z_m < 5 m) persisted about half a month longer in 1989.

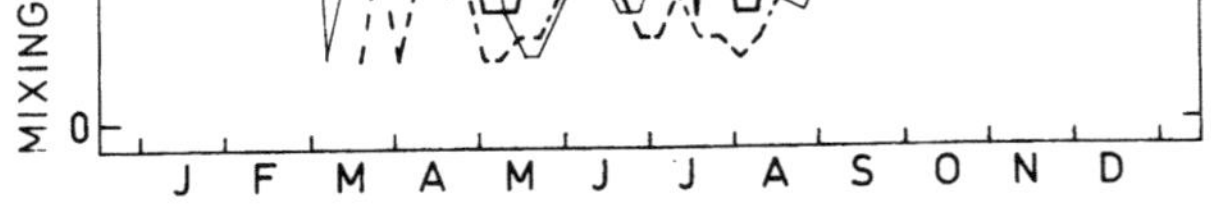

Fig. 1. Mixing depth of Behler See (1990) and Plu*ß*see (1989 and 1990). Values are truncated at the mean depth (13 m in Behler See and 10 m in Plu*ß*see).

Given that wind-induced mixing events during the stratified period can be major disturbances for phytoplankton succession (Reynolds, 1984, 1988) differences in the seasonal pattern of diversity between both lakes should be expected.

It was important for the selection of lakes that the same number of resources should become limiting during algal growth. Otherwise, differences in diversity indices would have been expected even under identical disturbance regimes (Tilman, 1982). A previous study on Plu*ß*see (Sommer, 1991b) had identified nitrogen and silicon (only for diatoms), but not phosphorus, as limiting nutrients by physiological methods (bioassays and measurements of cell quotas). Both lakes are quite similar in N and P availability. Total nitrogen concentrations before the start of the growing season were 40 μM in Plu*ß*see and 48 μM in Behler See, respectively. Total P concentrations at that time were 6.4 μM in Plu*ß*see and 5.6 μM in Behler See. During the entire study period TN:TP ratios were < 12; the same holds true for the ratio soluble reactive phosphorus:dissolved inorganic nitrogen ($NO_3 + NO_2 + NH_4$). Particulate C:N ratios of *ca.* 20:1 during the periods of maximal biomass in both lakes clearly suggest N-limitation; at the same time, N:P ratios in the particulate matter are < 15:1 (Goldman *et al.*, 1979; Sommer, 1989, 1991a, 1991b).

Unlike to N and P, there are great differences in the Si availability between both lakes. Before the onset of algal growth the concentrations of dissolved Si were only 25 μM in Plu*ß*see but 106 μM in Behler See. In both lakes they fall to strongly limiting levels during the spring maximum of diatoms (2 μM in Behler See; 0.2 μM in Plu*ß*see) and stay limiting for diatom growth during the entire summer period (< 12 μM in Behler See; < 5 μM in Plu*ß*see).

Thus, the same three essential resources may become limiting in both lakes: nitrogen, silicate (for diatoms) and probably also light. If the lakes were chemostats, an equilibrium coexistence of three phytoplankton species could be anticipated in each case.

The response of phytoplankton

The total phytoplankton biomass showed a bimodal seasonal pattern in either lake (Fig. 2). The spring-bloom started already before the onset of stratification and continued into the stratified period. In mid-May, there was a marked minimum of phytoplankton biomass in spite of nearly optimal growth conditions; e.g. high nutrient availability, long daylength, high irradiance. Such mid-season minima under otherwise good growth conditions are termed 'clear-water phases' and can be explained by excessive zooplankton grazing (Lampert, 1988). After the clear-water phase, there was an extended summer maximum of biomass which is a characteristic feature for eutrophic lakes (Sommer *et al.*, 1986). Despite a very similar trophic status of both lakes, maximal summer biomasses were 1½ to 2 orders of magnitude higher in Plußsee than in Behler See; carrying capacities calculated from total nitrogen concentration were quite similar in both lakes, however. The carrying capacity was reached only in Plußsee, during summer 1989; in summer 1990, the greatest attainment was 37% and, in Behler See, maximally 22%.

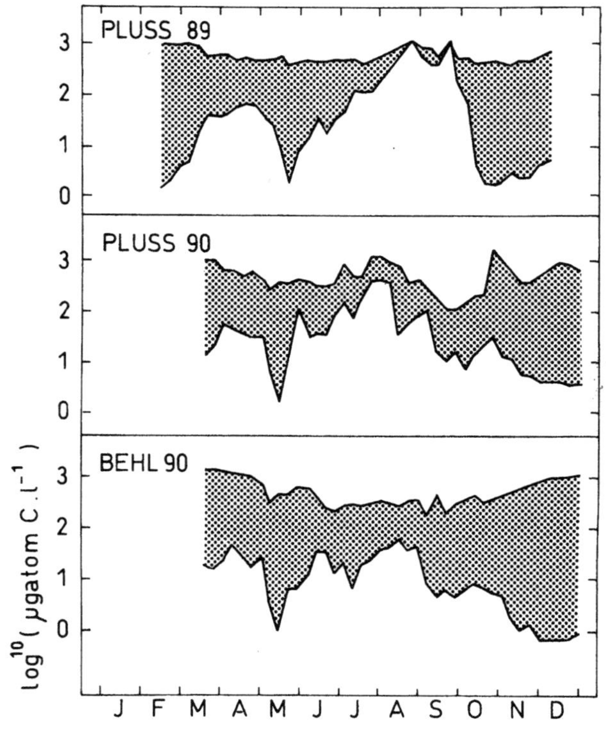

Fig. 2. Phytoplankton biomass (in µgatom C calculated according to Rocha & Duncan 1985; lower limit of stippled area) and carrying capacity (calculated as $27 \cdot N_{tot}$; upper limit of stippled area) in Behler See and Plußsee; logarithmic scale.

If resolved to the species level, the seasonal successions of phytoplankton in both lakes were quite different. However, if aggregated at a higher level of taxa (Fig. 3) and of functional groups (small, edible flagellates; large, inedible flagellates; small, edible immotile algae; large, immotile algae) the patterns were more similar and corresponded relatively well to the eutrophic standard model proposed by the plankton ecology group (Sommer *et al.*, 1986). The spring-bloom consists mainly of diatoms (*Stepanodiscus* spp. and *Asterionella formosa* Hass. in Plußsee; *Aulacoseira islandica* Simonsen, *Stepanodiscus* spp. and *Asterionella formosa* in Behler See) and Cryptophycean flagellates, which tend to become more important with the onset of stratification. After the clear-water phase phytoplankton species which are resistant against grazing became dominant, first colonial green algae (e.g. *Pandorina*, *Sphaerocystis*), the colonial Cyanophyta (mainly *Anabaena*, but also *Microcystis*) and dinoflagellates reached dominance. In all cases *Ceratium* spp. reached the highest values of p_i; *Ceratium hirundinella* Schrank reached 0.98 in Plußsee 1989 and 0.75 in Behler See 1990, *Ceratium furcoides* Langhans reached 0.90 in Plußsee 1990. In contrast to Plußsee, there were also some minor summer peaks of diatoms in Behler See. The greater importance of diatoms in Behler See can both been explained by higher Si:P or Si:N ratios (Sommer, 1983, 1985, 1989) or by longer mixing in spring and intermittent mixing events during summer (Reynolds, 1980, 1984, 1988).

If the biomass/capacity ratio is used as an operational, inverse measure of disturbance then not only wind induced mixing events but also internally driven reductions of biomass, such as the clear-water phase, qualify as disturbance. Figure 4 shows more or less reciprocal temporal changes of B/K and the diversity in both lakes. For the purpose of a cross-correlation analysis,

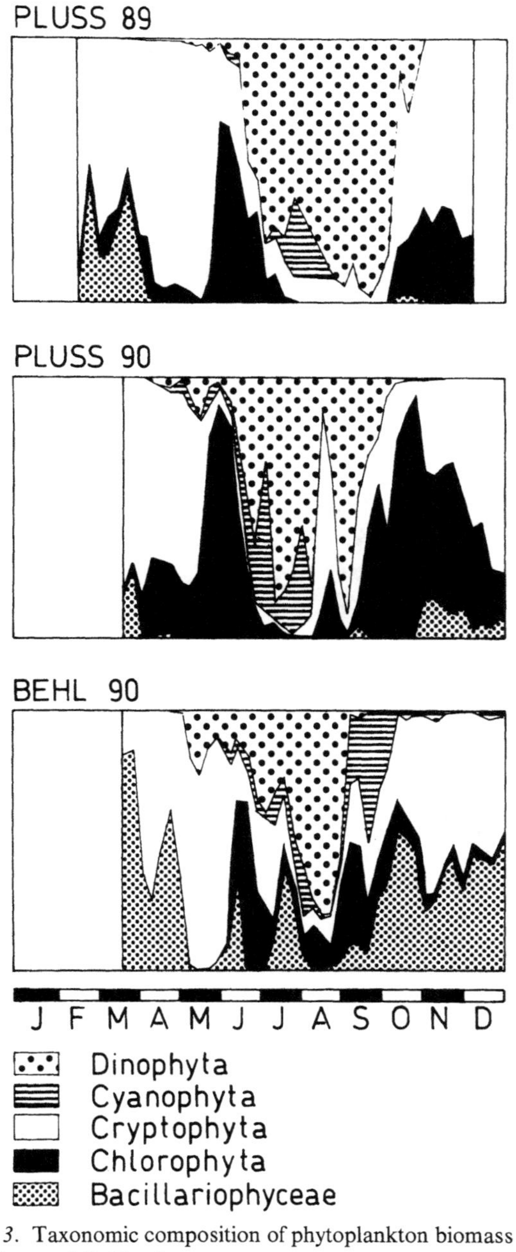

Fig. 3. Taxonomic composition of phytoplankton biomass in Plu*β*see and Behler See.

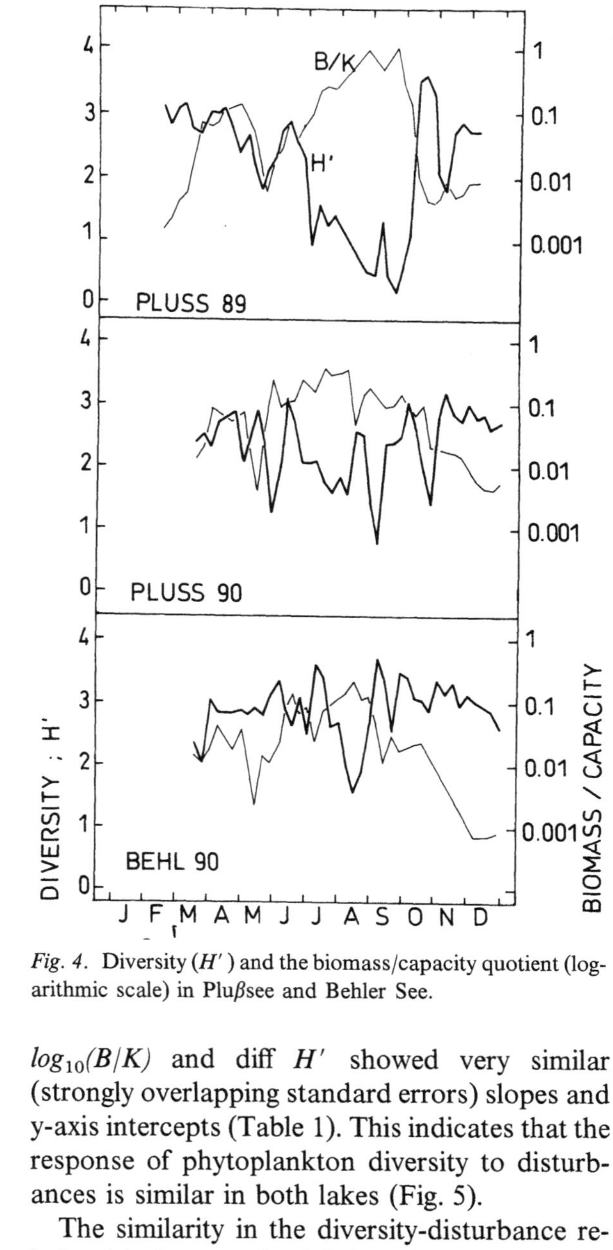

Fig. 4. Diversity (H') and the biomass/capacity quotient (logarithmic scale) in Plu*β*see and Behler See.

the time-series of $loq_{10}(B/K)$ and H' were differenced; i.e. the original values were replaced by the difference between two successive values (Chatfield, 1984). Diff $log_{10}(B/K)$ and diff H' were then used as input variables in a cross-correlation analysis. This analysis showed significant negative correlations for all three lake-years with no time-lag. A regression analysis between diff $log_{10}(B/K)$ and diff H' showed very similar (strongly overlapping standard errors) slopes and y-axis intercepts (Table 1). This indicates that the response of phytoplankton diversity to disturbances is similar in both lakes (Fig. 5).

The similarity in the diversity-disturbance relationship between both lakes also becomes evident from a scatter-plot of H' against log (B/C) (Fig. 5). When the biomass of phytoplankton reaches the carrying capacity diversity tends to become very low, the minimal recorded value being 0.16. Diversity begins already to decline when the biomass exceeds about 5%. If we accept the rationale of Hutchinson's (1961) paradox and of Connell's (1978) intermediate disturbance hypothesis, this means a surprisingly early onset

Table 1. Regression analysis of the week-to-week difference in diversity (y) on the weekt-to-week difference in $\log_{10}/BK$ (x) according to the model $y = a + bx$.

Lake/year	a + SE	$b \pm$ SE	r	p
Plußsee 1989	0.006 ± 0.091	-0.87 ± 0.26	−0.48	<0.002
Plußsee 1990	-0.055 ± 0.084	-0.67 ± 0.20	−0.52	<0.005
Behler S. 1990	0.086 ± 0.107	-0.59 ± 0.26	−0.43	<0.05
All data	0.007 ± 0.053	-0.72 ± 0.14	−0.48	<0.00001

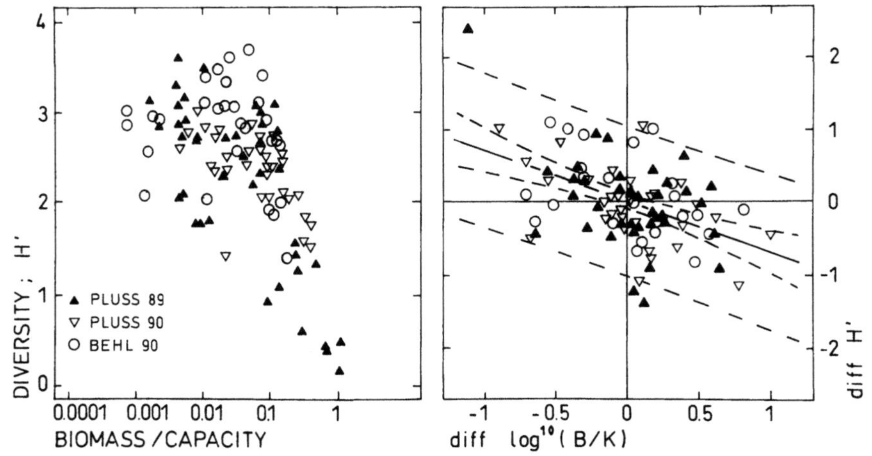

Fig. 5. Left: Diversity and B/K (logarithmic scale) in Plu*ß*see and Behler See. Right: Regression analysis of the week-to-week difference in H' on the week-to-week difference in log (B/K); the regression line and the 95% confidence intervals for the regression line and for the dependent variable refer to the entire data set (see Table 1).

of competitive exclusion while it takes 4 or 5 more doublings until competitive exclusion approaches completion. At B/C quotients <0.05 diversity fluctuates within relatively wide margins (H' = 1.5 to 3.7). There is a slight indication of a decline in diversity at very low B/K. This decline would be in agreement with the IDH, but the data are far from being convincing. Thus only one half of the IDH is fully supported: Diversity tends to decline, the longer the regime remains undisturbed.

References

Armstrong, R. A. & R. McGehee, 1976. Coexistence of species competing for shared resources. Theor. Pop. Biol. 9: 317–328.

Chatfield, C., 1984. The analysis of time series. 3rd edn. Chapman & Hall, London.

Connell, J., 1978. Diversity in tropical rain forests and coral reefs. Science 199: 1304–1310.

Droop, M. R., 1973. Some thoughts on nutrient limitation in algae. J. Phycol. 9: 264–272.

Ebenhöh, W., 1988. Coexistence of an unlimited number of algal species in a model system. Theor. Pop. Biol. 34: 130–144.

Gaedeke, A. & U. Sommer, 1986. The influence of the frequency of periodic disturbances. Oecologia 71: 98–102.

Goldman, J. C, J. J. McCarthy & D. G. Peavey, 1979. Growth rate influence on the chemical composition of phytoplankton in oceanic waters. Nature 279: 210–215.

Grenney, W. J., 1973. A theoretical approach to interspecific competition in the phytoplankton community. Am. Nat. 107: 405–425.

Grover, J. P., 1988. Dynamics of competition in a variable environment. Ecology 69: 408–417.

Grover, J. P., 1989. Effects of Si:P supply ratio, supply variability, and selective grazing in the plankton: An experiment with a natural algal and protistan assemblage. Limnol. Oceanogr. 34: 349–367.

Grover, J. P., 1990. Resource competition in a variable environment: Phytoplankton growing according to Monod's model. Am. Nat. 136: 772–789.
Grover, J. P., 1991. Non-steady state dynamics of algal population growth: experiments with two chlorophytes. J. Phycol. 27: 70–79.
Hutchinson, G. E., 1961. The paradox of the plankton. Am. Nat. 95: 137–145.
Lampert, W., 1988. The relationship between zooplankton biomass and grazing: A review. Limnologica 19: 11–20.
Levins, R., 1979. Coexistence in a variable environment. Am. Nat. 114: 765–783.
Petersen, R., 1975. The paradox of the plankton: An equilibrium hypothesis. Am. Nat. 109: 35–49.
Reynolds, C. S., 1980. Phytoplankton assemblages and their periodicity in stratifying lake systems. Holarctic Ecol. 3: 141–159.
Reynolds, C. S., 1984. Phytoplankton periodicity; the interaction of form, function and environmental variability. Freshwat. Biol. 14: 111–142.
Reynolds, C. S., 1988. Functional morphology and the adaptive strategies of freshwater phytoplankton. In C. D. Sandgren (ed.), Growth and Reproductive strategies of freshwater phytoplankton. Cambridge Univ. Press. 388–433.
Reynolds, C. S., 1981. Lake communities: An approach to their management for conservation: In F. Spellerberg, F. B. Goldsmith & M. G. Morris (eds), The scientific management of temperate communities for conservation. Blackwell, Oxford, pp. 199–225.
Richerson, P., R. Armstrong & C. R. Goldman, 1970. Contemporaneous disequilibrium, a new hypothesis to explain the 'paradox of plankton'. Proc. nat. Acad. Sci. 67: 1710–1714.
Robinson, J. D. & C. D. Sandgren, 1983. The effect of temporal environmental heterogeneity on community structure: a replicated experimental study. Oecologia 57: 98–102.
Rocha, O. & A. Duncan, 1985. The relationship between cell carbon and cell volume in freshwater algal species used in zooplankton studies. J. Plankton Res. 7: 279–294.
Sommer, U., 1983. Nutrient competition between phytoplankton in multispecies chemostat experiments. Arch. Hydrobiol. 96: 399–416.
Sommer, U., 1984. The paradox of the plankton: Fluctuations of phosphorus availability maintain diversity in flow-through cultures. Limnol. Oceanogr. 29: 633–636.
Sommer, U., 1985. Comparison between stedy state and non-steady state competition: Experiments with natural phytoplankton. Limnol Oceanogr. 30: 335–346.
Sommer, U., 1989. The role of competition for resources in phytoplankton succession. In: U. Sommer (ed.), Plankton ecology: succession in plankton communities. Springer, Berlin, 57–106.
Sommer, U. 1991a. The application of the Droop-model of nutrient limitation to natural phytoplankton. Verh. int. Ver. Limnol. 24: 791–794.
Sommer, U. 1991b. A comparison of the Droop-model and the Monod-model of nutrient limited growth applied to natural populations of phytoplankton. Funct. Ecol. 5: 535–544.
Sommer, U., Z. M. Gliwicz, W. Lampert & A. Duncan, 1986. The PEG-model of seasonal succession of planktonic events in fresh waters. Arch. Hydrobiol. 106: 433–471.
Tilman, D., 1977. Resource competition between planktonic algae: an experimental and theoretical approach. Ecology 58: 338–348.
Tilman, D., 1982. Resource competition and community structure. Princeton Univ. Press, Princeton.
Turpin, D. H. & P. J. Harrison, 1980. Cell size manipulation in natural marine planktonic diatom communities. Can. J. Fish. aquat. Sci. 37: 1193–1195.
Washington, H. G., 1984. Diversity, biotic and similarity indices. Wat. Res. 18: 653–694.

Hydrobiologia **249**: 67–92, 1993.
J. Padisák, C.S. Reynolds & U. Sommer (eds), Intermediate Disturbance Hypothesis in Phytoplankton Ecology.

Importance of intermediate disturbances for the species composition and diversity of phytoplankton in two very different Berlin lakes

Ingrid Chorus & Gertrud Schlag
Institute for Water-, Air-, and Soil-Hygiene of the Federal Health Agency, Corrensplatz 1, D-1000 Berlin 33, Germany

Key words: phytoplankton species composition, diversity, intermediate disturbances, seasonal succession, phytoplankton community structure, physical forcing

Abstract

Species composition and diversity of phytoplankton were studied for several years in two lakes which differ with respect to mixing conditions and nutrient limitation: Schlachtensee regularly stratifies very stably. In contrast, size and wind-exposure predispose Lake Tegel to deeper mixing; additionally, stratification is artificially destabilized by aeration. As the duration of aeration was varied, the study period includes interannual changes in mixing conditions. For both lakes, it also covers trophic change due to restoration; this was especially pronounced in Schlachtensee.

Results show that mixing conditions affect species composition on two levels: on a superordinate level, lake morphology or hydrology govern stability of stratification and susceptibility to perturbation, and hence the extent to which motile species can develop. In Schlachtensee, species with some means of actively seeking preferred depths usually dominated during summer stratification: *Planktothrix agardhii* during the hypertrophic phase, and flagellates since restoration. In contrast, in Lake Tegel deeper mixing as a generally prevailing condition favored non-motile species. Their seasonal pattern was remarkably constant from year to year. Although changes in the extent of mixing were pronounced during the four years studied (1987–1990), these were within a range that affected species composition only slightly: in summer, cyanobacteria and diatoms represented 'climax species' whose dominance was not offset by additional, weather-induced increases of turbulence.

On a subordinate level, and within the constraints set by nutrient limitation as well as by grazing pressure, small-scale changes in mixing conditions caused by meteorological cycles were shown to strongly affect species composition and in consequence diversity: Results for the fouryear post-restoration study period at Schlachtensee show that considerable interannual variations of species composition and diversity can be attributed to variations in the frequency of meteorological changes. In accordance with the 'intermediate disturbance hypothesis' (IDH), diversity was lowest during 1989, the year with the longest and most pronounced cycles of fair weather (14 to 27 fair days on end). However, the mechanism for this was rarely a decline of diversity caused by competitive exclusion within single long phases of stable conditions, as conceived by the 'intermediate disturbance hypothesis'. Instead, diversity responded to changes in mixing conditions with a variety of patterns – often with low values during phases of increased mixing and with high values under quiescent conditions, especially during the first calm days just after increased mixing. Thus, not disturbance as such, but rather the rate of change between phases of disturbance and quiescence appears to determine the frequency of high diversity indices.

In Lake Tegel, high diversity indices were somewhat more frequent in 1989, the year during which thermal stratification was most stable. For species adapted to frequent or continuous mixing, interjected calm phases with unusually high stability of thermal stratification may represent a 'disturbance' of accustomed conditions. Thus, in turbulent Lake Tegel, meteorological cycles appear to act in reversal to the IDH, in a sense which may be termed 'intermediate quiescence hypothesis'.

Phosphorus limitation due to successful restoration was found to decrease winter and vernal diversity in Schlachtensee. Presumably, without nutrient constraints, new populations could grow more rapidly in response to the rapid changes of physical parameters during this season. In contrast, restoration has increased summer diversity, as phosphorus concentrations no longer allow the virtual monocultures of *Planktothrix agardhii* which prevailed previously. In Lake Tegel, the reduction of phosphorus concentration down to 60 μg/l P in 1989 limited biomass, but this level was still too high to significantly alter species composition or diversity.

Introduction

The role of disturbance in the maintenance of diversity in phytoplankton communities is regarded as well established, especially under experimental conditions (Sommer *et al.*, 1993); however, the operation of this mechanism in lake ecosystems is difficult to verify. For an understanding of the driving forces of seasonal periodicity, Reynolds (1988) differentiates between 'allogenic, environmental variability' as opposed to 'autogenic, equilibrium dynamics' of a community. The question is: to what extent is species composition of a given lake governed by biogenically driven seasonal succession in the sense of the PEG-model (Sommer *et al.*, 1986), and to what extent is it determined by random external disturbances? In this context, distinction between disturbance caused by grazing (one of the mechanisms of autogenic succession), from allogenic events (such as storms or hydrological changes) is important. Thus, during the workshop dedicated to the 'intermediate disturbance hypothesis' ('IDH') held in Baja in July 1991, the following definition of 'disturbance' was derived:

> 'Within the context of seasonal change in phytoplankton species composition, disturbances are primarily non-biotic and stochastic events that result in distinct and abrupt changes in competition and which interfere with internally driven progress towards ecological equilibrium (sensu Odum). Such events operate through (eg.) the medium of weather and at the frequency scale of 1–10 algal generation times.'

Also of interest, on a larger time scale, is the effect of thermal stratification and turbulence within the epilimnion (as determined by site and morphometry of a lake) on species composition and hence on diversity.

Data were collected on phytoplankton biomass and species composition from two very different lakes in Berlin in the course of different research projects, in Lake Tegel over a time span of 4 years (1987–1990) and in Schlachtensee during two years before restoration success (1982/1983; Chorus, 1989) and during 4 years since (1987–1990). In the following, we attempt to evaluate these data with respect to three questions:

1. Do overall differences in the mixing conditions of both lakes affect diversity?
2. Can cyclical weather changes during summer stratification be traced as effects on temperature profiles, species composition, and diversity; are they discernible as 'intermediate disturbances' of seasonal succession?
3. How do trophic state and disturbance interact to affect diversity?

Description of the two lakes studied

Both the morphometry (Table 1) and the trophic state of the two lakes are very different. Schlachtensee is small, elongated and rather well sheltered by high banks and surrounding woods. Motorboats are not permitted. Thus, it regularly establishes stable thermal stratification from late April or early May until well into September. Mixing

Table 1. Morphometric data and nutrient concentrations

	Schlachtensee	Lake Tegel
area	0.42 km^2	4 km^2
mean depth	4.6 m	6.6 m
maximum depth	9.6 m	15 m
volume	1.97 mio. m^3	24.6 mio. m^3
retention time	0.5 a	0.3–0.25 a
thermal stratification	pronounced	temporary, slight
total phosphorus (anual means)		
before restoration	800 $\mu g\ l^{-1}$ P	800 $\mu g\ l^{-1}$ P
1987	27 $\mu g\ l^{-1}$ P	170 $\mu g\ l^{-1}$ P
1988	27 $\mu g\ l^{-1}$ P	81 $\mu g\ l^{-1}$ P
1989	22 $\mu g\ l^{-1}$ P	60 $\mu g\ l^{-1}$ P
1990	29 $\mu g\ l^{-1}$ P	89 $\mu g\ l^{-1}$ P
total nitrogen (1987–1990)	0.8–3.7 mg l^{-1} N	4.8–12.6 mg l^{-1}

depth then ranges between 2 and 4 m, depending on weather conditions.

In contrast, Lake Tegel is large, wind-exposed, and rather shallow in relation to its area. Freight ships, tourist liners, and motorized sports boats of all sizes cause considerable wave action, especially on fair and sunny days during which perturbation of the water column would otherwise be slight. Additionally, 15 aerators were installed in 1979, originally for compensation of oxygen deficiency threatening because of the inflow of scarcely treated sewage. Data from several years before 1979 show that without aeration Lake Tegel usually developed stable stratification, but mixing depths ranged between 5 and 8 m. Since aeration was commenced, Lake Tegel has scarcely stratified. As sewage has been diverted and treated since 1985, an attempt to employ the aerators only upon demand (decline of hypolimnetic oxygen concentrations) was begun in 1989.

Both lakes are being restored, chiefly by drastic reduction of their phosphorus input through treatment of their main inflow, Schlachtensee since autumn 1981 and Lake Tegel since autumn 1985 (Klein, 1988). Total phosphorus concentrations initially around 800 $\mu g\ l^{-1}$ P (annual mean) decreased exponentially in both lakes Table 1). Phytoplankton biomass in Schlachtensee showed no reaction until total phosphorus concentrations fell below 60 $\mu g\ l^{-1}$ P in 1985; then species composition switched and biomass declined steeply (Chorus & Wesseler, 1988 and Klein & Chorus, 1991). This situation has been stable (at annual means of total phosphorus around 30 $\mu g\ l^{-1}$ P) since 1986. Thus, for Schlachtensee the effect of trophic state on diversity can be compared to the effect of cyclic weather events. At Lake Tegel, the decline of phosphorus concentrations followed a very similar curve, from annual means around 800 $\mu g\ l^{-1}$ P before 1985 down to 60 $\mu g\ l^{-1}$ P in 1989, but concentrations have increased again because of remobilisation of sedimented phosphorus (very likely enhanced by artificial mixing through aeration). Phytoplankton density has reacted more gradually than in Schlachtensee and reached a minimum in 1989 (Klein & Chorus, in prep.).

Nitrogen was never limiting in either of the lakes during the period presented here. Data for silicate by Ripl (pers. comm., kindly supplied by the Inst. for Limnology of the Technical University of Berlin and by the Gesellschaft für Gewässerbewirtschaftung mbH) for Schlachtensee in 1988 show that silicate may become limiting to diatoms by late spring with values below 0.35 mg l^{-1}; in Lake Tegel silicate concentration did not fall below 0.5 mg l^{-1} in 1990.

Methods

Meteorological data were kindly supplied by the Meteorological Institute of the Free University of Berlin. Measurements were taken within the city (Berlin Dahlem) on a slightly elevated site within 7 km of Schlachtensee and 14 km of Lake Tegel. This distance to the lakes will rarely seriously affect daily means of temperature, irradiance or precipitation, but windspeed for Lake Tegel may be biased: values there may be up to twice as high as measured in Dahlem. However, day-to-day trends observed in Dahlem will also apply for Lake Tegel, and absolute values are less relevant for the following evaluation, as their effect on mixing of the lake cannot be quantified anyhow. Data for temperature, irradiance, and precipitation are true daily means or integrals, respectively. Daily means for windspeed were calculated from 10 min. measurements performed every three hours.

Water temperature was measured with electrodes (WTW, SYLAND) or with a thermometer (differentiating 0.1 °C).

During thermal stratification, phytoplankton samples were taken from the euphotic zone (as determined by measurements of rates of photosynthesis) in Lake Tegel. In Schlachtensee, this was also the case in 1987 (sampling by Gervais, 1991) and in 1988. In 1989 and 1990, photosynthesis was no longer measured in Schlachtensee, so samples from 2 m depth were taken to represent the euphotic zone. On a few occasions in 1989, only integrated samples over 0–7.5 m depth were available (May 5th, 11th, 16th and 26th, June 6th and 12th, and August 28th) for Schlachtensee. These were included in the evaluation of the data: chlorophyll profiles show that this underestimates euphotic biomass by at most 40%. In order to avoid a bias of diversity calculations, photoautotrophic bacteria inhabiting the hypolimnion in June were not included. Very few other sources of bias are to be expected for the diversity calculations in May and June, because at low biomass density, the euphotic zone covered most of the water column, and for August 28th because the water column was well mixed down to 5 m depth and the chorophyll profile shows that scarcely any phytoplankton occurred beneath this. For the winter months, data from samples integrated over the entire water column or from 2 m depth are included. No bias is to be expected from this, as the water column was well mixed and vertical profiles of chlorophyll-*a* show homogeneity.

Phytoplankton was counted in 2 to 25 ml chambers using an inverted microscope. Usually, two diagonals through the chamber (at right angles) were counted; if results between these diverged by more than 20%, additional diagonals were counted. With the exception of *Actinocyclus normannii* Hust. and in some cases *Cyclotella spp.*, centric diatoms were counted by size classes. Cell volume was determined by measuring 10 to 20 cells (depending on the variability of dimensions), at Lake Tegel whenever species were important. At Schlachtensee, measurements taken 1988 were used for calculations of biovolume in 1989 and 1990; new measurements were only taken for species not measured previously.

Diversity was calculated using the Shannon function to the basis of 2 (Shannon, 1948):

$$H' = -\sum \frac{BV_n}{BV_{tot}} \log_2 \left(\frac{BV_n}{BV_{tot}}\right)$$

where BV_n is the biomass of a given species, and BV_{tot} is the total biomass.

Results and discussion

Meteorological conditions

Irradiance, temperature, precipitation and windspeed were chosen for evaluation as key parameters which affect the warming of the water column and stability of thermal stratification. Figure 1a shows monthly means of the first three of these parameters in relation to long-term mean values; panel b shows the deviations of these means from the long-term values. 1987 was an exceptionally cold year, already beginning with an extremely cold winter. This was the only one of

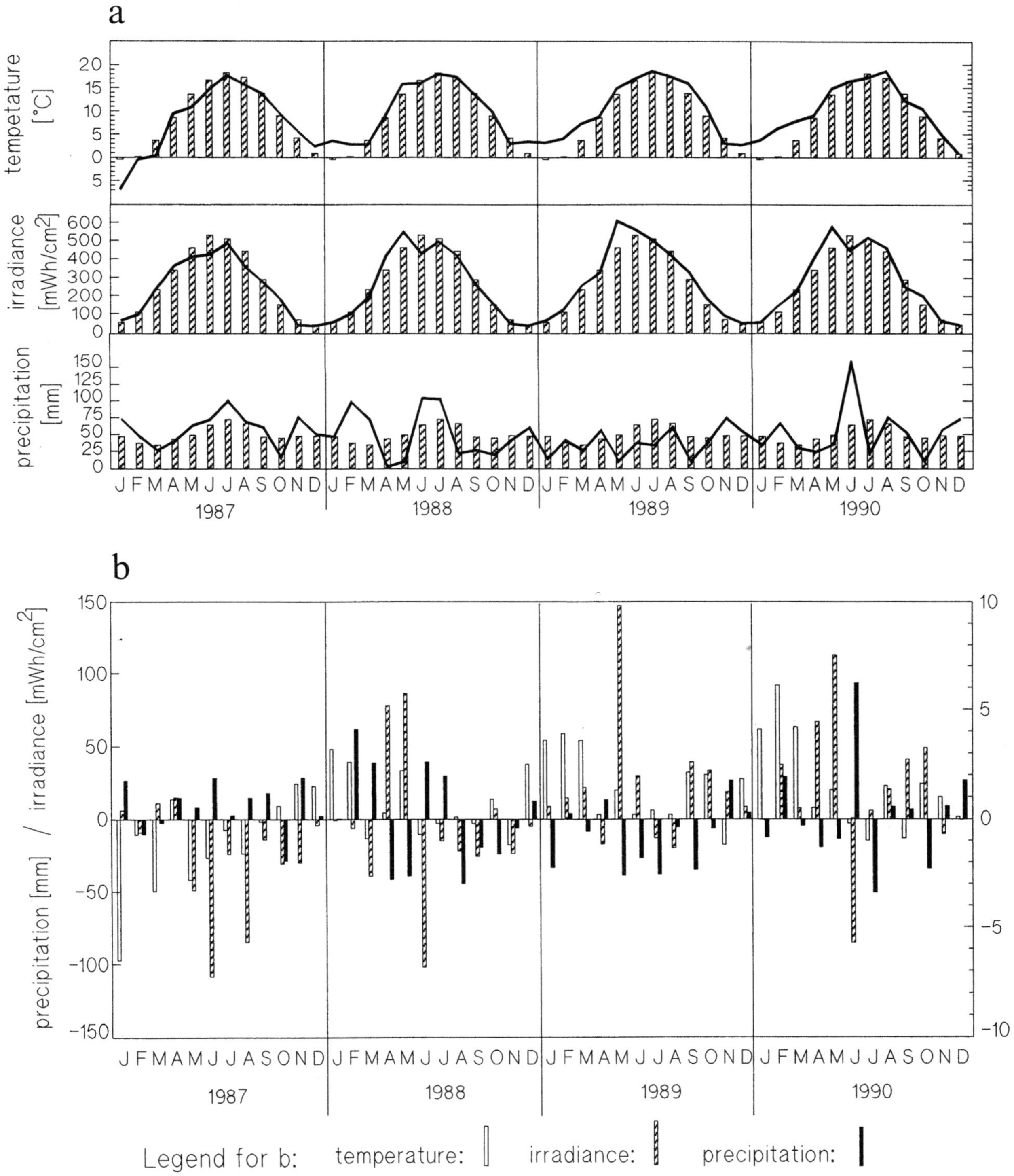

Fig. 1. (a) Monthly means (curves) of temperature, irradiance, and precipitation 1987–1990 as related to monthly long-term mean values (bars). Long-term means (bars) range from 1909–1969 for temperature, from 1965–1977 for irradiance and from 1963–1972 for precipitation. (b) Deviations of monthly means of temperature, irradiance, and precipitation from monthly long-term means over the time spans indicated in the legend for panel a.
(Data kindly supplied by the Meteorological Institute of the Free University of Berlin).

the four years presented here with a long period of ice cover. Also, monthly means for temperature and irradiance were far below long-term means from May through August, while means for precipitation were higher than average. In contrast, the winters of the following three years were exceptionally warm, and irradiance as well as temperature were often well above the long-term means. This also applies to the spring months. The above-average solar energy input especially during May 1988, 1989 and 1990 was important for the establishment of stable stratification in the lakes.

In order to assess disturbances caused by meteorological events on the time scale relevant for intermediate disturbances of phytoplankton succession, these parameters are shown on a day-to-day basis in Fig. 2 for the time span of thermal stratification (from May to mid September of each year). Cyclic patterns of quiescence and disturbance can be recognized with some regularity only for 1988 and 1989, when phases with high irradiance and temperature alternated with phases of low irradiance and/or low temperature, usually combined with precipitation (during the first 2–3 weeks of July 1988, rainfall occurred chiefly at night and thus had little effect on irradiance and temperatures). The frequency of alternation between these phases was higher in 1988: 'fair spells' lasted for 14 to 18 days, whereas (with one exception) they lasted for 18 to 27 days during the summer 1989. In 1987, the time scale of disturbances usually was much smaller, with three to five sunny days alternating with rainy or overcast days. Thus, the data for 1987 scarcely show cyclic patterns. In 1990, all of May was fair, with one slight disturbance around the 10th, but June, August and the first half of September were very variable.

Temperature, thermal stratification and water column stability

Temperature curves for both lakes show similar trends which reflect the differences in weather cycles between the years investigated, but these effects were more pronounced in Schlachtensee than in Lake Tegel (Fig. 3).

During the cold year 1987, warming of the water column occurred more slowly than in the following years, and maximum temperatures were lower. Fortnightly measurements show that water temperatures increased rather regularly up to the end of June. In both lakes, the long period of rainy and cool weather from mid July to mid August is reflected by a decrease of epilimnetic temperatures, whereas in deeper strata, temperatures continued to increase steadily. In Lake Tegel, this destroyed the minor temperature gradient that had developed between May and mid-July. During August, increase of temperatures, somewhat reduced precipitation, and several consecutive days of higher irradiance caused another warming, in Schlachtensee of the epilimnion and in Lake Tegel of the whole water column.

During 1988, similarities between temperatures of both lakes are less conspicuous, due to a lower frequency of sampling in Lake Tegel (measurements were often taken at weekly intervals in Schlachtensee), but essentially the same steep incline occurred during May, and some of the surface temperature peaks found in Schlachtensee (in early June, late July and mid-August) correspond to those observed in Lake Tegel. Characteristically, the fair spell in October caused some renewed stratification in Schlachtensee, but in Lake Tegel, it merely slowed down the temperature decline of the whole water column. Small-scale responses of Lake Tegel's water temperatures to weather changes were generally damped (as compared to Schlachtensee), because mixing distributed effects over a larger water column.

In spring and early summer 1989 and 1990, when Lake Tegel was allowed to stratify more stably before aeration was begun, and when sampling frequencies were similar in both lakes, temperatures followed quite similar patterns: maxima were registered in the upper layers of both lakes in May, June and July 1989 and in late June 1990; minima were observed in both during early June 1989 and in mid-June and mid-July 1990. From August onward, temperature patterns differed, as artificial mixing became the overriding effect in

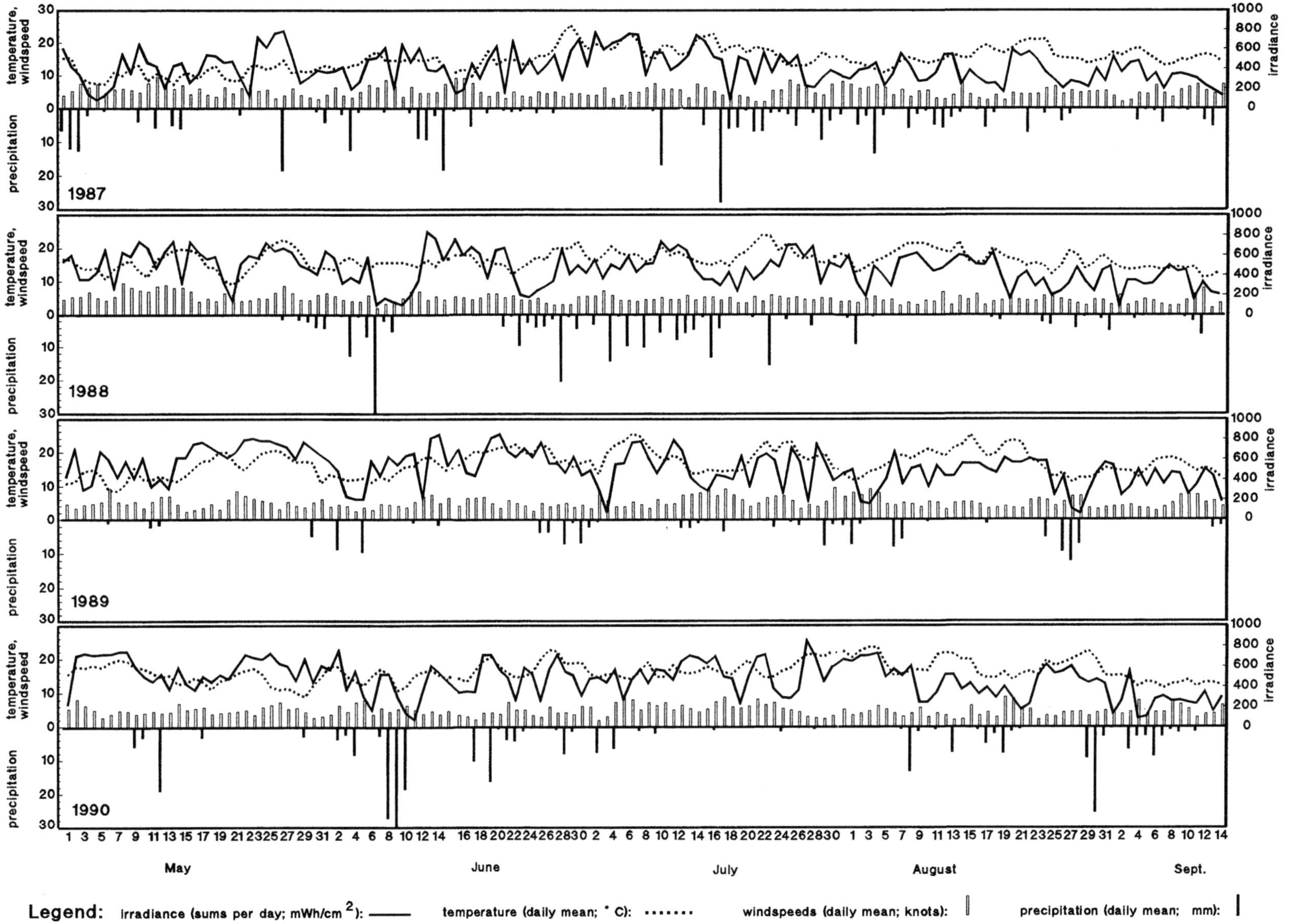

Fig. 2. Daily values for May to mid-September of irradiance (solid curves), temperature (dotted curves), wind speed (unshaded bars) and precipitation (shaded bars).

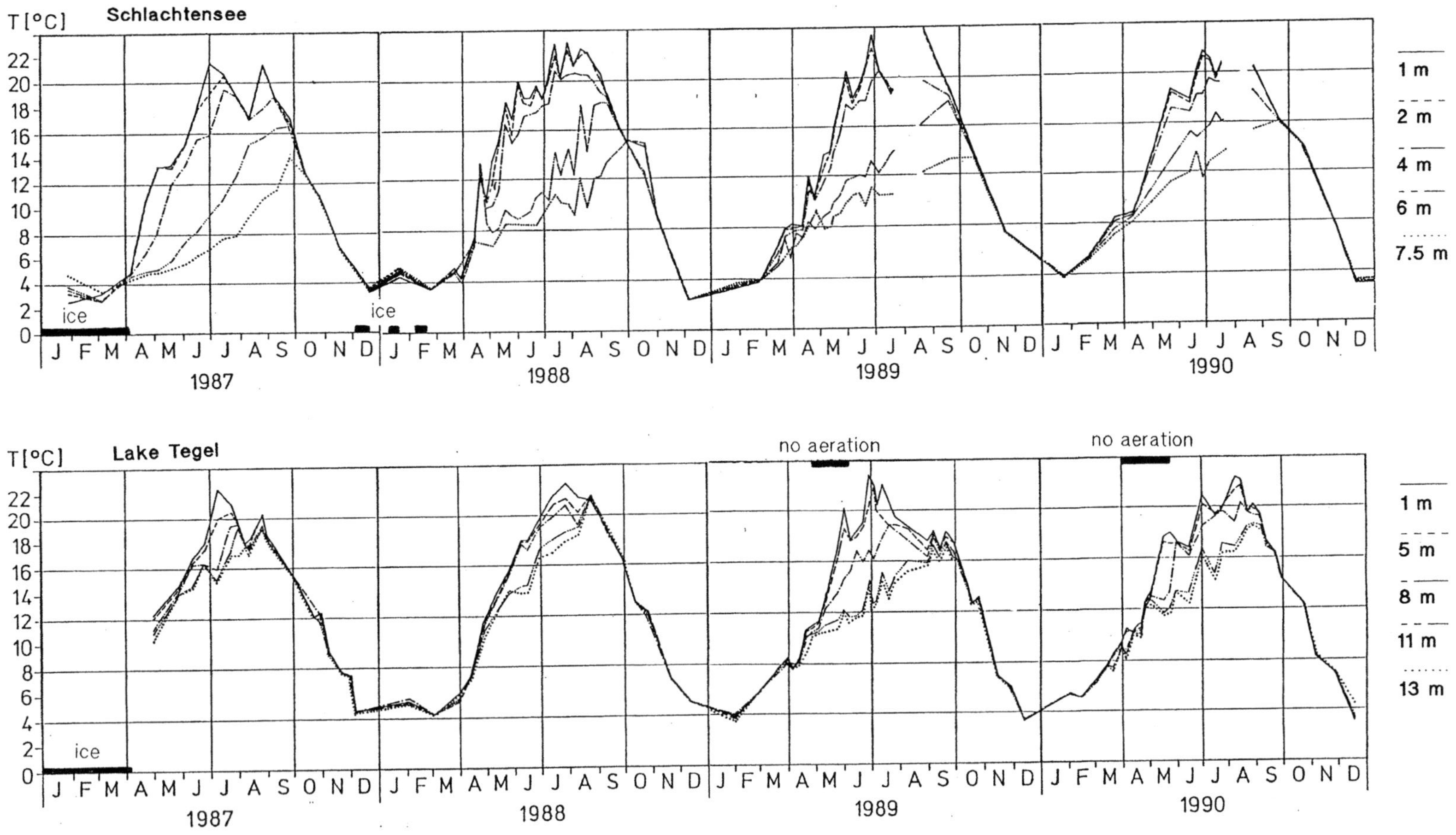

Fig. 3. Temperature in Schlachtensee and in Lake Tegel (Data for Schlachtensee 1987 are taken from GERVAIS 1989).

Lake Tegel. However, the parallels during spring and early summer confirm the strong effect that fluctuations of weather conditions exercise upon stratification and turbulence within the epilimnion. In detail, this can be traced in Figs 7–9. These show that irradiance often is the primary factor for stability or turbulence within the epilimnion. For example, in May 1989, temperatures in both lakes scarcely increased during the first 12 days of the month, as irradiance was low, but increased steadily during the rest of the month, as irradiance was constantly high (Figs 8 and 9). Likewise, the following drop in irradiance during the first days of June was reflected by a decrease of surface temperatures in both lakes; also, water temperature curves show closer proximity for the upper depths measured, and thus indicate increased mixing within the epilimnion. Conspicuous effects of increased windspeeds were scarce between May and September: one example is Schlachtensee from May 9th to 17th, where 9 consecutive days of above-average windspeeds transported heat down to 4 m, causing a steep temperature increase at that depth (Fig. 7).

In spite of the obvious reaction of both lakes to meteorological conditions, effects in Lake Tegel occur on a larger spatial scale: Whereas temperature curves for Schlachtensee show close proximity only at 1 m and 2 m depth (the 4 m curve often being distinctly lower), a similar degree of proximity can frequently be noted for the 1 m and 5 m temperature curves of Lake Tegel. Mixing depth is at least twice that of Schlachtensee, and this has significant implications for light supply to entrained phytoplankton. This deeper mixing within the epilimnion is primarily a consequence of the lake's morphology. Additionally, aeration caused turbulence in the bottomnear layers. Thus, perturbation from the surface and from the bottom reduced temperature differences between the epilimnion and the hypolimnion and – especially before 1989 – often precluded thermal stratification altogether. The lower stability of stratification as compared to Schlachtensee enhanced nutrient recycling in Lake Tegel.

Phytoplankton species composition

For Schlachtensee's hypertrophic phase, data on phytoplankton biomass and species composition are available from January 1982 to June 1983. During this time, phytoplankton biomass reached maxima of 40 cm^3 m^{-3} (Chorus, 1989). The seasonal pattern of species composition in 1982 was typical for a hypertrophic lake, with virtually a monoculture of *Planktothrix agardhii* (Gom.) Anagnostidis & Komárek during late summer and autumn (Table 2). A storm on August 8th caused a temporary decline of its biomass to 1/3 of the density measured on the previous sampling occasion (July 19th), but no change in species composition. During the following winter and spring, cyanobacteria remained dominant (chiefly *Limnothrix redekei*, Van Goor; Meffert), but during March large pennate diatoms also temporarily attained high population densities.

Since 1986, nutrient concentrations in Schlachtensee scarcely changed, and the predominant variable among environmental conditions was the extent of vertical mixing, caused by weather. Patterns of phytoplankton development were remarkably variable (Fig. 4 and Table 2). They showed regular, characteristic features only in late winter and early spring: in all four years, diatoms developed high population densities of 10 cm^3 m^{-3} or more during vernal overturn after ice-out, as in 1987, or already during late winter in years without much ice cover (1988–1990). However, even among these diatoms, species composition differed considerably between years, as did the share of non-diatom species. This was already the case during the lake's eutrophic phase. In every year, the diatom populations declined steeply as thermal stratification developed. Summer phytoplankton populations varied strongly from year to year, largely between species that have some means of motility: *Uroglena sp.*, *Dinobryon sp.*, *Cryptomonas spp.*, *Chrysochromulina parva* Lackey, *Ceratium spp.*, *Peridinium willei* Huitfeldt-Kaas, *Planktothrix agardhii*, *Limnothrix redekei*, *Chroococcus sp.*, but sometimes also *Chlamydocapsa planktonica* (W. et G. S. West) Fott or large, chiefly pennate Diatoms.

Table 2. Seasonal change of species composition in Schlachtensee before and after success of restoration, and in Lake Tegel in the course of restoration (presentation as by Gervais, 1989, altered).

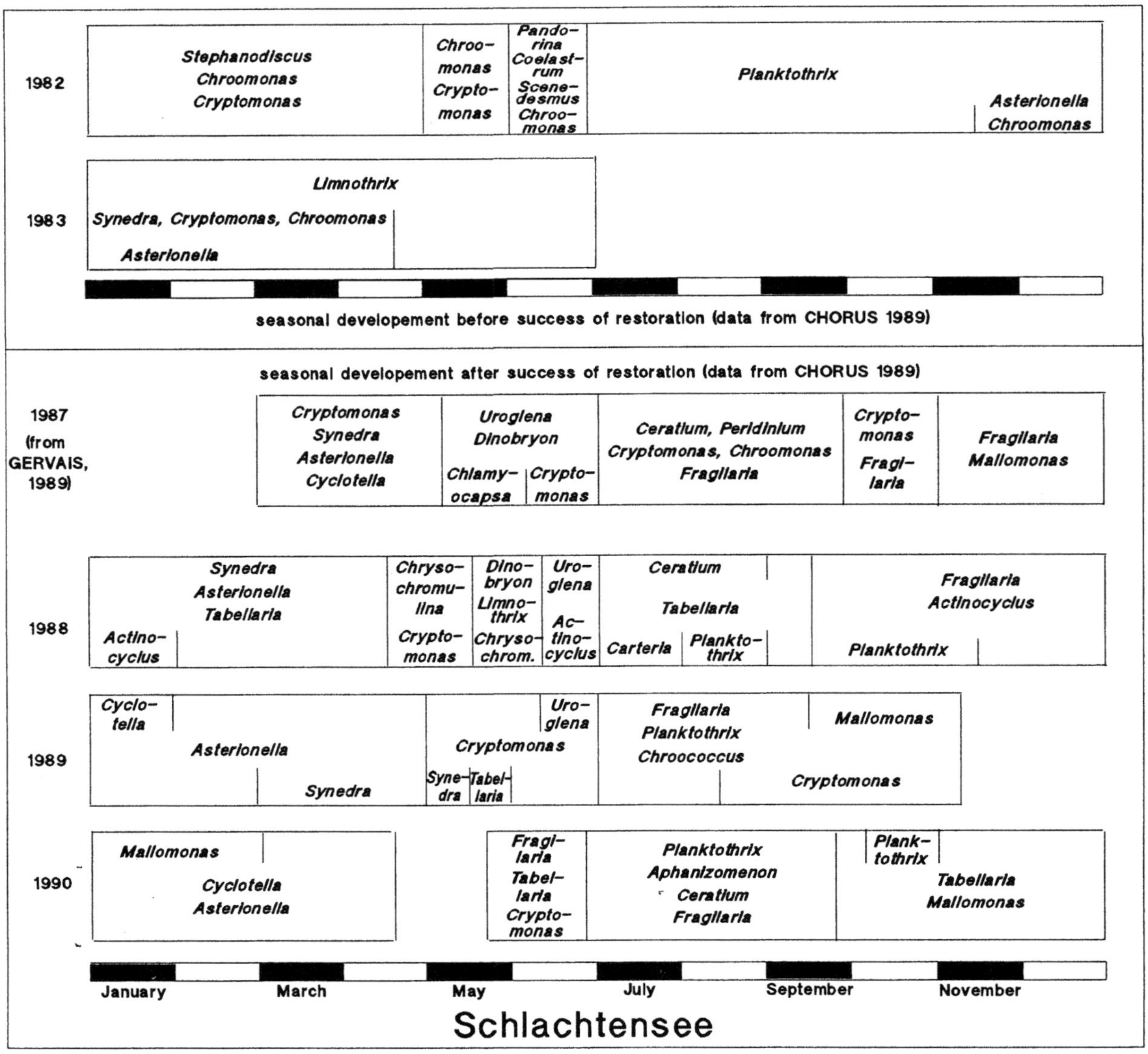

In Lake Tegel, maxima of phytoplankton biomass density were similar to those during the hypertrophic phase of Schlachtensee, but their duration was shorter. In 1989, the spring and summer maxima were limited by phosphorus. The general seasonal pattern of phytoplankton development in Lake Tegel contrasts to the variable pattern of Schlachtensee by showing a surprising similarity between years, in spite of the pronounced differences in phosphorus concentration as well as in mixing conditions which occurred from 1987 to 1990. Also, in contrast to species composition in Schlachtensee, most of the significant species in Lake Tegel were not motile; the turbulence of the lake appears to favour coccal forms. During all four years, the diatom species *Actinocyclus normannii* Hust. and *Cyclotella spp.* dominated from autumn to spring; during winter, up to April, they were associated with *Stephanodiscus spp.* and in spring 1989 and 1990 also with *Asterionella formosa* Hass. and *Synedra acus* Kütz. A distinct clear water phase was noted every year

Table 2. (Continued)

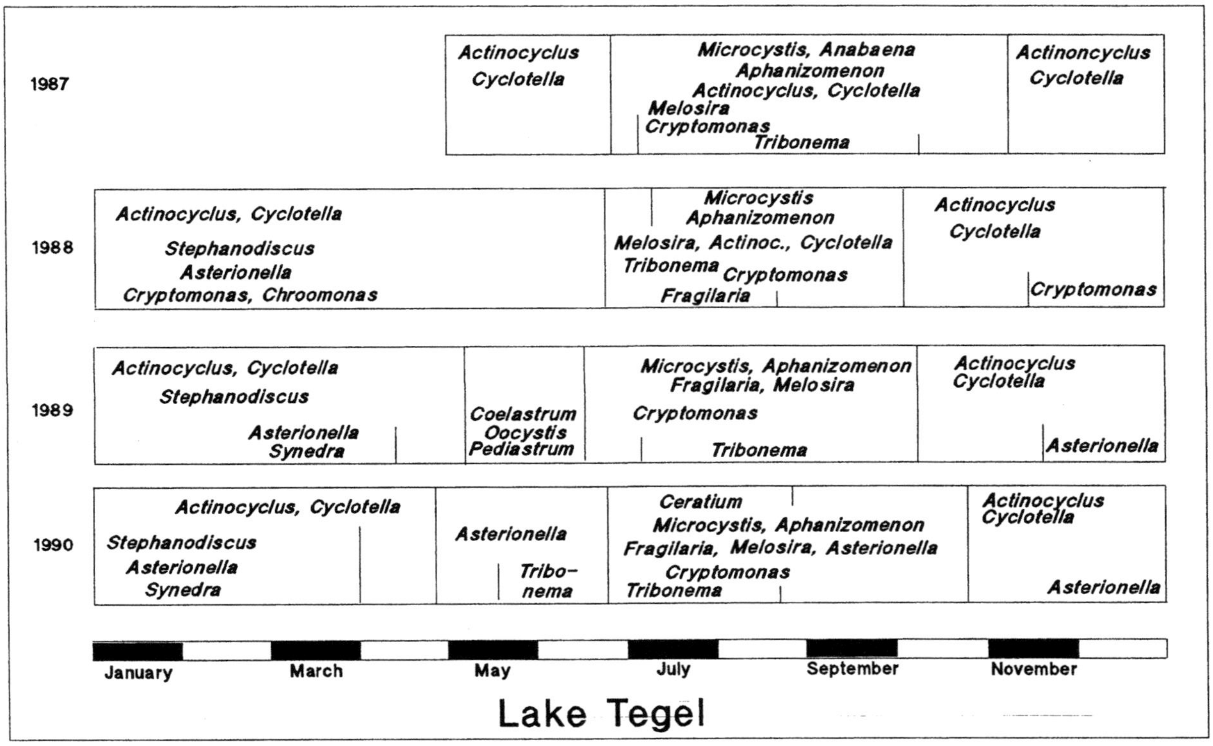

during late spring. During summer of all four years, the cyanobacteria *Microcystis spp.* and *Aphanizomenon flos-aquae* Ralfs. dominated with high biomass densities. These two species form remarkably large colonies in Lake Tegel; diameters of up to 5 cm were observed for *Microcystis* in 1989. This may be a mechanism by which they increase their buoyancy and thus are able to reduce their entrainment into the turbulences of Lake Tegel's epilimnion. Only in August 1990 was their density briefly exceeded by *Ceratium spp.* Populations of *Aulacoseira spp.*, *Tribonema affine* West and *T. vulgare* Pasch. as well as some Cryptophycean species also developed in every one of the four summers. Variability occurred among the summer diatoms – all of which were large or colonial: centric diatoms prevailed in 1987 and 1988, but *Fragilaria crotonensis* Kitt. and *Asterionella formosa* were the chief species in 1989 and 1990. This is probably a reaction to reduced mixing, as cylindrical cells have higher sedimentation losses than the bands of *Fragilaria crotonensis* or the star-shaped colonies of *Asterionella formosa* (Reynolds, 1986). Also, a short maximum of green algae, chiefly Chlorococcales, occurred only in early June 1989, and *Ceratium spp.* became dominant only during August 1990. The latter may be due to two spells of very warm weather (end of July and end of August, Fig. 4) which warmed the upper 5–8 m of the lake (Fig. 1) and caused more pronounced thermal stratification than had existed during August of any of the previous years.

Thus, species composition and seasonality of phytoplankton were very different in both lakes. During summer stratification, physical conditions in Schlachtensee favored motile species with some means of actively seeking a favorable depth. Among these, the subset which prevailed while total phosphorus concentrations exceeded 40 $\mu g\ l^{-1}$ P was different from that which presently occurs under phosphorus limitation of total biomass. In Lake Tegel, species dominate which would probably have excessive sedimentation losses in Schlachtensee during stratification.

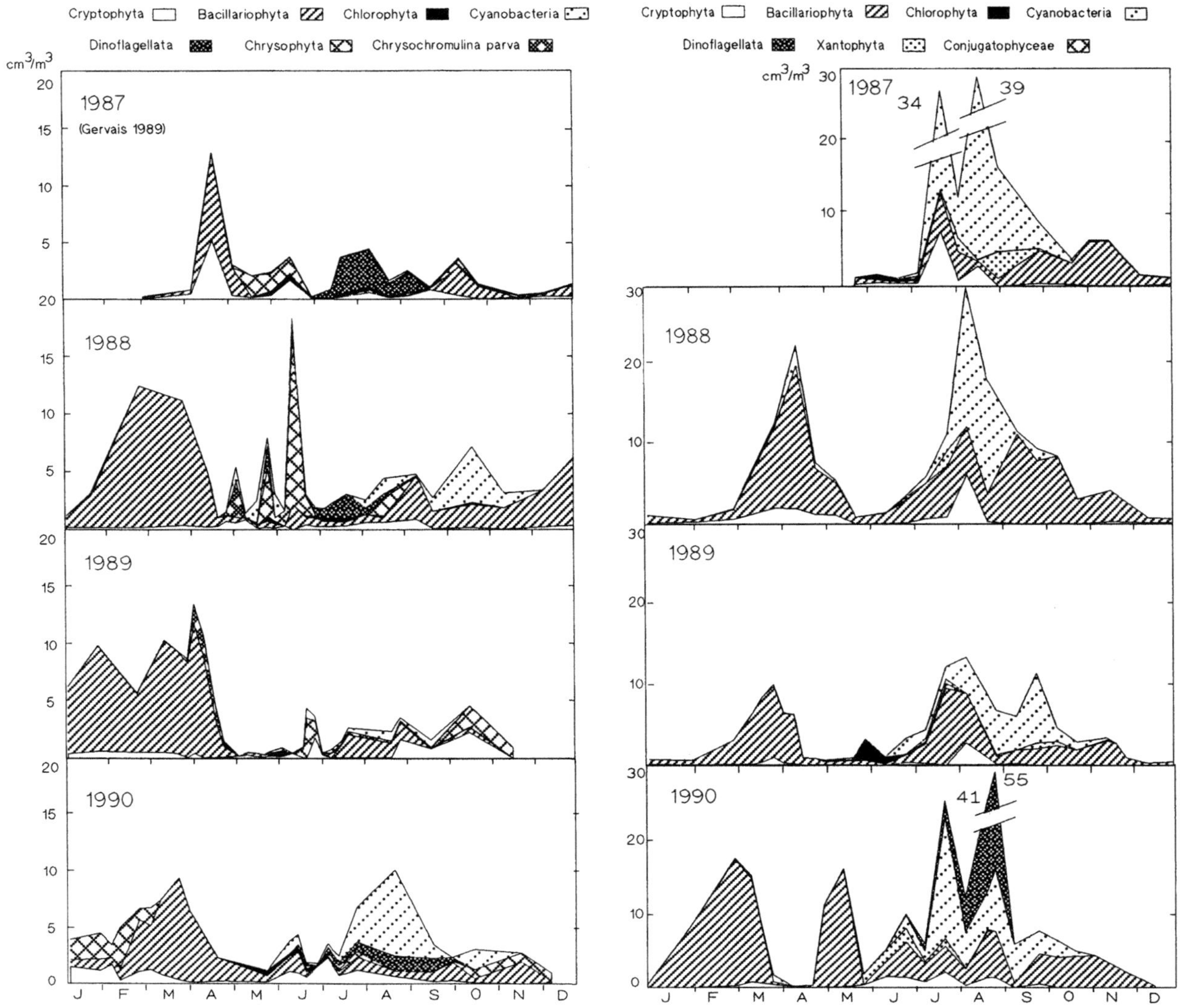

Fig. 4. Composition of phytoplankton in the euphotic zone of Schlachtensee and Lake Tegel. (Data for Schlachtensee 1987 are taken for 2 m depth from Gervais, 1989; note different biomass-scales for both lakes!)

Diversity and disturbance

Schlachtensee before restoration success

The data collected during Schlachtensee's hypertrophic phase in 1982/83 show a clear seasonal pattern for diversity as measured by the Shannon function: values were high in winter, spring and early summer, but extremely low during late summer and autumn (Fig. 5a). During spring, high diversity was not the result of meteorological disturbances of intermediate frequency. Instead, the rapid change of phytoplankton species composition was governed by directional resource developments typical for the season: ice-off, increasing stratification and sedimentation. In 1982, the biological response to these parameters followed the pattern described as seasonal succession by the PEG-model (Sommer *et al.*, 1986) quite closely; species change was caused by sedimentation losses, presumably by silicate depletion, and also by increasing grazing pressure (Chorus, 1989). In 1983, a mild winter with only brief ice

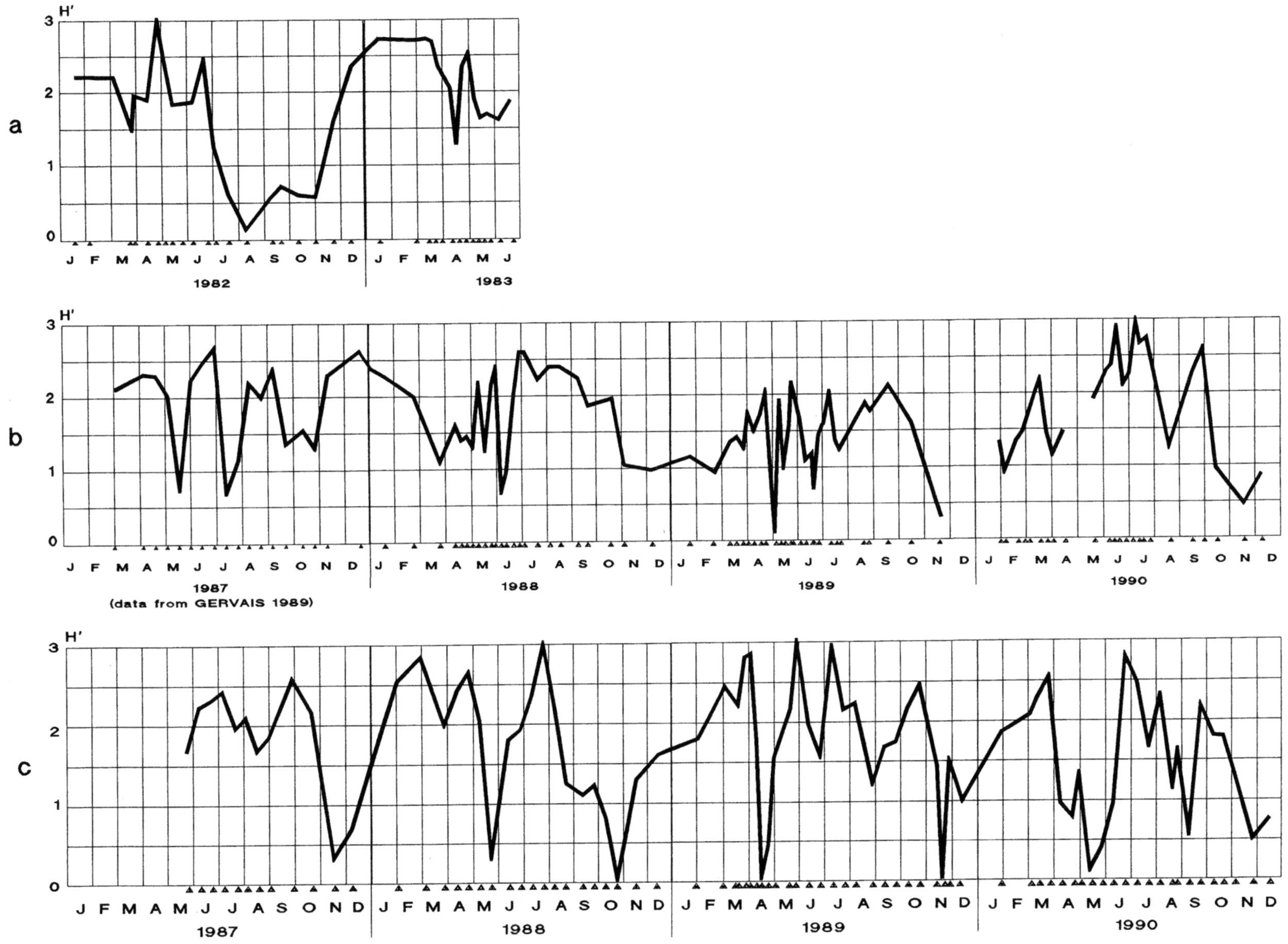

Fig. 5. Diversity as calculated by the Shannon function for
a: Schlachtensee before reaction of phytoplankton to restoration b: Schlachtensee 1987–1990, after successful restoration
c: Lake Tegel 1987–1990. Triangles on the time axis indicate sampling occasions.

cover allowed the long-term survival of a 'climax species' (*Limnothrix redekei*) and large pennate diatoms, whose population densities changed only slowly. During the spring of both years, the lack of limitation by phosphorus or nitrogen was important for the maintenance of high diversity: as one species declined, the next species could already grow.

Low diversity during summer 1982 reflects the pronounced dominance of *Planktothrix agardhii* over a period of several months during a very warm and calm summer. This is an example for competitive exclusion by a 'climax species' under stable physical conditions and limitation by only one resource: *P. agardhii*, once dominant, causes a high degree of turbidity (secchi depths ranged between 0.4 and 0.5 m), in which other species could no longer unfold sufficiently rapid growth rates to outcompete *P. agardhii.* Berger, (1987) and Berger & Sweers (1988) have shown that this dominance may be stable for months and be terminated only by severe and prolonged changes in mixing depth, which reduce light availability even below the level required by mass developments of *P. agardhii.* The duration of the storm which occurred on August 8th 1982 was too short to accomplish this; as described by Reynolds (1988), this summer climax species was resilient towards such brief physical disturbances. Only prolonged decrease of irradiance and increasing mixing depth towards winter favored *Limnothrix redekei*, which gradually replaced *Planktothrix agardhii* and remained dominant up to July 1983. However, such a stable dominance of *P. agardhii* requires two conditions: one is the lack of nutrient limitation (otherwise sufficiently high population densities cannot be achieved, and turbidity is not high enough to preclude the growth of other species). The second condition is that mixing depth does not exceed 2.5 m, because entrainment in deeper mixing would reduce the trichomes' residence time within the euphotic zone to a point where biomass would decline because of light limitation (Berger, 1987); a reduction of the biomass of Planktothrix would, however, increase light availability and thus give other species a chance to grow. This second condition was also fulfilled during the unusually warm and fair summer 1983, during which mixing depth rarely exceeded 2.5 m (Chorus, 1989).

Schlachtensee after successful restoration

After successful restoration, the seasonal pattern of diversity in Schlachtensee has been quite variable (Fig. 5b), as is to be expected from the variability of phytoplankton species composition described above. In contrast to the hypertrophic phase of the lake, diversity was frequently low in winter and spring, with values declining to 1 or less during the strong dominance of *Synedra acus* (1988), *Asterionella formosa* (1989), or *Mallomonas caudata* IWAN. (1990). This may have been due to phosphorus limitation: at concentrations of soluble reactive phosphorus within detection limits (0.3 $\mu g\ l^{-1}$) and 1 $\mu g\ l^{-1}$, which prevailed during spring from 1987 to 1990, liberation by the declining species or its consumers may be necessary, before another species can begin to grow, whereas without limitation, the ascending species may form a considerable population density before the previous species has strongly declined.

Also in contrast to the hypertrophic phase of the lake, post-restoration summer diversity values have usually been high. Phosphorus concentrations no longer allowed the extremely high biomass levels which had previously secured the 'climax dominance' of *Planktothrix agardhii.* Meteorologically induced disturbances of intermediate frequency may have been the key mechanism maintaining high diversity and precluding stable late summer climax populations. One indication for this are the low values for the Shannon diversity in 1989, the year with the longest phases of steady meteorological conditions. Not only did maxima for H′ never exceed 2.3, but low values also occurred more frequently than in other years (Fig. 6). This appears to correspond to the prediction of the 'intermediate disturbance hypothesis': longer undisturbed phases should allow the process of competition to run further towards the point of exclusion, which would cause a decline of diversity.

In order to test whether this actually was the mechanism responsible for diversity in Schlach-

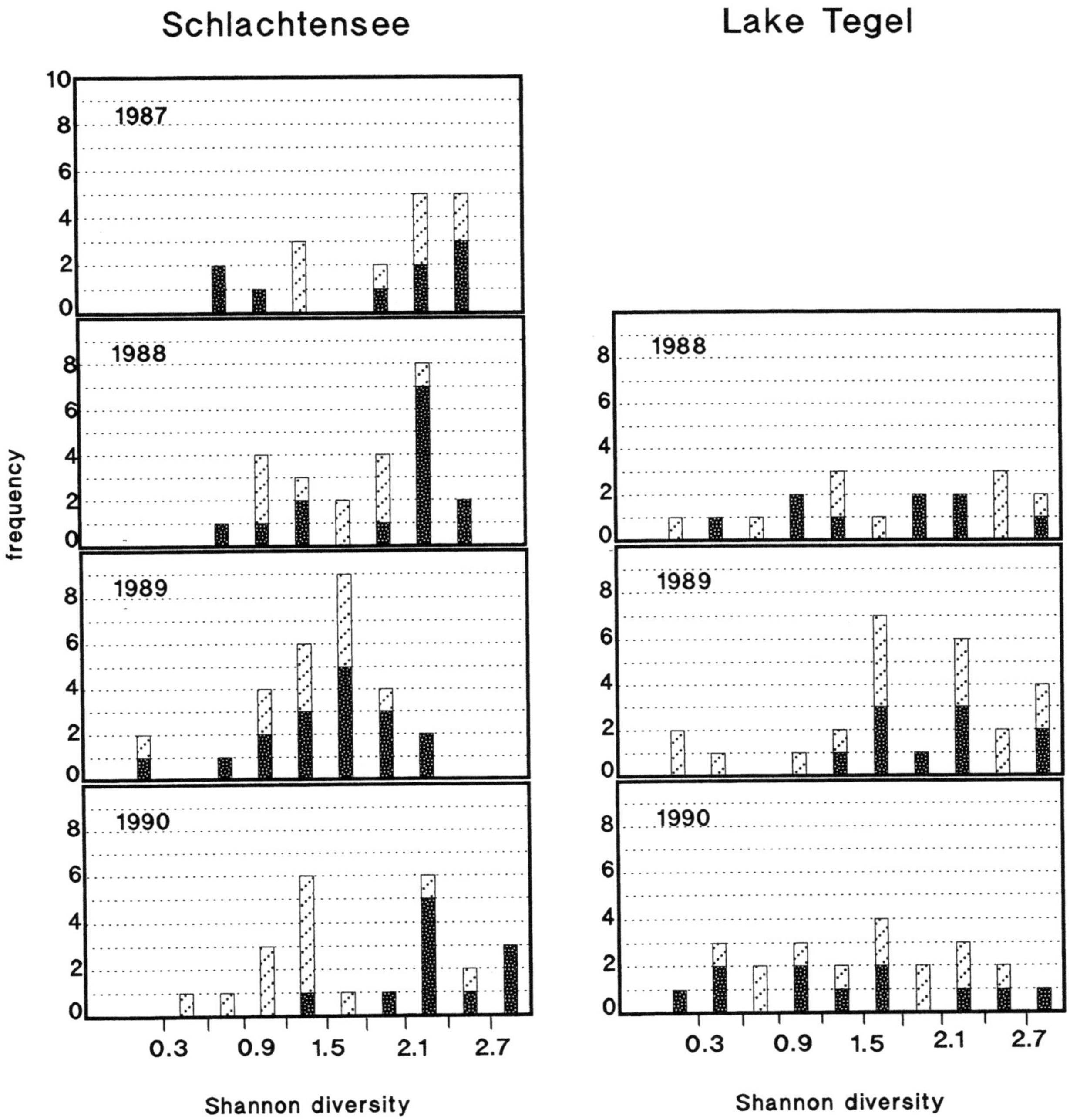

Fig. 6. Frequency distribution of diversity ratings as calculated by the Shannon funktion; dark shading: values during thermal stratification, between May 1st and Sept. 15th; light shading: values during mixed or scarcely stritified periods (before May or after Sept. 15th).

tensee, population development must be compared to physical forcing, as is discernible from meteorological parameters and lake temperature profiles during thermal stratification. In the following, such an evaluation is undertaken for the stratified period from May to mid-September using the examples 1988 and 1989, as both years showed pronounced, but different cycles of meteorological disturbances.

Spring. During May of both years, large and mainly non-ingestible phytoplankton species prevailed. This may be attributed to seasonal succession in the sense of the PEG-model (Sommer, 1986), as herbivorous crustacean populations were large and growing during this time (Ewald, 1991). However, pronounced differences in the prevalence of these phytoplankton species occurred between 1988 and 1989, and these differences can be attributed to interannual variations of water column mixing and stability:

During May 1988, situations with low irradiance and sometimes also high windspeeds alternated with fair weather at a frequency of a few days, and this was traced by an irregular increase of water temperatures. Phytoplankton reflected this high frequency of disturbances by a rapid progression of dominant species (Fig.7), which caused high diversity. *Limnothrix redekei* dominated during periods with higher water column stability in mid-May and towards the end of the month. Increased perturbations between May 17th and 25th (marked by a decrease in temperature) is likely to have promoted the growth of *Dinobryon sp*., as turbulence would assist in keeping the colonies in suspension.

In 1989, meteorological conditions were characterized by two distinct phases: during the first two weeks of May, low irradiance and high windspeeds caused increased mixing within the upper 4 m of the water column (which had already begun to develop during April, Fig. 3). This was followed by a long phase of steady fair weather and stable stratification from May 14th to the end of the month (Fig. 8). According to the expectation of the 'intermediate disturbance hypothesis', progression of phytoplankon species composition was considerably more steady than during the previous year, and as a consequence, diversity was lower. During the turbulent first phase of the month, diatoms (*Synedra acus* and *Tabellaria spp.*) reappeared. Diversity briefly increased, but rapidly declined again by May 16th, as *Synedra acus* became increasingly dominant. During the following quiescent phase, the 'internally driven progress towards ecological equilibrium' (see definition of 'IDH') was remarkably slow: the diatom population declined only gradually, and it took more than one week for other species (*Ankyra* sp., *Koliela longiseta, Dinobryon sp.* and cryptophytes) to begin growing. As these additional species were growing, diversity increased. Their population density remained low, and the stable period apparently was not long enough for competition to proceed towards the exclusion of any of these species. Thus, although the period of stable conditions lasted for more than two weeks, diversity failed to fulfil the prediction of the 'intermediate disturbance hypothesis' within the time available.

Early summer. Differences in the early summer situation between both years were more extreme, and followed a pattern explicable by mixing conditions only in one year:

During June 1988, diversity was low during a period of meteorological disturbance and increased turbulence within the epilimnion. However, phytoplankton species composition calls for doubt as to whether physical forcing was a relevant cause: although perturbed conditions are unlikely to favor motile species, the colonial flagellate *Uroglena sp.* became dominant by June 6th. Increased mixing induced a slight come-back of diatoms, but the mass development of *Uroglena* outweighed these by far. *Uroglena* remained dominant well into the beginning of the following steady and fair spell (June 12th to 24th), but nonetheless declined rapidly after June 13th, instead of maintaining its position in the sense of a 'climax species'. Diversity increased, not because of the ascent of new species, but because of increasing equitability due to this decline. Thus the pattern was reversed to the predictions of the 'intermediate disturbance hypothesis', and shows that factors other than disturbance were overriding in governing the development.

In 1989, the month of June also began with a few days of weather-induced increased vertical mixing, but this quickly favored non-motile species: the density of *Dinobryon sp.* increased further, and *Oocystis sp.* briefly contributed a rather high share of total biomass (51%). This suggests that perturbation may very rapidly select species

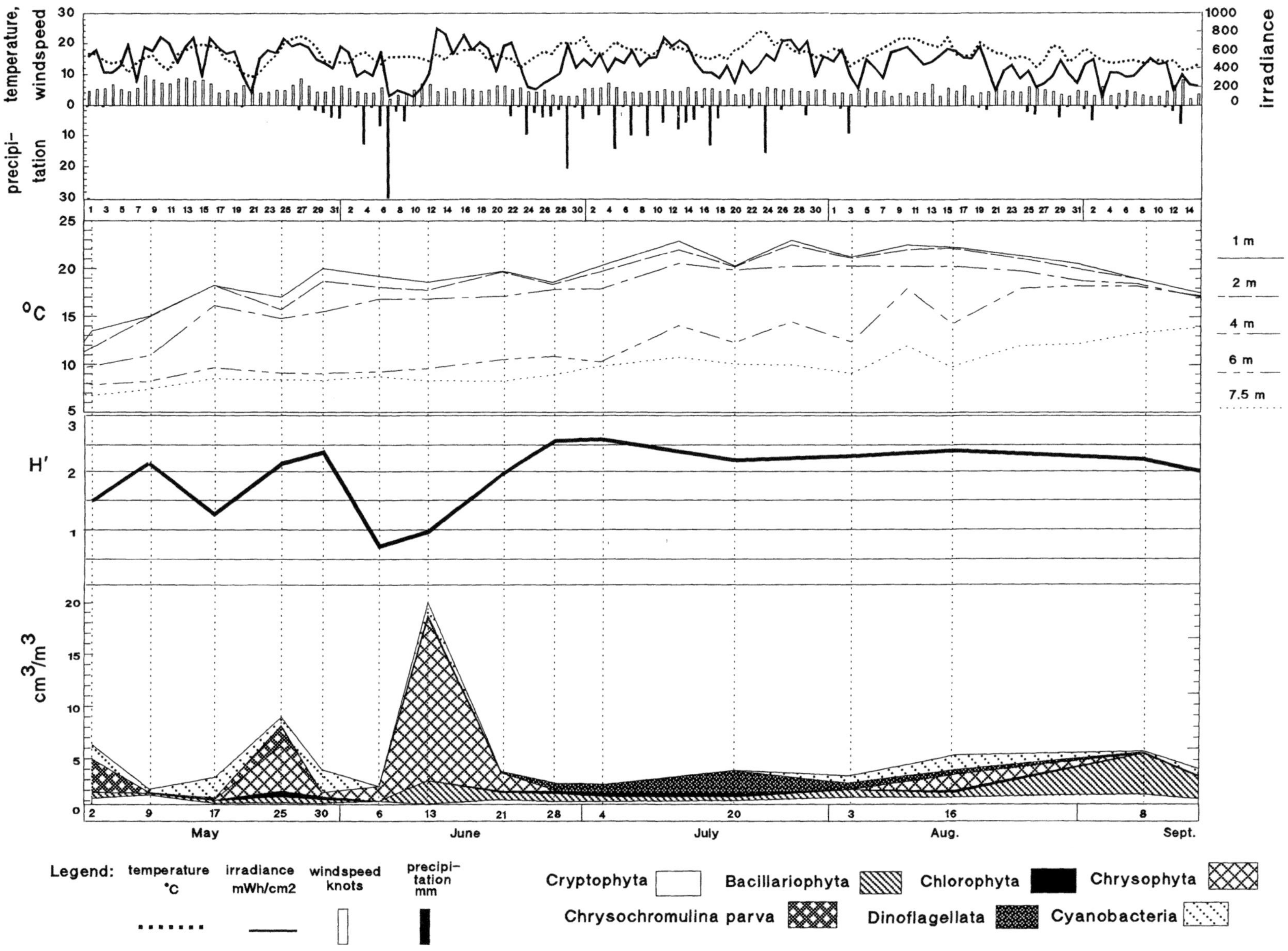

Fig. 7. Meteorological conditions, water temperature, diversity, and species composition during summer stratification in Schlachtensee 1988.

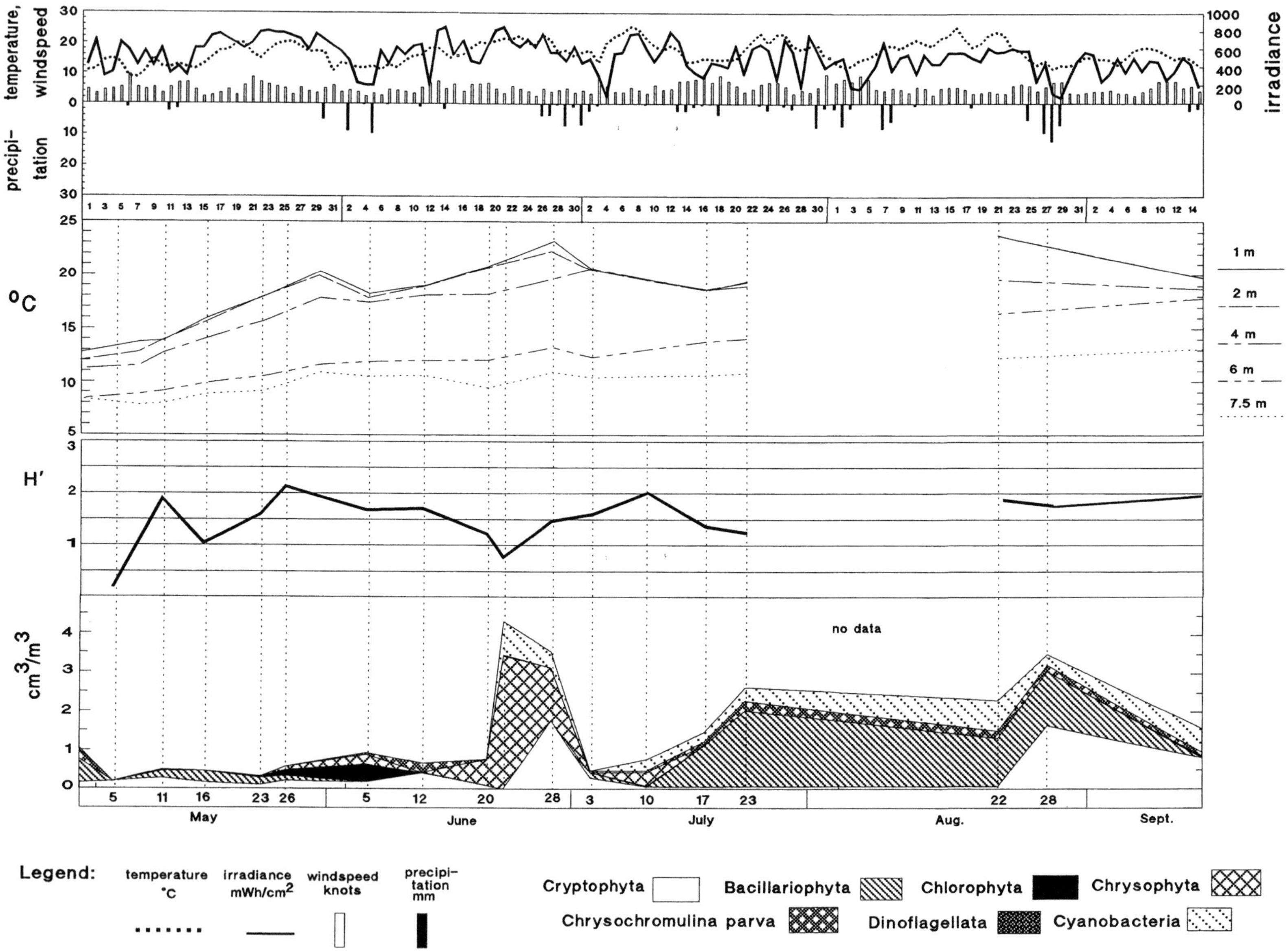

Fig. 8. Meteorological conditions, water temperature, diversity, and species composition during summer stratification in Schlachtensee 1989.

favored by a increased intensity and frequency of epilimnetic mixing. In contrast, progression was gradual during the three-week fair spell that followed from June 6th to the end of the month; as expected, the extremely stable stratification (due to water temperatures up to 23 °C) favored flagellates, first *Chrysochromulina parva* together with crypophytes. These were followed by species better protected against grazing: during the second half of the month, increasing dominance of *Uroglena* sp. together with a smaller population of *Planktothrix agardhii* caused diversity to decline. This sequence of species follows the pattern of competitive exclusion under stable allogenic conditions very well; also in agreement with the 'intermediate disturbance hypothesis', this situation appears to have been disrupted by increased mixing (mediated by weather) at the end of the month. However, the development of a significantly larger *Uroglena*-population during the preceding year inspite of disturbance represents a serious objection to such an interpretation. Apparently, more information on the ecology of *Uroglena* is necessary for an understanding of the factors governing its dominance.

Late summer. From July to mid-September, diversity was remarkably constant in both years, but clearly higher during 1988. As for May of both years, the differences in population development are likely to be a result of weather-induced differences in mixing of the epilimnion.

In late June 1988, a few days of meteorological disturbance (from the 24th to 28th) were followed by three weeks of high irradiance. Some precipitation was also registered during this time, chiefly at night, and temperature curves show that this disturbed epilimnetic water column stability only slightly (Fig. 7). Phytoplankton responded with the development of an assemblage of large flagellates (*Peridinium willei, Carteria sp.*, and *Ceratium spp.*, treated as one species for diversity calculation). Among motile algae, these species appear to be rather 'powerful swimmers': in Schlachtensee, all three have proven to migrate vertically throughout the aerobic depths (Chorus & Speck, 1991). Thus, they can probably cope with the slight turbulence caused by precipitation, although increased mixing will terminate their dominance (as observed in 1987 by Gervais, 1991). Their coexistence is probably not linked to disturbance, but rather due to niche separation by vertical migration. Subsequently, towards the end of July and during the first 3 weeks of August, conditions of high irradiance were more constant than during July, and temperature curves indicate that stratification within the epilimnion, between 2 m and 4 m depth, was rather pronounced. This probably favored less powerful swimmers among the motile species: *Planktothrix agardhii* and *Uroglena sp.*; these replaced the large flagellates very gradually. Thus, in this case not disturbance, but the environmental change towards increased quiescence is likely to have promoted the change in species composition. In contrast, by September 8th, disturbance in the sense of increased mixing depth had allowed three species of diatoms to outcompete *Planktothrix* and *Uroglena* – a typical situation for autumn.

In 1989, the fair period during the first 10 days of July is a good example of succession after a 'setback' (sensu Reynolds, 1988) presumably induced by increased mixing: after the decline of *Uroglena*, competition started anew between the same species which had previously been dominant – *Uroglena sp., Planktothrix agardhii, Chroococcus sp.*, and cryptophytes. Characteristically, diversity was high because biomass was rather evenly distributed between them. Again, competition was a slow process which showed little change within the week from July 3rd to 10th. However, by July 17th, reduced irradiance and increasing wind speeds very rapidly induced growth and dominance of diatoms, chiefly *Fragilaria crotonensis.* They outcompeted *Uroglena* and prevented the blue-greens from increasing their stock. This is one more example of reduced diversity because of rapid dominance of one species under conditions of perturbation. Subsequently, the calm period from August 9th to 22nd allowed for the population growth of Cyanobacteria (chiefly *Planktothrix agardhii*), but was not long enough for them to outcompete *Fragi-*

laria; consequently, diversity increased due to increased evenness.

Consequences for disturbance – diversity mechanisms in Schlachtensee

These results show that in Schlachtensee higher diversity in 1988 as compared to 1989 was indeed linked to weather events, but the causal connection is not as straightforward as previously assumed:

1. The cases in which long phases of stable fair weather conditions actually led to reduced diversity because of competitive exclusion were rare: the data show clear evidence for this only during the summer of 1983, under hypertrophic conditions. The only other example is the *Uroglena*-maximum in June 1989, but its interpretation primarily as a result of stable physical conditions was shown to be somewhat questionable. Usually, the process of population growth and competition was too slow to reach the point where one species clearly outweighed the others, even during long phases of quiescence.
2. Low diversity may just as well be the result of perturbation after phases of quiescence (such examples are the dominance of *Fragilaria* on July 17th 1989 and of *Oocystis* on June 5th 1989); in such situations, turbulence is especially favorable for one species.
3. In some cases, such as August 1988, not disturbance, but the gradual increase of quiescence within the epilimnion lead to a gradual trade-off of populations – a slow process which also maintained high diversity.

Thus, the detailed comparison of population development and physical parameters for 1988 and 1989 shows that the rate of chance between perturbed and quiescent conditions is a key factor controlling species trade-off and as a consequence also diversity.

Lake Tegel

In face of the very different mixing conditions in Lake Tegel as compared to those in Schlachtensee, the parallels as well as the differences in diversity patterns are informative. Similarities with Schlachtensee under hypertrophic conditions occurred especially during winter and spring: in Lake Tegel diversity was also high during spring, especially at the end of the vernal phytoplankton maximum. This confirms the observation at Schlachtensee, that a lack of severe nutrient limitation may enable a more rapid exchange of dominant species. In late spring of every year observed, diversity in Lake Tegel declined to a first minimum: in 1988 and 1989 during the clear water phase, when phytoplankton consisted only of a few centric diatoms, and in 1990 two weeks after the clear water phase, because *Asterionella formosa* caused a second, single-species diatom maximum in May.

Summer diversity was high in Lake Tegel in all four years observed, in contrast to Schlachtensee under hypertrophic conditions. Dominance of cyanobacteria was not due to only one species, but to a rather even share of *Microcystis spp.* and *Aphanizomenon flos-aquae*; also, diatoms and either *Cryptomonas spp.* or *Tribonema spp.* were present. The lack of competitive exclusion by one species can be attributed to the special mixing conditions of the lake, which cause fluctuations of resource availability: variations in frequency and depth of mixing of the upper layers alternate the degree of limitation by light and the possibilities for *Microcystis* to float near the surface. Further, aeration-induced turbulence near the lake bottom entrains some of the silicate which accumulates above the sediment and transport this upward. To a variable extent, it reaches the epilimnion via eddy diffusion and resupplies the diatoms, enabling the growth of their summer populations. The same applies to phosphorus: especially in 1989 and 1990, concentrations of soluble reactive phosphorus in the upper strata of the lake frequently amounted to only 1 $\mu g\, l^{-1}$ and thus were low enough to limit uptake rates and to affect competition between species, but turbulence resupplied these strata with phosphorus accumulated in bottom-near layers. Thus, during summer three resources – light, phosphorus, and silicate – may have interchanged in limiting growth of phytoplankton in Lake Tegel, which by theory (Sommer, 1989) would allow for the coexistence of three species. Mixing conditions in Lake Tegel therefore were important in maintaining high summer diversity, not only via their direct effects on sedimentation or motility of phytoplankton, but also by affecting nutrient supply.

Further, the 'intermediate disturbance hypothesis' predicts a decline of diversity at increasingly strong disturbances, beyond an optimum (Sommer *et al.*, 1993). For the verification of this part of the hypothesis, the situation in Lake Tegel during autumn is of interest: towards the end of all four years, diversity decreased to a minimum as *Actinocyclus normannii* became dominant. However, the decline of diversity in Lake Tegel occurred in October or November, several months after complete vertical mixing had begun. Thus, overturn alone does not qualify as disturbance strong enough to cause the predicted decline of diversity. Only together with the annual minima of irradiance (Fig. 1) in November and December did deep mixing strongly restrict the number of species capable of coping with such severe light limitation.

Notably, in Lake Tegel summer diversity showed more fluctuations during 1989 and 1990, the two years with a higher stability of thermal stratification. The analysis of frequency distribution of diversity ratings (Fig. 6) shows a tendency towards higher ratings in 1989 – the year with the most pronounced thermal stratification and longest phases of calm weather. The phytoplankton species behind this diversity show that in addition to the general, regular pattern of the lake, some new populations appeared during these years with more phases of quiescence: high water column stability enabled the growth of a population of Chlorococcales in late May 1989, *Anabaena spp.* in June 1989, and the mass development of *Ceratium sp.* in August 1990 (Fig.4). This reflects a slight increase in the dynamics of phytoplankton population development, which probably is due to the increased amount of chance in the mixing conditions in the epilimnion (Fig. 3). Thus, Lake Tegel shows that the 'intermediate disturbance hypothesis' may apply in a reversed sense which could be termed 'intermediate quiescence hypothesis': in a system adapted to perturbation, interjected phases of quiescence may increase diversity and represent the actual 'disturbance' .

The outstanding feature of phytoplankton species composition and diversity in Lake Tegel however, is the high degree of interannual regularity, especially when compared to Schlachtensee. This shows that the seasonal course of physical forcing is a primary factor governing the phytoplankton assemblage in this shallow lake.

On the other hand, in face of the basic differences between mixing conditions and species composition in both lakes, details in the time course of diversity show a surprising amount of congruence of diversity; numerous maxima and minima occurred simultaneously: in 1987 at the end of June and August, 1988 in March, April, May, July, and September, 1989 in late March, in late May, in early July and in August, 1990 in March, in June, towards the end of July, in August, and towards the end of September in 1990. Similarity of the seasonal patterns of diversity in these two very different lakes was most striking during 1989, when stratification in Lake Tegel was more stable than in any other year, and cycles of weather changes were most pronounced. This is further strong evidence for the effect of meteorological cycles on the wax and wane of species, even in a system adapted to turbulence. An explanation of such similar effects in so different systems requires a detailed investigation of population development, together with the physical conditions. For this purpose, data from Lake Tegel for 1989 are evaluated and discussed in comparison to the results for Schlachtensee:

Towards the end of May and in early June, high diversity was due to the development of Chlorococcales in both lakes (Fig. 8 and Fig. 9), and in both, these were favored by the same mechanisms: high irradiance at low biomass and thus turbidity allowed enough light penetration for these green algae to unfold high growth rates. However, in the usually more turbulent epilimnion of Lake Tegel, the relatively low extent of vertical mixing ('intermediate quiescence') during the fair period was probably also an important requirement for their development, as entrainment

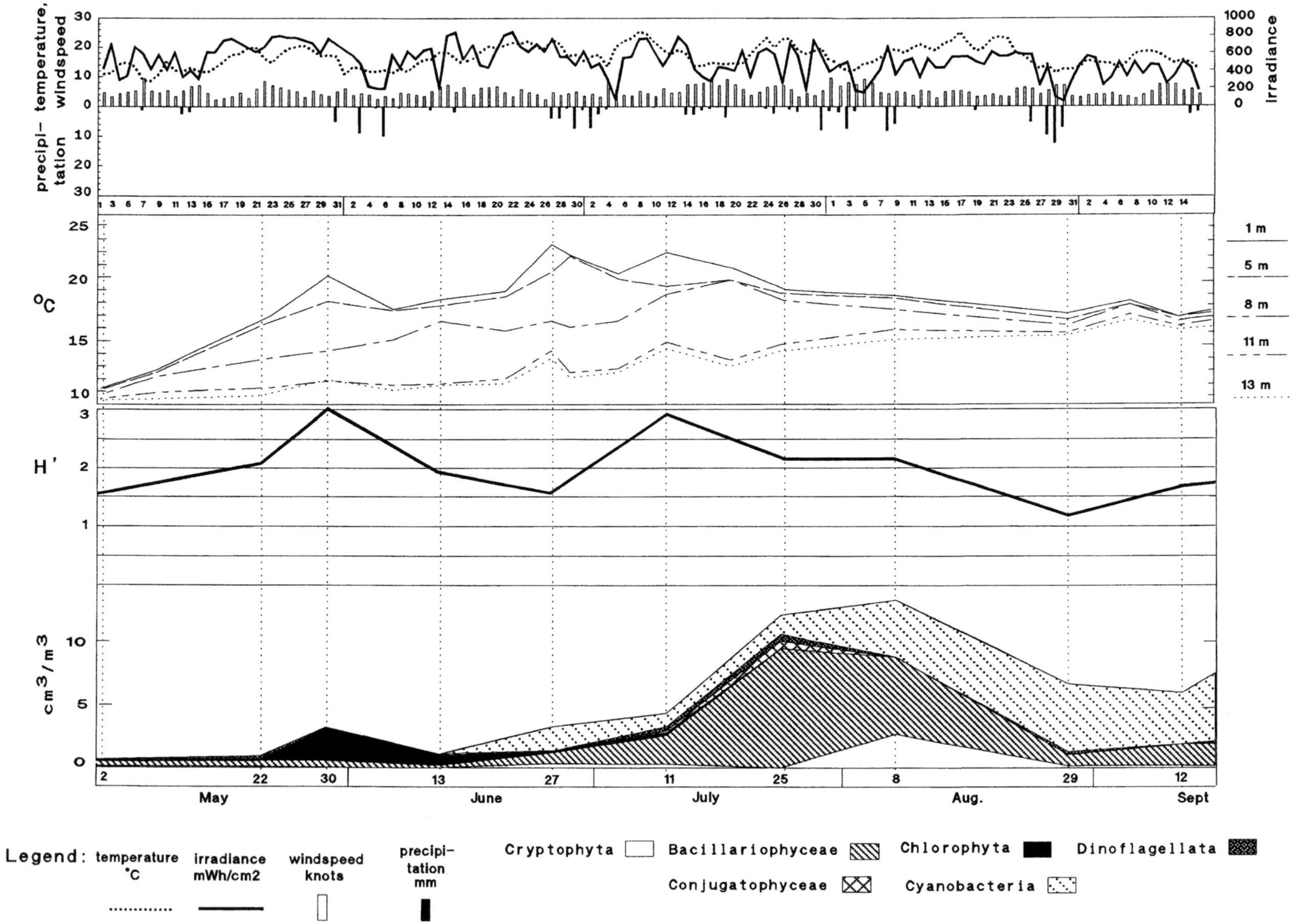

Fig. 9. Meteorological conditions, water temperature, diversity, and species composition during summer stratification in Lake Tegel 1989.

by the usual deeper mixing would have reduced light availability.

During the quiescent and well stratified phase from June 6th to the end of the month, diversity declined to a minimum in both lakes. Although very different species dominated (*Anabaena spp.* in Lake Tegel, and *Uroglena sp.* in Schlachtensee) both of these are motile and probably favored by calm conditions. However, in Lake Tegel aerator-induced hypolimnetic turbulence ensured higher diversity: aeration was begun partially on June 8th and accelerated June 20th, which can be traced in hypolimnetic temperatures (Fig. 9). Subsequently, silica concentrations increased again (from a minimum of 0.47 $mg\,l^{-1}$ on May 15th to 3.5 $mg\,l^{-1}$ by June 12th), and simultaneously diatoms (*Asterionella formosa)* reappeared.

Cooling of surface temperatures and increased mixing during the first days of July appears to have terminated the dominance of the motile species in both lakes, and diversity attained a maximum. In Lake Tegel it reached 2.9 by July 11th, as numerous new species (*Microcystis spp.*, *Aphanizomenon flos-aquae*, two species of *Tribonema*, three species of diatoms, *Chlamydocapsa planktonica*, *Mougeotia sp.*) began growing during these first days of quiescence. Again, the higher diversity maximum in Lake Tegel was due to the high share of diatoms.

Phytoplankton reaction to increased turbulence towards the end of July was also similar to the response in Schlachtensee: by July 25th, diatoms (*Aulacoseira spp.* and *Fragilaria crotonensis*) dominated and diversity declined. Subsecuently, the fair weathter during August and September enhanced the growth of cyanobacteria and reduced the share of diatoms.

Thus, although species composition was different, on a level of higher taxa or functional ecological groups ('blue-greens', 'diatoms', or 'motile forms' rather than individual species), much of the phytoplankton reaction to changes in mixing conditions in Lake Tegel paralleled that in Schlachtensee. At a first glance this appears to contradict the result stated previously, that physical conditions are a dominant factor in Lake Tegel and select almost totally different species. This contradiction may be resolved by regarding the 'hierarchical sequence of factors ... ultimately conditioning the structure of phytoplankton assemblages', as proposed by Reynolds (1989), which move 'from interactions involving factors that are primarily physical ... to chemical ... and then to biotic' (p. 46). Within such a sequence, physical factors act on two levels: First, they define the overall conditions of a given lake (general susceptibility to stratification or mixing; in some cases also turbidity) and of seasonality (irradiance, vernal overturn, monomixis, etc.). Within this frame, nutrient limitation may impinge upon phytoplankton, setting limits to biomass and influencing species composition. Within this further constraint, biogenic succession can develop, for example favoring species protected against grazing losses. Physical factors come into action once more upon a level beneath all these, and on a much smaller temporal as well as spatial scale: here they are minor (and often transient) changes in mixing conditions caused by meteorological cycles. This adds a fourth level to the sequence and means that within the set of major, comprehensive constraints, meteorologically induced intermediate 'disturbances', 'quiescences', or simply 'changes', are a further, subordinate governing factor: For example, in Schlachtensee they do not interfere with selection for species protected against grazing, or abolish nutrient limitation. Likewise in Lake Tegel, intermediate phases of quiescence allow transient populations of motile species such as *Ceratium* and *Anabaena*, or of Chlorococcales, but do not change the regular pattern of seasonality (diatoms – *Microcystis*/*Aphanizomenon* – diatoms) which appears to be the lake's basic response to overall mixing and nutrient conditions.

Summary and conclusions

Although mixing conditions in Schlachtensee and Lake Tegel are quite different, the approach to assessing the effect of allogenic 'disturbances' using diversity as a summarizing parameter has shown only small differences between them: The general level of diversity was similar, and temporal patterns showed numerous parallels. However, the reaction of both lakes to a reduced frequency (and in Lake Tegel also a reduced intensity) of 'disturbances' in 1989 showed a slightly opposite trend: in well-sheltered and stratified Schlachtensee, diversity decreased somewhat, whereas in the more turbulent Lake Tegel, unusually calm circumstances in 1989 caused an increase of diversity. This suggests that in a basically turbulent system interjected calm phases are the actual disturbance; an 'intermediate quiescence theory' may be proposed for such systems. Better yet, meteorologically induced changes of mixing conditions on an intermediate time scale should be regarded as key factor determining diversity.

For an understanding of the mechanisms behind this relation of diversity and physical factors, the 'mediators' of diversity – the individual phytoplankton species – must be regarded with respect to their morphological adaptations to mixing or stratification. This shows that physical factors act on two different levels.

On an superordinate level, the general susceptibility to perturbation because of lake morphology, wind exposure, hydrology, or artificial mixing determines which functional groups of species can inhabit a given lake. For Schlachtensee, this often favors motile species with some means of adjusting their depth within the water column – flagellates or cyanobacteria with gas vacuoles. This applied regardless of trophic state, although within the set of motile species, different subsets dominated under hypertrophic conditions and after restoration. In turbulent Lake Tegel, non-motile species dominated; the only significant exceptions were *Microcystis* and Cryptophyta, but these also prevailed only as long as some stratification existed.

The next level in the sequence of governing factors beneath this general frame set by a lake morphology (and sometimes also by hydrology or by anthropogenic measures), is nutrient limitation (Reynolds, 1989). A further question pursued in the evaluation of the phytoplankton data of the two lakes was how physical factors relate to nutrient limitation in affecting diversity. This was found to differ seasonally:

For winter and vernal overturn, results show that with little or no phosphorus limitation, diversity was high in both lakes, whereas after successful restoration in Schlachtensee, winter and vernal diversity decreased. Apparently, the rapid changes of physical parameters (increasing light availability, decreasing perturbation) during this season may be answered more rapidly by population development of adapted species, if nutrient constraints are low.

During summer, physical conditions appear to be decisive for diversity in eutrophic lakes: results for Schlachtensee show that in a stably stratified system without limitation by phosphorus or nitrogen, a summer 'climax' population can establish a stable dominance quite resistant to changes and not easily offset even by a storm (as in August 1982); thus, diversity may be very low. This can also apply to continuously mixed lakes if they are very shallow, such as the IJsselmeer polders described by Berger, (1987) and Berger & Sweers (1988). In such situations, due to self-shading, light becomes the only limiting resource. In contrast, summer diversity can be quite high in somewhat deeper eutrophic lakes in which stratification is weakly developed, such as Lake Tegel. There, intensity and depth of mixing within the upper layers in Lake Tegel was high enough to maintain diatoms in suspension, but not too high for the buoyant blue-green species *Microcystis* and *Aphanizomenon*. Additionally, perturbations of the sediment-near layers contributed to diversity by resupplying silicate. Although this latter process was especially enhanced in Lake Tegel through artificial mixing induced by aeration, it may be important in any perturbed lake with larger shallow areas. Thus, in such lakes physical factors may increase the number of limiting resources.

Under pronounced nutrient constraints, summer populations even of potential 'climax species' can no longer establish maximal biomass densities. For *Planktothrix agardhii*, biomass reduction by phosphorus limitation has been shown to destabilize dominance and to increase susceptibility to changes in mixing conditions. Thus, under conditions of stable stratification, nutrient limitation interacts closely with 'disturbances' in maintaining diversity (see below).

The third level of factors determining phytoplankton species composition and hence diversity is biogenic seasonal succession in the sense of the PEG-model (Sommer *et al.*, 1986). In calm and stably stratifying Schlachtensee, succession can be a major driving force (Chorus & Wesseler, 1988). However, in relation to physical parameters it is important to note that these biogenic mechanisms can select only among those species favored by the mixing conditions of a given lake – in the case of Schlachtensee during summer stratification among motile species. In generally perturbed systems such as Lake Tegel, physical parameters may be more important than biogenic succession for structuring the biocoenosis.

On a level subordinate to these other governing factors discussed above, transient changes in weather conditions act upon phytoplankton by affecting the intensity of mixing within the epilimnion, and allowing or precluding diurnal microstratification. In Schlachtensee, these were shown to be very important as a further factor selecting among species which can prevail under the given nutrient concentrations and grazing pressure. Detailed evaluations of population development showed that at least four mechanisms of phytoplankton reaction to such meteorologically induced changes in physical conditions are important:

1. Diversity decreases after conditions have changed from turbulent to quiescent because competition runs towards the dominance of one species and comes close to exclusion of the others before the calm phase is terminated. Such responses were rare; evidence for this was found only in one or two cases in Schlachtensee: under hypertrophic as well as quiescent conditions in 1982/83, as *Planktothrix agardhii* dominated, and (with some uncertainty) also during 1989, as *Uroglena* spp. dominated. Generally, the process of competitive exclusion under stable conditions was too slow to reach the point of one species actually 'winning the race' and ousting the others. Therefore, even prolonged calm periods rarely caused reductions of diversity.

2. Diversity decreases during phases of increased epilimnetic mixing because turbulence selects adapted species very rapidly. In Schlachtensee, populations of diatoms or Chlorococcales could develop within one week (see mid-June 1988 and early May, early June and the second half of July 1989) and obviously very rapidly outcompeted motile species which had dominated during the preceding calm phase.

3. Diversity increases at low biomass densities during the first calm days after a disturbance, as the decline of species promoted by turbulence overlaps the beginning development of species requiring more stability of stratification. This was a frequent response of phytoplankton in Schlachtensee (see late May and early July 1988, or early July 1989) but could also be observed in Lake Tegel in late May 1989.

4. Diversity remains constant, although dominant species change almost entirely, but the quantitative relationship between the new species is similar to that between the previous species. This was also observed in Schlachtensee during spring of 1989.

In Schlachtensee almost every meteorological cycle during the stratified period of each year could be traced in temperature curves, and these changes in stratification stability nearly always induced marked changes in species composition. This usually prevents competition from reaching the point of exclusion, and thus confirms the assumption proposed by the 'intermediate disturbance hypothesis' that diversity is maintained by such changes of physical conditions. However, the results show that there is no direct relation between diversity and disturbance: community reaction to changes in physical conditions may cause increases as well as decreases of diversity, or even leave it unchanged inspite of significant population changes. Thus, the Shannon function is insufficient as a tool for assessing the effect of disturbances. For this, investigation must include the response of the prevailing phytoplankton species.

Acknowledgements

We would like to express very hearty thanks to all those who helped collect the data in the course of the years covered: Mr. Wolf (sampling), Mrs. Pawlitzky (sampling, figures and computerizing data), Mrs. Laskus (phytoplankton counts), Mrs. Ewald (cooperation and help in our project 1982 and 1983), Mr. Gervais (cooperation in 1987), and Mr. Klein for enabling this research. Equally, we wish to thank Mrs. Padisák and Mr. Schmidt for focussing the IAP workshop in 1990 on the intermediate disturbance hypothesis, and the participants for their impulses in the discussion there. Thanks are also due Mr. Wintermeyer for advice as to the choice and interpretation of meteorological data, and to the Meteorological Institute of the Free University of Berlin for supplying these. Last but not least, hearty thanks go to C. Reynolds, U. Sommer, J. Padisák, and F. Gervais for helpful comments and kind advice on the manuscript.

References

Berger, C., 1987. Habitat en ecologie von *Oscillatoria agardhii* Gomont. Van Zee tot Land 55. Rijksdienst voor de IJsselmeerpolders, Lelystad (with an extensive English summary.): 1–233.

Berger, C. & H. E. Sweers, 1988. The IJsselmeer and its phytoplankton – with special attention to the suitability of the lake as a habitat for *Oscillatoria agardhii* Gom. J. Plankton Res. 10: 579–599.

Chorus, I., 1989. Phytoplankton und Primärproduktion im Schlachtensee 1982/83 zu Beginn der Sanierung. WaBoLu Hefte 3: 1–17.

Chorus, I. & E. Wesseler, 1988. Response of the phytoplankton community to therapy measures in a highly eutrophic urban lake (Schlachtensee, Berlin). Verh. int. Ver. Limnol. 23: 719–728.

Chorus, I. & C. Speck, 1990. Bedeutung der Sedimentationsverluste für das Phytoplankton des Schlachtensees. Extended abstracts of the annual congress of the German Society for Limnology: 101–106.

Ewald, S., 1991. Long-term changes of crustacean plankton during successful restoration of Lake Schlachtensee (Berlin-West). Verh. int. Ver. Limnol. 23: 866–872.

Gervais, G., 1989. Das Phytoplankton des Schlachtensees im Jahre 1987 – sechs Jahre nach dem Beginn der Sanierungsmaβnahmen. Diploma thesis at the Free University of Berlin, Inst. für Systematische Botanik und Pflanzengeographie: 1–149.

Gervais, G., 1991. Which factors controlled seasonal and spatial distribution of phytoplankton species in Schlachtensee (Berlin, F.R.G.) 1987 Arch. Hydrobiol. 121: 43–65.

Klein, G., 1988. Seensanierung in Berlin. Wasserkalender 1988. Erich Schmidt Verlag, Berlin: 46–74.

Klein, G. & I. Chorus 1991. Nutrient balances and phytoplankton dynamics in Schlachtensee during oligotrophication. Verh. int. Ver. Limnol. 23: 873–878.

Reynolds, C., 1988. The concept of ecological succession applied to seasonal periodicity of freshwater phytoplankton. Verh. int. Ver. Limnol. 23: 683–691.

Reynolds, C., 1989. Physical determinants of phytoplankton succession. In U. Sommer (ed.): Plankton ecology – succession in plankton communities. Springer-Verlag, Berlin Heidelberg New York London Paris Tokyo. ISBN 3-540-51373-6: 9–56.

Shannon, C. E., 1948. A mathematical theory of communication. Bell Syst. tech. J. 27: 623–656.

Sommer, U., 1989. The role of competition for resources in phytoplankton succession. In U. Sommer (ed.): Plankton ecology – succession in plankton communities. Springer-Verlag, Berlin Heidelberg New York London Paris Tokyo. ISBN 3-540-51373-6: 57–106.

Sommer U., J. Padisák, C. S. Reynolds & P. Juhász-Nagy, 1993. Hutchinson's heritage: The diversity-disturbance relationship in phytoplankton. In J. Padisák, C. S. Reynolds & U. Sommer (eds), Intermediate Disturbance Hypothesis in Phytoplankton Ecology. Developments in Hydrobiology 81. Kluwer Academic Publishers, Dordrecht: 1–7. Reprinted from Hydrobiologia 249.

Sommer, U., Z. M. Gliwicz, W. Lampert & Duncan, 1986. The PEG-model of seasonal succession of planktonic events in fresh waters. Arch. Hydrobiol. 106: 433–471.

Hydrobiologia **249**: 93–100, 1993.
J. Padisák, C.S. Reynolds & U. Sommer (eds), Intermediate Disturbance Hypothesis in Phytoplankton Ecology.

Seasonal succession of phytoplankton and its diversity in two highly eutrophic lakes with different conditions of stratification

Irina Trifonova
Institute of Lake Research, Sevastjanova 9, 196199 St. Petersburg, Russia

Key words: lake phytoplankton, algae, diversity, succession, disturbance

Abstract

Seasonal succession of phytoplankton biomass, its diversity and its photosynthetic activity in two highly eutrophic lakes have been compared. In order to test the intermediate disturbance hypothesis, the lakes have been chosen with almost the same level of trophy but different conditions of stratification, through two ice-free periods of open water with different weather conditions.

High phytoplankton diversity throughout the period of investigation was characteristic for the shallower Lake Lobardzu.The number of species here was usually more than 30 and the Shannon diversity changed from 1.2 to 4.2. Owing to the frequent external disturbances, periods characterized by autogenic succession with establishing dominance and declining diversity alternated with periods of biomass reduction and rises of diversity and photosynthetic activity. In the warmer summer of 1983, with more intense warming of bottom layers and predominance of blue-greens, phytoplankton biomass was higher and diversity lower than in the cold summer of 1982.

In stratified Lake Rudusku, phytoplankton diversity and number of species were usually much lower. During the long summer stratification up to three-four dominant species of blue-greens and dinoflagellates become established and competitive exclusion leading to low diversity advanced. Some changes in biomass and diversity, were caused by zooplankton activity.

Introduction

A number of surveys of phytoplankton seasonal succession has advanced our understanding of this process as being mainly an autogenic successional development (Reynolds, 1980, 1984; Sommer, 1986; Sommer *et al.*, 1986). However, the question of diversity as a measure of community structure and stability has remained one of the most controversial issues. High phytoplankton diversity in many eutrophic lakes contradicts the competitive exclusion principle (Hardin, 1960). This dilemma, also known as Hutchinson's 'paradox of the plankton', has been explained by non-equilibrium fluctuation of the environment (Hutchinson, 1961). The relation of phytoplankton diversity to changes in environmental conditions, including weather and mineral nutrition, has been shown also by others (for example Fedorov, 1971). The intermediate disturbance hypothesis, adapted to phytoplankton ecology by Reynolds (1988), has suggested new interpretation of seasonal succession as the interaction between true successional development and intermediate disturbance.

Good support for the role of intermediate disturbance in the maintenance of phytoplankton diversity comes from the fact that high diversity of phytoplankton is usually related to shallow well-mixed lakes, not only eutrophic (Padisák *et al.*, 1988) but even highly eutrophic examples (Trifonova, 1990). The different patterns of phytoplankton seasonal change in shallow and stratified lakes has been shown in the PEG-model (Sommer *et al.*, 1986). The relation of seasonal succession type to the lake trophic type and its change in the process of eutrophication has also been established (Reynolds, 1984; Sommer, 1986; Trifonova, 1986, 1989, 1990). This is why the intermediate disturbance hypothesis can be best tested by comparing of lakes of similar trophic type but different conditions of stratification.

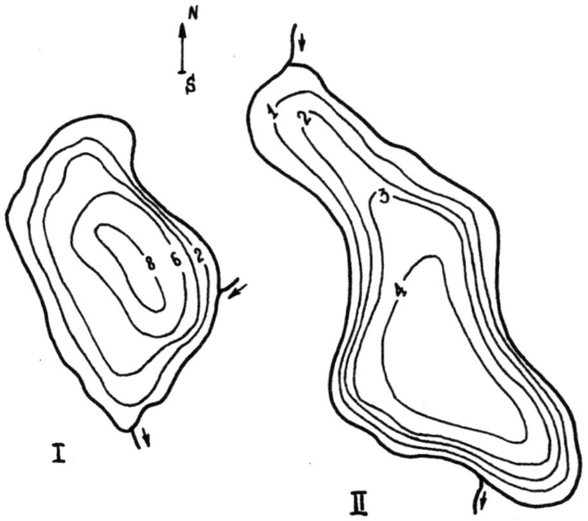

Fig. 1. Maps of the lakes with depth contours (m). I – L. Rudushku, II – L. Lobardzu

Characteristics of the lakes

Here, in order to test the intermediate disturbance hypothesis, we have chosen two highly eutrophic lakes situated in the Latgalian Elevation. This lake district is in the eastern part of Latvia, near the border with Russia at about 56° N, in which lakes have originated by glacial activity. The bedrocks of the elevation are composed mainly of carbonate rocks, which influence the high mineral content of lake water (total ionic concentration 300–400 mg l^{-1}). About 80% of the area is arable. Lake catchments are cultivated and heavily fertilized. Intensive anthropogenic eutrophication results in a high phosphorus content in many lakes.

A detailed description of both lakes and their phytoplankton has been given in Drabkova & Pritkova (1988). Only the main limnological characteristics of the lakes are summarised in Fig. 1 and Table 1.

Both lakes are small with areas < 1 km^2. Lake Rudushku is the deeper and has a smaller ratio of surface area to mean depth. Its morphometry permits stable thermal stratification: the lake is dimictic and even monomictic in some years. A stable thermocline forms at a depth of 4–6 m and a very thick (1–3 m) anoxic zone with H_2S develops during summer and winter.

The more shallow and more open Lake Lobardzu, with a larger ratio of area to depth, is polymictic with persistent mixing during the ice-free period, which is about 200 days from April to November. Another characteristic feature of Lake Lobardzu is its high water exchange. Together with the persistent mixing it creates unstable conditions in this lake.

Table 1. General characteristics of the lakes

Characteristic L.	Rudushku	L. Lobardzu
Water area (km^2)	0.06	0.09
Depth (m) max	8.2	4.0
mean	4.2	2.5
Water exchange rate (yr^{-1})	0.86	5.2
Transparency (m)*	1.7	0.9
pH	7.0–8.4	7.1–8.6
Total ions (mg l^{-1})	242–296	269–361
P_{tot} (mg l^{-1})**	0.072	0.080
P_{min}/P_{tot}(%)	60	29
N_{tot}(mg l^{-1})**	1.15	1.58
N_{min}/N_{tot}(%)	45	35
N_{tot}:P_{tot}	16	20
Chl*a* (mg m^{-3})*	15.0	28.0
Phytoplankton biomass (g m^{-3})*	8.2	12.0
Primary production (g C m^{-2} $year^{-1}$)	251	326

* Mean for the season (April-October)
** Mean for the year

Both lake waters have rather high mineral and nutrient content, especially for phosphorus and nitrogen. The concentration of total phosphorus was especially high in the hypolimnion of Lake Rudushku – up to 400 μg l^{-1}.

Inorganic (soluble-reactive) phosphorus was practically absent in the epilimnion during the period of stratification, yet usually very high in the hypolimnion. In Lake Lobardzu, nutrients were more uniformly distributed in the water column and the inorganic forms were completely depleted by algae. The content of Fe was also rather high in both lakes, reaching 1.7–4.6 mg l^{-1} in the bottom layers.

Regarding trophic status, both lakes can be considered highly eutrophic. According to the chlorophyll content, phytoplankton biomass and primary production the trophic status of Lake Lobardzu is even the higher in spite of its lower nutrient content.

Methods

Both lakes were studied over two years, 1982 and 1983. Samples were taken in the deepest parts of the lakes with a Ruttner sampler every 10 days in 1982 and biweekly in 1983. They were fixed with 40% formalin, concentrated by sedimentation and counted in a Najotte chamber (Guseva, 1959). The biomass was determined as algal volume for each lake, and converted to fresh weight assuming a specific gravity of 1 g cm^3. We considered algae with cell dimensions < 30 μm to be nanoplankton. Photosynthetic intensity was measured by the oxygen light and dark bottle method. For evaluation of diversity the Shannon function (H) based on biomass was used:

$$H = \sum \frac{B_i}{B} \log_2 \frac{B_i}{B} \quad \text{(Odum, 1971)}$$

where B_i is the biomass of the ith species and B is the total phytoplankton biomass.

Results

Lake Lobardzu

The two ice-free periods were quite different in weather conditions and thermal regime (Fig. 2a). In 1982, the spring was rather cold with very slow warming of the water after ice-break. Only in the beginning of June did the water temperature reach 15 °C and very weak stratification begin. During summer, two periods of rapid cooling and full mixing were observed – in mid June and late July. Autumnal cooling and mixing began at the end of August and was very rapid. In early September, water temperature was already 13 °C under full homothermy. In 1983, the spring was warmer. In mid May the water temperature reached 15 °C

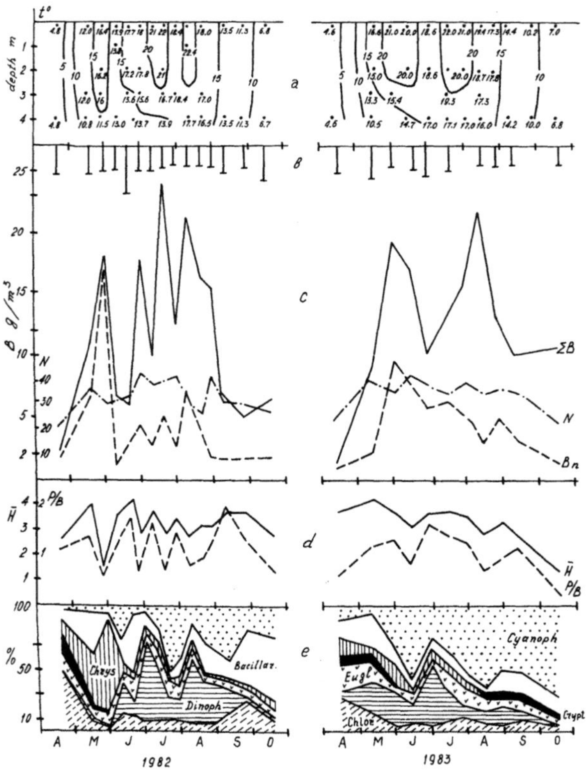

Fig. 2. Seasonal changes in Lake Lobardzu of: a. water temperature-time-depth disposition of isotherms (°C); b. Secchi-disc transparency (m); c. total phytoplankton fresh weight biomass (B), nanoplankton biomass (B_n) and number of species (N) in samples of surfacewater; d. Shannon diversity (H) and photosynthetic activity (P/B); e. Percentage contribution of the main algal groups to total phytoplankton biomass.

and at the end of May stratification was rather strong. Only one period of disturbance at the end of June, with cooling and full mixing, took place during summer. The decrease of temperature with depth was rather small, and full mixing promoted the warming of the bottom layers. As a result the temperature near bottom sediments was higher than in 1982. Autumnal cooling and mixing began also in late August but was more slow.

Total phytoplankton biomass in the surface layer during the two open water periods ranged from 2 to 25 g m^{-3} (Fig. 2c). The mean biomas for the season (April–October) was higher in 1983 due to mass development of blue-greens.

The spring growth of phytoplankton biomass in both years began in April immediately after ice-break and continued to the end of May. Spring phytoplankton was dominated by small nanoplanktonic algae, mainly chlamydomonads and cryptomonads. Most numerous were *Chlorogonium maximum* Skuja, *Chlamydomonas mucosa* (Korsch.) Pasch. and *Cryptomonas erosa* Ehr. In the middle of May these species were replaced by the chrysophytes and diatoms: *Dinobryon sertularia* Ehr., *D. divergens* Imh., *Stephanodiscus hantzschii* Grun., *Asterionella formosa* Hass. and *Synedra acus* Kütz. At first the number of species, diversity and photosynthetic activity of phytoplankton increased parallel with biomass (Fig. 2d). By the end of the month however only two species of *Dinobryon* contributed 85% of the spring crop biomass in the end of May (Fig. 2e). The diversity was at a minimum, although the number of species recorded hardly changed. This spring crop coincided with a first peak of zooplankton caused by Rotifera. As a consequence of zooplankton grazing, the phytoplankton biomass decreased rapidly, mainly at the expense of small agae. This fall of biomass also coincided with a period of rapid cooling and full mixing, and seems to be the result of the combined effect of autogenic and physical factors. In spite of an increase of transparancey, this period could not be considered as a real 'clear-water phase' since the phytoplankton biomass was still higher than 5 g m^{-3}.

Parallel with this decrease of phytoplankton biomass, its diversity and photosynthetic activity increased. About 10 species contributed 80% of biomass. Among the dominants were *Oscillatoria agardhii* Gom., *Peridinium aciculiferum* (O.F.M.) Ehr., *Glenodinium elpatiewskii* (Ostenf.) Schiller, *Trachelomonas hispida* (Perty) Stein and the diatoms *Fragilaria crotonensis* Kitt. and *Aulacoseira granulata* (Ehr.) Sim. Indirect confirmation of intermediate disturbance was also the appearance in the plankton of large benthic colonies of *Pediastrum duplex* Meyen and *P. boryanum* (Turp.) Menegh.

Biomass increase began at the end of June after onset of stratification. Diatoms and chrysophytes were replaced by dinoflagellates. *Peridinium cinctum* and *Glenodinium elpatiewskii* contributed about 70% of biomass. The diversity and photosynthetic activity decreased, although the number of species increased. A small decrease of biomass in the beginning of July was probably caused by zooplankton grazing, this time mainly by *Bosmina longirostris* (O.F.M.). It was followed by increase in diversity and photosynthetic activity. The role of dinoflagellates in the total biomass became smaller and, under conditions of intensive warming and strong stratification, the mass development of blue-greens began. The summer maximum of biomass in mid-July was determined primarily by *Oscillatoria agardhii* and *O. redekei* van Goor. Together with *Glenodinium elpatiewskii* and *Trachelomonas hispida*, they contributed more than 90% of biomass.

Another disturbance of stratification due to the cooling and mixing in late-July caused a further small decrease of biomass; at the same time diversity and photosynthetic activity increased. *Oscillatoria agardhii* and *Glenodinium elpatiewskii* contributed only 40% of biomass. Among the dominants, diatoms, Chrysophyceae and Eugleonophyceae appeared again – mainly *Synedra actinostroides* Lemm., *Dinobryon sertularia* and *Trachelomonas hispida*.

This very short intermediate disturbance was followed by another period of intensive warming and increase of phytoplankton biomass particularly of small algae. The decrease of diversity was rather weak, but some 53% of biomass was contributed by *Glenodinium elpatiewskii*. Subdominants were diatoms and blue-greens. In spite of the beginning of autumnal mixing at the end of August, the decrease of biomass was rather weak. The dominance of 4 species – *Glenodinium elpatiewskii*, *Stephanodiscus hantzschii*, *Oscillatoria agardhii* and *O. redekei* – continued. An autumnal peak of zooplankton, mainly *Daphnia cucullata* Sars and *Bosmina longirostris*, caused a decrease of small algal biomass. Diversity and photosynthetic activity increased.

Rapid biomass decrease resumed at the beginning of September. *Oscillatoria agardhii* and *Glenopdinium elpatiewskii* continued to be dominant, but diatoms and green algae became relatively more important – mainly *Synedra acus*, *Asterionella formosa*, *Aulacoseira ambigua* (Grun.) Sim. and *Chlorogonium maximum*. Only in October, parallel with the rise of phytoplankton biomass (mainly diatoms), did diversity and photosynthetic activity began to decrease.

In 1983, the spring phytoplankton composition was similar to that in 1982 although the contribution of Chrysophyceae was much smaller and that of bluegreens more intensive. *Oscillatoria agardhii*, *Anabaena scheremetievii* Elenk., *Anabaena spiroides* Kleb. and *Glenodinium elpatiewskii*, together with *Dinobryon sertularia* and *Trachelomonas hispida*, dominated during May. The diversity was higher than in the previous year. The zooplankton peak was also greater. As a result of grazing the biomass of small algae, mainly diatoms, Chrysophyceae and Euglenophyceae, was lower. The biomass of blue-greens continued to increase and the end of May they contributed, together with dinoflagellates, more than 8% of biomass. Phytoplankton diversity slightly decreased.

The weak decrease of biomass in late July was caused by the combined effect of zooplankton grazing and disturbance of stratification. This intermediate disturbance was the only one during the summer, and resulted in a weak rise of diversity and photosynthetic activity. In early-July the phytoplankton biomass again began to increase in parallel with the water warming. The summer peak of biomass at the end of July was mainly caused by blue-greens. This time *Oscillatoria redekei* and *O. agardhii* contributed about 7% biomass. Subdominants were *Glenodinium elpatiewskii* and *Trachelomonas hispida*. The phytoplankton diversity decreased.

At the end of August, in parallel with autumnal mixing and water cooling, a decrease of biomass began. Dinoflagellates and Euglenophyceae were substituted by diatoms, mainly *Synedra acus*. Blue-greens continued to be dominant. In conditions of full homothermy the competitive equilibrium of 3 species – *Oscillatoria agardhii*, *O. redekei* and *Synedra acus* – was ended. Phytoplankton diversity continued to decrease and was lower in October, when *Oscillatoria redekei* became absolutly dominant, contributing 8% of biomass.

Lake Rudushku

In the stratified Lake Rudushku the difference in thermal regime between the two years was less pronounced (Fig. 3a). In 1983 the stratification onset was earlier – in early May instead of the end of May as in 1982 but the thickness of the epilimnion was less. Only in July did it increase to 5 m. Autumnal overtun began in September in both years, but full homothermy only in October. All intermediate disturbances that were connected with wind mixing and water cooling during summer affected only the epilimnion. The temperature of the bottom layer was almost the same (5–7 °C) during both summer periods.

Total phytoplankton biomass during the two ice-free periods changed from 1.7 to 19 g m^{-3} (Fig. 3c). The mean biomass for the season was a little higher in 1983. Because of the very complex vertical distribution of algae in Lake Rudushku (Hindák & Trifonova, 1989; Trifonova, 1990), average data for the layer 0–2 m have been analysed in this paper.

The spring growth of phytoplankton biomass in both years started in April after ice-break, and continued during May parallel with the increase in diversity and photosynthetic activity. In 1982, small algae contributed about 90% of the spring biomass. Chlamydomonads and dinoflagellates dominated in April, especially *Chlorogonium maximum* and *Peridinium aciculiferum*. During May they were replaced by Chrysophyceae and dia-

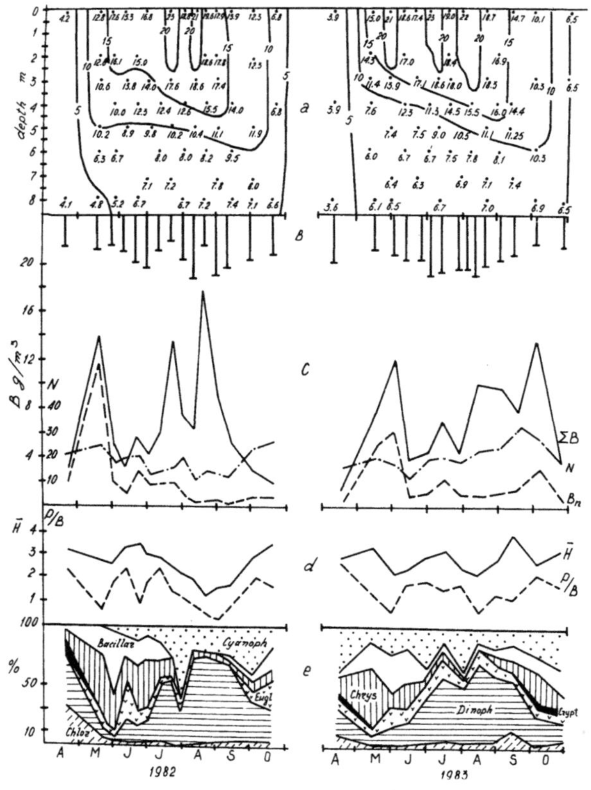

Fig. 3. Seasonal changes in Lake Rudushku (legends as in Fig. 2).

toms. At the end of May the crop mainly comprised *Dinobryon sertularia, Stephanodiscus hantzschii, Asterionella formosa* and *Synedra acus*. These four species together contributed about 80% of biomass. The spring peak of zooplankton, mainly Rotifera, caused a rapid decrease of phytoplankton biomass in the beginning of June. It was followed by increase of transparency, phytoplankton diversity and photosynthetic activity (Fig. 3b,d). This typical 'clear water phase' (Sommer *et al.*, 1986) continued till the middle of June. As a result of zooplankton grazing the biomass of small algae – especially *Dinobryon* and *Stephanodiscus* – decreased.

The next rise of biomass began from mid June. This small crop was again grazed by zooplankton. *Dinobryon, Stephanodiscus* and *Trachelomonas* disappeared from the plankton. The biomass of blue-greens continued to rise, parallel with the decrease of diversity and photosynthetic activity. *Oscillatoria agardhii* and *Ceratium* contributed about 90 % of biomass in late July. The weak decrease of biomass at the beginning of August also coincided with enhanced zooplankton grazing, this time mainly by *Daphnia cucullata* and *Bosmina longirostris*. As result the biomass of blue-greens decreased, but that of *Ceratium* continued to increase and contributed about 70 % of the summer peak biomass at the end of August. The phytoplankton diversity and photosynthetic activity were then low.

The dominance of two species, *Oscillatoria* and *Ceratium hirundinella*, continued even after the beginning of water cooling and the decrease of biomass in September. It was disturbed only in mid September, when the depth of mixing exceeded the mean depth of the lake. Diatoms, Chrysophyceae and Euglenophyceae then appeared in the plankton, and its diversity began to increase parallel with the number of species and photosynthetic activity.

The general pattern of phytoplankton periodicity in 1983 was almost the same as in 1982. The main difference was in more intensive development of blue-green algae, mainly of *Oscillatoria agardhii* whose growth began just after the icebreak in April. *O. agardhii* was among the dominants throughout the growing season April–October.

The spring crop at the end of May was derived from the same algae as in 1982 except *O. agardhii*, but the diversity of phytoplankton was higher. In mid June, presumably as a result of zooplankton grazing, a rather long 'clear water phase' was observed. Small diatoms and Chrysophyceae were grazed and from the end of June larger algae dominated. Some 3–4 species, mainly *O. agardhii, Ceratium hirundinella, Peridinium cinctum* and *Trachelomonas hispida,* contributed more than 80% of biomass. The diversity decreased but not so rapidly as during 1982.

The increase of phytoplankton diversity began at the end of August, parallel with the increase of the mixing zone and the increase of small algal biomass. Among dominants there appeared diatoms, chrysophyceae and chlamydomonads.

Discussion

In both lakes vernal phytoplankton community developed in the successional pattern predicted by the PEG-model (Sommer *et al.*, 1986) and in correspondence to the 'first stage' of Margalef's formulation of succession (Margalef, 1958) was dominated by small fast-growing nanoplanktonic algae. Immediately after the icebreak, they were represented mainly by chlamydomonads and cryptophytes. Though they were later substituted by chrysophytes and diatoms. Usually the diversity increased in parallel with the phytoplankton biomass but than declined during the spring crop in the late May (Fig. 2,3).

The difference in successional changes of summer community structure between two lakes was more pronounced.

High diversity of summer phytoplankton was characteristic of the shallow, highly eutrophic Lake Lobardzu. The number of species in its phytoplankton was usually higher than 30 and the Shannon diversity ranged from 1.2 to 4.2. High diversity is a consequence of unstable environmental conditions, determined first by the persistent wind mixing and constant circulation of nutrients. Due to the frequent disturbances short periods of autogenic succession alternate with periods of biomass reduction and increase of diversity and photosynthetic activity. There were no real 'clear-water' phase since the phytoplankton biomass was always higher than 5 g m^{-3}. True 'clear-water' phase is not characteristic of shallow-eutrophic lakes with rapid nutrients turnover (Trifonova, 1990). In the warmer summer of 1983, due to the more stable conditions and stronger warming of bottom layers, blue-green algae dominated and phytoplankton diversity was lower than in 1982.

In the stratified and highly eutrophic Lake Rudushku, phytoplankton diversity was rather lower. The number of species represented was less than 20 and Shannon diversity ranged from 1.2 to 3.5. In both periods of investigation, phytoplankton was dominated by the motile flagellates and blue-greens typical of stratified lakes (Reynolds, 1980). During summer stratification, the majority of them became concentrated in the metalimnion. It is known that, in stratified lakes, the thermocline zone plays the part of a 'second bottom' (Gliwicz, 1979) restricting the loss of nutrients from the epilimnion. Zooplankton also concentrates in the epilimnion and its excretion together with thermocline erosion, is the main source of nutrients for phytoplankton growth.

Phytoplankton changes from the spring crop until the autumnal overturn represented the process of autogenic succession with all its typical stages, including a classic 'clear-water-phase'. The autogenic succession led to the equilibrium of three-four species and, perhaps as a result of competitive exclusion, to low diversity. Some changes in phytoplankton activity, its grazing and nutrient excretion. In the warmer summer of 1983 the abundance of zooplankton was higher than in 1982. Probably, the higher phytoplankton diversity was the result of this activity, especially the more intensive turnover of nutrients. The complete disturbance of autogenic succession and increase of diversity took place only after autumnal overturn when the depth of mixing exceeded the mean depth of the lake.

It was generally very difficult to distinguish the role of different factors in intermediate disturbances of succession. A zooplankton peak very often coincided with the stratification onset, so that grazing loss was combined with that by sedimentation. Sometimes enhanced zooplankton grazing coincided with water cooling and mixing. Usually intermediate disturbance was a result of a combined effect.

Thus the different character of phytoplankton periodicity in shallow and stratified lakes of the same trophic type has been shown. Support has also been obtained for an important role of intermediate disturbances in the maintenance of phytoplankton diversity. In almost all instances, the increase of biomass connected with more stable conditions was followed by the decrease of phytoplankton diversity and photosynthetic activity. *Vice versa*, the rises in diversity and photosynthetic activity took place when biomass decrased. Only under conditions of persistent autumnal mixing and cooling of water in shallow Lake

Lobardzu did phytoplankton diversity decrease in parallel with the biomass decrease. It may be the confirmation of the hypothesis that diversity should decrease under too frequent disturbances (Reynolds, 1988).

Generally, however, disturbance of autogenic succession can be an explanation of the high phytoplankton diversity in many shallow and highly eutrophic lakes. Besides, it seems that this hypothesis can be applied not only to the seasonal succession but also to the general succession of phytoplankton communities in lake development (Trifonova, 1986; 1990).

Acknowledgements

I am very grateful to Dr. Judit Padisák for having invited me to submit this paper during the Workshop at Baja as well as to Dr. J. F. Talling and Dr. C. S. Reynolds for editing the manuscript, critical review and many valuable remarks.

References

Drabkova, V. G. & M. J. Pritkova, (eds), 1988. Izmenenija a structuri ozernich ecosystem v uslovijach uvelichenija biogennoi nagruzki (Changes of lake ecosystems structure in conditions of nutrient loading increase). Nauka Publ. House, Leningrad, 311 pp (in Russian).

Hardin, G., 1960. The competitive exclusion principle. Science 131: 1291 -1297.

Hutchinson, G. E., 1961. The paradox of the plankton. Am. Nat. 95: 137–146.

Hindak, F. & I. Trifonova, 1989. Morphology and ecology of three Lymnothrix species (Cyanophyta) from the hypolimnion of a highly eutrophic lake in Latvia. Biologia (Bratislava) 44: 3–11.

Fedorov, V. D., 1971. Svjaz vidovogo raznoobrazija a phytoplanktona s izmeneniem uslovii mineralnogo pitanija (Relation of phytoplankton diversity to changes in the conditions of mineral nutrieiton). Hydrobiol. J., Kiev 9: 21–24 (in Russian).

Gliwicz, Z. M., 1979. Metalimnic gradients and trophic state of lake epilimnia. Mem. Ist. ital. Idrobiol. 37: 121–143.

Guseva, K. A., 1959. Metodi podscheta phytoplanktona. (Methods of phytoplankton counting). Tr. Inst. Biol. Vodochr. Borok 2: 44–51 (in Russian).

Margalef, R., 1958. Temporal succession and spatial heterogeneity in phytoplankton. In: Perspectives in Marine Biology. Berkeley-Los Angeles: 323–349.

Odum, E. P., 1971. Fundamentals of ecology. Saunders, Philadelphia.

Padisák, J., L. G. Tóth & M. Rajczy, 1988. The role of storms in the summer succession of the phytoplankton community in a shallow lake (Lake Balaton, Hungary). J. Plankton Res. 10: 249–265.

Reynolds, C. S., 1980. Phytoplankton assemblages and their periodicity in stratifying lake systems. Holarctic Ecol. 3: 141–159.

Reynolds, C. S., 1984. The ecology of freshwater phytoplankton. Cambridge University Press, Cambridge, 384 pp.

Reynolds, C. S., 1988. The concept of ecological succession applied to seasonal periodicity of freshwater phytoplankton. Verh. int. Ver. Limnol. 23: 683–691.

Sommer, U., 1986. The periodicity of phytoplankton in Lake Constance (Bodensee) in comparison to other deep lakes of central Europe. In M. Munawar & J. F. Talling (eds), Seasonality of Freshwater Phytoplankton – a Global Perspective. Developments in Hydrobiology 33. Dr W. Junk Publishers, Dordrecht: 1–7. Reprinted from Hydrobiologia 138.

Sommer, U., Z. M. Gliwicz, W. Lampert & A. Duncan, 1986. The PEG-model of seasonal succession of planktonic events in fresh waters. Arch. Hydrobiol. 106: 433–471.

Trifonova I. S., 1986. Sezonnaja i osnovnaja successija ozernogo phytoplanktona (Seasonal and general succession of lake phytoplankton). Hydrobiol. J., Kiev. 23: 21–28 (in Russian).

Trifonova, I. S., 1989. Changes in community structure and productivity of phytoplankton as indicators of lake and reservoir eutrophication. Arch. Hydrobiol. Beih. Ergebn. Idmnol. 33: 363–371.

Trifonova, I. S., 1990. Ecologia i successija ozernogo phytoplanktona (Ecology and succession of lake phytoplankton). Nauka Publ. House, Leningrad. 178 pp (in Russian).

Hydrobiologia **249**: 101–109, 1993.
J. Padisák, C.S. Reynolds & U. Sommer (eds), Intermediate Disturbance Hypothesis in Phytoplankton Ecology.

Seasonal fluctuations in the diversity and compositional stability of phytoplankton communities in small lakes in upper Bavaria

Roswitha Holzmann
Zoologisches Institut der Universität München, Limnologische Station Seeon, Seidlstr. 25, DW-8000 München 2, Germany

Key words: diversity, phytoplankton, community structure, intermediate disturbance, Principal Component Analysis

Abstract

A study of diversity and compositional stability of phytoplankton communities during one vegetation period was carried out in small lakes in upper Bavaria. Shannon-Weaver diversity index was calculated on the base of number of individuals and on the base of biomass. On average, the diversity (annual mean) was highest in mesotrophic lakes. A comparison of three morphologically different (esp. exposure to wind, surface area and mean depth) lakes (Pelhamer See, Thalersee and Kautsee) sought to find out how the phytoplankton community structure reacts to events of intermediate disturbance, in terms of diversity- or biomass changes. Principal Component Analysis was used to measure the persistence of the phytoplankton association. The examples given in this paper led to these conclusions: High diversity or increase in diversity occur in compositionally instable communities, in high wind-stress events, with small algae and with high grazing. Low diversity or decrease in diversity occur in compositionally stable periods, when conditions select few species, as large forms dominate and as grazing by zooplankton takes effect. Gradual seasonal changes are observed in structurally stable lakes. In lakes exposed to frequent disturbances, seasonal changes may be dominated by intermediate responses.

Introduction

Much ecological research in the field begins with a list of species together with their relative abundance. The two measures combined give some indication of ecological diversity. Diversity is dependent on the number of species present in the community and on the distribution of individuals amongst those species (eveness) (Pielou, 1966). Since the diversity index merely is a spot measure of the community structure in a dynamic system, diversity index measurement over a vegetation period would probably be a more satisfactory measure. Numbers of individuals were used to calculate diversity indices (Harris, 1986; Elber & Schanz, 1989); however, the introduction of an index based on biomass is indicated. This index emphasizes the influence of bigger size classes (having larger volume and hence more biomass). If cell numbers are used, a massive bloom of larger algae could be obscured by huge masses of small algae. Therefore, diversity measurement based on biomass seems to give a realistic view of the true situation in a lake.

A network of factors determines the composition and hence the diversity of a phytoplankton community (Reynolds, 1984; Sommer, 1989). An alteration of any of the factors has an effect on the community itself and, depending on the scale of alteration (severity and frequency), the successional state of the community, and its species composition will remain more or less stable. The term 'stability' refers to the qualitative and quantitative constancy of composition of the ecosystem during a long period of time (constancy of the system) and it can also mean a dynamic component that turns to another state of stability after the occurrence of changes in the environment (Gigon, 1983; Elber & Schanz, 1989). Harris & Piccinin (1980) and Kaneta *et al.* (1985) used Principal Component Analysis to determine compositional changes in the phytoplankton community structure.

According to Reynolds (1988), seasonal changes in a natural phytoplankton community are best interpreted as alterations between true succession and intermediate disturbances. Intermediate disturbances are allogenic, including events such as floods, episodes of increased wind mixing or they are biotic, including factors like grazing. Therefore, it may be expected that: structurally stable lakes follow rather gradual seasonal changes. However, such seasonal changes may be overridden in lakes subject to intermediate disturbances. As there is some indication that temporally variable disturbances increase diversity (Gaedeke & Sommer, 1986; Padisák *et al.*, 1988) it will be the aim of this paper to compare morphologically different lakes and to make an attempt to investigate the impact of disturbances on the phytoplankton community structure, in terms of diversity or biomass changes.

Description of sites studied

The lakes are situated in two small nature reserves in upper Bavaria, near Chiemsee. The lakes came into being as the result of the last glaciation (Wrobel, 1989). They lie close together and are shallow to moderately deep small lakes with different trophic states (see Table 1). Three lakes are subject to a closer analysis in this paper.

Pelhamer See is characterized by a mean depth of 9.5 m and a relatively large, wind exposed surface area (A_0 71.4 ha). The lake is eutrophic; the rich nutrient conditions owing to intensive agriculture in the catchment area of the lake.

Thalersee has a mean depth of 4.2 m and a small surface area (A_0 3.8 ha). It is sheltered from windstress. This highly eutrophic lake is the main site of nutrient import into the whole lake system due to anthropogenic imports into the lake.

Kautsee has a mean depth of only 3.2 m, a small surface area (A_0 16.5 ha) and is exposed to westerly winds. It is situated in the center of the nature reserve. The absence of direct anthropogenic influence may be the reason for the relatively low nutrient content (mesotrophic).

Table 1. Shannon-Weaver diversity index H′ (based on biomass) in lakes with different nutrient conditions (trophy). Annual averages in 1988 and 1989 ± SD.

Lake	Trophic state	1988	1989
Kautsee	mesotrophic	2.1 ± 0.4	2.0 ± 0.5
Klostersee	mesotrophic	2.0 ± 0.6	2.0 ± 0.6
Hartsee	mesotrophic	2.1 ± 0.5	2.0 ± 0.6
Kesselsee	mesotrophic	2.2 ± 0.3	2.5 ± 0.2
Langbürgner See	mesotrophic	1.3 ± 0.5	1.6 ± 0.5
Brunnsee	oligo.-meso.	1.9 ± 0.3	1.9 ± 0.5
Schloßsee	eutrophic	1.8 ± 0.4	1.3 ± 0.4
Pelhamer See	eutrophic	1.8 ± 0.4	1.7 ± 0.4
Thalersee	highly eutr.	1.5 ± 0.7	1.8 ± 0.4
Einbessee	dystrophic	1.2 ± 0.4	0.7 ± 0.4
Blassee	dystrophic	0.7 ± 0.4	1.6 ± 0.3

Materials and methods

The investigations were carried out during the vegetation periods of 1988 and 1989. In these years the general sequence of species dominance in the lakes was nearly identical. Therefore, the results of 1988 presented in this paper can be seen as representative for both vegetation periods. Each of the lakes was sampled a total of 13/14 times, biweekly in 1988 from April to September and once in October and November. Samples of the aerobic zone were taken with a Ruttner Sampler at intervals of one or two metres (depending on the depth of the lake) at two sampling points near to the deepest point of each lake. Aliquots of the mixed samples were taken for phytoplankton counts (fixed with Lugol's iodine) and-spectrophotometric Chl-*a* determination, following the procedure of Marker *et al.* (1980).

Environmental measurements included Secchi-depth, temperature, dissolved oxygen, pH, conductivity and mixing depth (according to maximum of relative thermal resistance). Information about zooplankton was derived from parallel samples. It was collected in vertical hauls, using a 30 μm plankton net, fixed with 4% sucrose-formalin and counted under a dissecting microscope.

For quantitative phytoplankton determination, Utermöhl's method (Utermöhl, 1958) in sedimentation chambers was used. Counting units were individual cells, filaments or colonies, depending on organisation of algae. Biomass was calculated according to the proposal of Rott (1981) as fresh-weight.

Shannon-Weaver diversity function (Shannon & Weaver, 1949) was calculated on the base of numbers of individuals (ind l^{-1}) and on the base of biomass (g m^{-3}), in each case, picoplankton was excluded.

$$H' = - \sum_{i=1}^{S} p_i \log^2 p_i ,$$

where H' is the number of bits per individual; S is the number of species and p_i is the relative importance of species i, either relative abundance (N_i/N_{tot}) or relative biomass (B_i/B_{tot}).

Compositional stability was measured by the use of Principal Component Analysis (PCA) (Harris & Piccinin, 1980; Kaneta *et al.*, 1985). The PCA ordination method used is included in a computer package called Mulva-4 (Wildi & Orloci, 1990). Before ordination, original biomass data are calculated due to chord dissimilarity index (Wildi & Orloci, 1990). This index emphasizes dominant species (high biomasses) and comes close to the often-recommended average linkage sorting (Clifford & Stephenson, 1975).

Results

Regarding the average diversity for 1988/1989, there seems to be a tendency among the mesotrophic lakes of the series to give the highest diversity values (exception Langbürgner See which had a *Planktothrix agardhii* (Gom.) Anagn. et Kom. bloom over a long period in both years). The oligo-mesotrophic Brunnsee and the eutrophic lakes generally have lower diversities than the mesotrophic lakes; the lowest values occurred in the dystrophic lakes (see Table 1). An F test for difference of variance was applied to the diversity data, and significant differences ($P < 0.025$) were found between the mesotrophic lakes and the eutrophic and dystrophic lakes in 1988 and 1989; the oligo-mesotrophic Brunnsee however, differed from the mesotrophic lakes only in 1988.

Pelhamer See

The markers seen in Fig.1 describe the coordinates of the biomass data after Principal Component Analysis. The PCA summarizes the biweekly changes in the phytoplankton associations of a lake. The clusters of ordination markers cor-

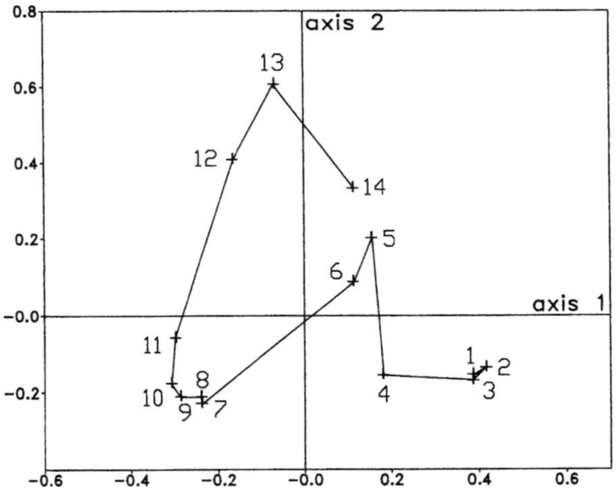

Fig. 1. A plot of the results of a biweekly ordination of the biomass of phytoplankton species in Pelhamer See along the first two axes (axis 1 explains 40% and axis 2 explains 22% of the total variance). The sequential samples are joined and numbered. Samples from April to November, 1988.

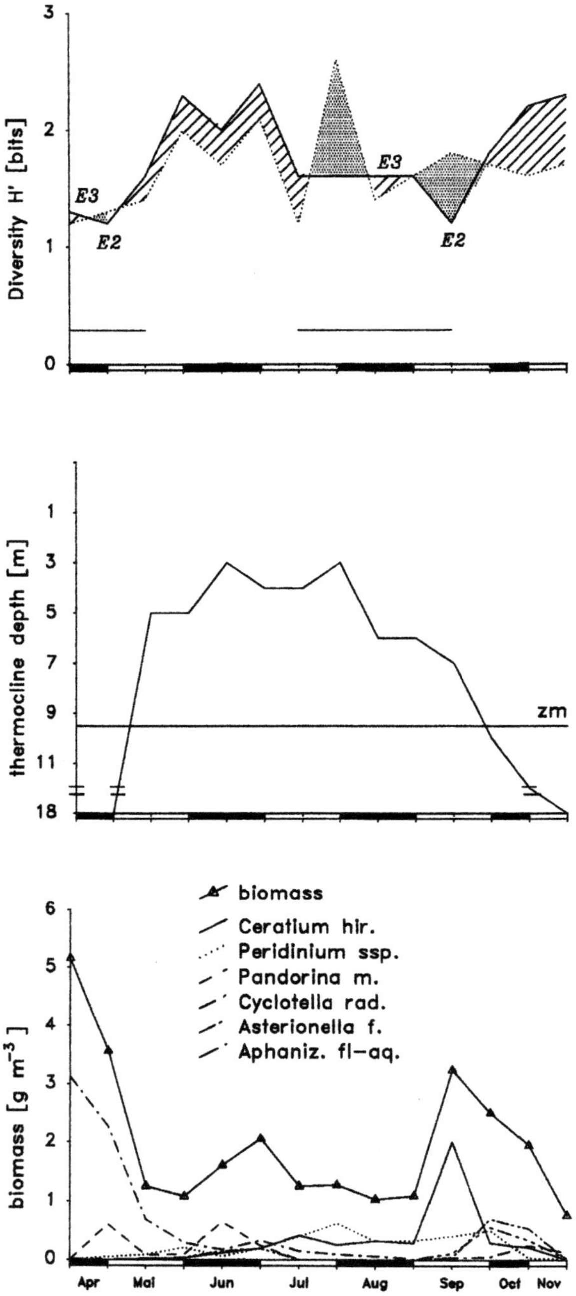

Fig. 2. (a) Shannon-Weaver diversity index based on biomass (g m^{-3}; solid line) and based on numbers of individuals (ind l^{-1}; dotted line) in Pelhamer See, 1988. Hatching indicates the difference between the indices (for full explanation, see text). E# stands for samples with 1, 2 or 3 taxa forming 80% of total biomass. Bars represent the stable periods of the ordination plot (see Fig. 1). (b) Depth of thermocline (m). Mean depth (zm). zm exceeds depth of thermocline most of the year; indicating stable conditions. (c) Total biomass (g m^{-3}) and biomass (g m^{-3}) of important species.

respond to groups of species which occur together in the seasonal succession of species.

In the ordination of Pelhamer See two periods of compositional constancy (1–3/7–11) and two periods of change (4–7/12–14) can be distinguished. The grouping is the same at spring turnover, early summer, summer stagnation and decrease of stagnation till turnover in autumn.

In Figs 2a, 4a, 6a, the difference between the graphs of the differently-calculated diversity indices is shown. When the index based on biomass is higher than the index based on number of individuals, then small to middle-sized classes (vol. $< 10^4\ \mu m^3$) dominate the phytoplankton community, i.e. the striped area. The dotted hatches indicate big size classes (vol. $> 10^4\ \mu m^3$) are dominating. The change in diversity during the vegetation periods can also be inferred from the graphs.

If the grouping of the ordination (Fig. 1) is compared to the diversity graphs (Fig. 2a), it may also be deduced that, on the basis of biomass, the periods of compositional stability feature lowered diversity and, to a large extent, large size classes are dominant. In both periods with constancy of phytoplankton community, the biomass comprised populations made up by only three species. The spring phytoplankton was dominated by large *Cyclotella radiosa* Lemm., *Aulacoseira* ssp. and *Cryptomonas erosa* Ehr. or *C. ovata* Ehr. During stagnation, *Peridinium gatunense* Nygaard, *P. willei* Huitf.-Kaas and *Ceratium hirundinella* Schrank were predominant.

The highest diversity index in Pelhamer See was soon after the onset of stagnation (see Fig. 2b). Obviously, the selective advantage moved from *Cyclotella radiosa* to other species co-occurring during this period (Fig. 2c).

The second period of the species' compositional change was also characterized by high diversity, and at the same time it was a period of decreasing thermal stability of the water column, until complete autumnal turnover (see Fig. 2b). Taxa number increased from 11 to 18.

In October, the phytoplankton consisted mainly of *Asterionella formosa* Hass., *Aphanizomenon flos-aquae* Ralfs ex Born. et Flah.,

Fragilaria crotonensis Kitton, *Stephanodiscus minutulus* (Kütz.) Cleve et Möller and, in November, of *Oscillatoria* sp., *Closterium acutum* var. *variabile* W. Krieg and *Aulacoseira subarctica* Haworth. The correspondence among compositional instability, high diversity and the dominance of small and middle-sized phytoplankton groups is remarkable; yet the figures of thermocline and biomass of Pelhamer See follow a typical gradual seasonal change. The phytoplankton community and diversity react according to the given pattern.

Thalersee

The ordination plot in Fig. 3 opens with two similar spring samples 1–2. Another comparatively long period of constancy is obvious between 5–9. A third though short period occurred in autumn. Similarly to Pelhamer See, three distinct periods of community change were evident (3–5/10–12/13–14). Again the compositionally stable periods correspond to a large extent to low diversity or decreases in diversity and the periods of change correspond to high diversity or increase in diversity, as shown in Fig. 4a. There is a pronounced

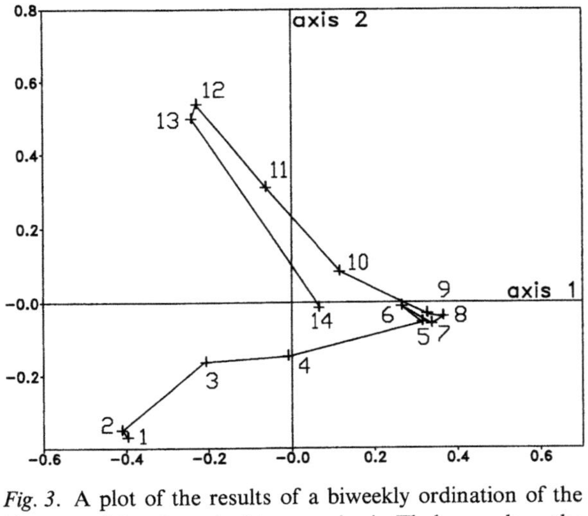

Fig. 3. A plot of the results of a biweekly ordination of the biomass of the phytoplankton species in Thalersee along the first two axes (axis 1 explains 37% and axis 2 explains 24% of the total variance). The sequential samples are joined and numbered. Samples from April to November, 1988.

difference between minima and maxima of diversity (0.4 bits – 2.4 bits) in Thalersee.

A closer look at Fig. 4b/c helps to explain the strong variance in the diversity in Thalersee. An important fact is that, depending on wind stress and solar heating a second, more or less stable, thermocline in the upper column of the lake may develop, see Fig. 4b. In May/June, there was a decrease of total biomass due to a decrease of dominant *Synedra acus* Kütz., following the onset of thermal stagnation (Fig. 4c/b). As a result, diversity increased. An increase in the number of taxa, from 13 to 20, was also observed. Simultaneously, *Bosmina longirostris* O.F. Müller increased to its maximum (832 ind l^{-1}). Diversity was lowered when *B. longirostris* left *Planktothrix agardhii*, which is highly resistant to Cladoceran grazing (Sterner, 1989), as the overwhelming dominant.

A fresh increase in diversity occurred simultaneously to the destruction of the upper thermocline by windstress. However, other factors may have contributed, for example, increased grazing pressure.

A period of warm and calm weather followed, inducing a new second layer of thermocline (see Fig. 4b). This favoured *Planktothrix agardhii*, which would have been able to stratify in the lower part of Thalersee. The biomass curve (Fig. 4c), however, indicates that the main growth period was soon after the wind episode (last week in June). These data agree with the results of Reynolds *et al.* (1984). Afterwards the biomass remained approximately constant, indicating a period of equilibrium (90% of biomass, coexistence one month, biomass not significantly increasing). This shows that resistance against grazing and low loss rate is advantageous in this situation (Padisák *et al.*, 1988).

Again, a period of increased wind mixing in August destroyed the upper layer (see Fig. 4b). Although the windstress was not strong enough to cause complete turnover, the mixing of the upper layer stimulated the renewed increase of *Cyclotella comensis* Grunow biomass, while the biomass of *Planktothrix agardhii* decreased. From this point, further disturbances were observed,

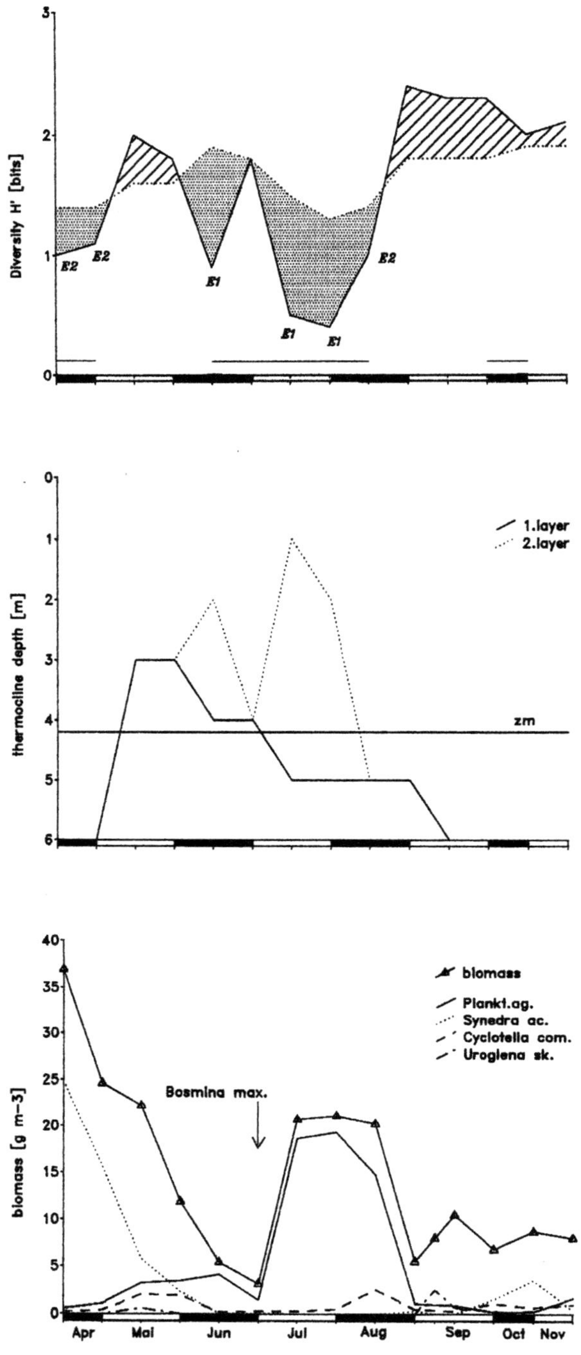

Fig. 4. (a) Shannon-Weaver diversity index based on biomass (g m^{-3}; solid line) and based on numbers of individuals (ind l^{-1}; dotted line) in Thalersee, 1988. Hatching indicates the difference between the indices (for full explanation, see text). E# stands for samples with 1, 2 or 3 taxa forming 80% of total biomass. Bars represent the stable periods of the ordination plot (see Fig. 3). (b) Depth of thermocline (m). Solid line: First layer; broken line: Second layer (further expl., see text). Mean depth (zm). (c) Total biomass (g m^{-3}) and biomass (g m^{-3}) of important species.

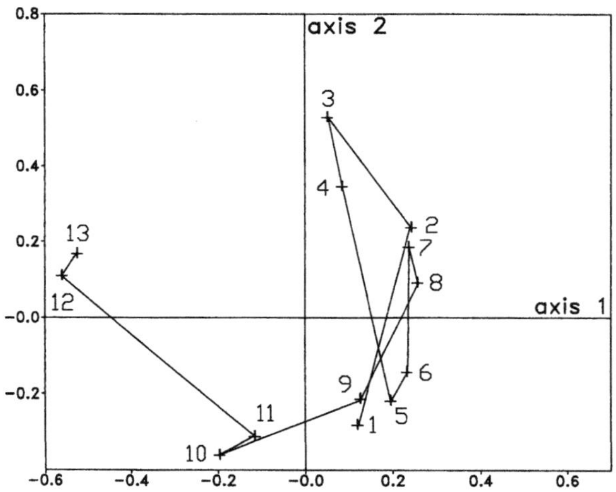

Fig. 5. A plot of the results of a biweekly ordination of the biomass of phytoplankton species in Kautsee along the first two axes (axis 1 explains 37% and axis 2 explains 24% of the total variance). The sequential samples are joined and numbered. Samples from April to October, 1988.

until complete turnover in November (see Fig. 4b; thermocline exceeding mean depth). During this period, the increase in taxa number from 13 to 20–23 underlines the higher diversity. However, the number of taxa at complete turnover in November was only 13. This fact, together with the low diversity index observed during spring turnover, may be a hint that with high-frequency disturbance only those tolerant taxa are able to increase, so diversity declines.

Kautsee

The disturbances observed in Kautsee are of similar nature, though stronger. In Fig. 5, the ordination plot shows strong oscillation, indicating no persistent stable period of the phytoplankton assemblage; there are only short periods with similar algae, moving rapidly on to another community structure.

Most of the year the depth of the thermocline exceeds the mean depth of Kautsee. There is no

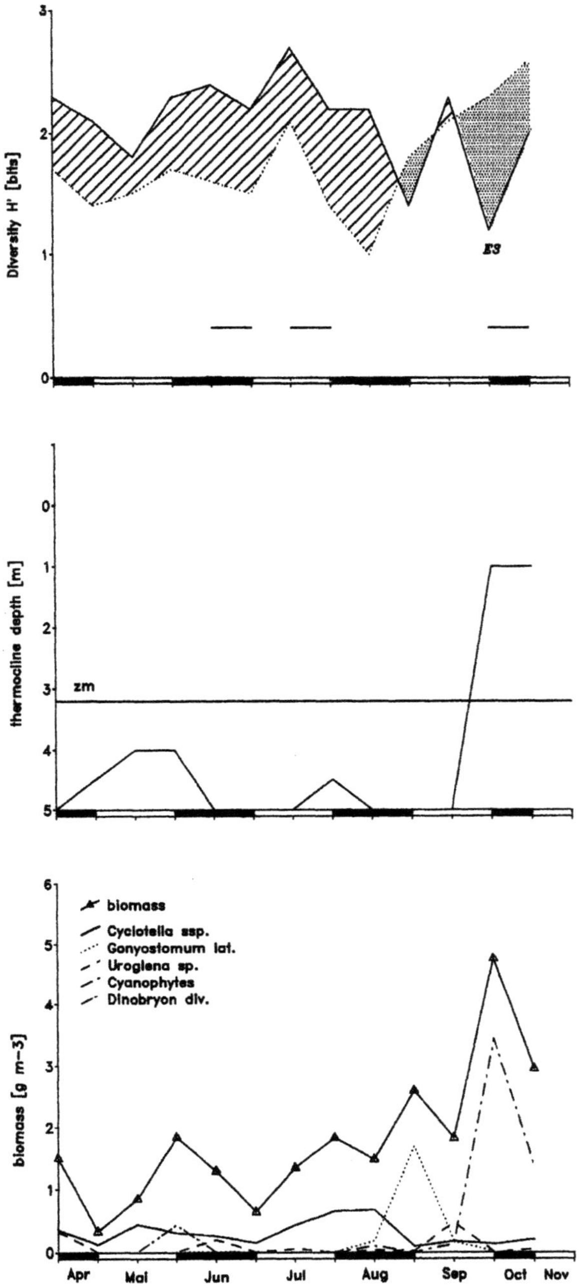

Fig. 6. (a) Shannon-Weaver diversity index based on biomass (g m^{-3}; solid line) and based on numbers of individuals (ind l^{-1}; dotted line) in Kautsee, 1988. Hatching indicates the difference between the indices (see text). E# stands for samples with 1, 2 or 3 taxa forming 80% of total biomass. Bars represent the stable periods of the ordination plot (see Fig. 5). (b) Depth of thermocline (m) and mean depth (zm) (further expl., see text). (c) Total biomass (g m^{-3}) and biomass (g m^{-3}) of important species/species groups.

strong stratification and, therefore, the possibility of complete mixing through long periods of the year (Harris, 1986) is given. (see Fig. 6b). A result of this behaviour could be the high diversity until August. Evidence of a disturbance was the sudden occurrence of periphyton species after strong winds in the last week of June. As those forms were not registered for biomass calculation they did not contribute to the risen diversity value. The interventions of increased wind mixing tended to restore the biomass of *Cyclotella* ssp. (see Fig. 6c) . Maximum diversity was observed in the middle of July when 31 taxa were counted. Further, smaller sized classes of algae dominated till the end of August, which is later than in the other lakes mentioned.

In the middle of August, windy weather again caused complete turnover. A bloom of *Gonyostomum latum* Iwanoff then led to a low diversity (see Fig. 6a). A warm period in September induced stratification right beneath the surface (see Fig. 6b). This environment provided an advantage for colony-forming cyanophytes (*Snowella lacustris* Kom. et Hind., *Aphanothece clathrata* W. et G.S. West, *Radiocystis geminata* Skuja) as total biomass proves. Diversity in Kautsee fell to a minimum at this time.

In Einbessee, a strongly stratified, dystrophic lake (see Table 1), an increase of diversity coincided with the peak of a *Bosmina longirostris* population. They grazed on the dominant alga, *Stichococcus minor* Skuja (Sterner, 1989).

Discussion

This analysis of phytoplankton changes in three upper Bavarian lakes is consistent with the view that there is a tendency to higher diversity indices in mesotrophic lakes than in lakes of other trophic states. The coexistence of more species than in extremely nutrient-poor or nutrient-rich environment is possible (Harris, 1986; Sommer, 1989; Reynolds, 1993).

In the present study high diversity or increase of diversity were found to be associated with the following factors: periods of compositional change of phytoplankton communities, especially at the onset of summer stagnation (Pelhamer See, Thalersee, Kautsee). The high diversity is also connected with lakes where the stability of stratification is intermittent (Kautsee), also with periods of increased wind mixing, resulting in more fluctuation of resources, disturbance (Kautsee) and thinning out of dominant phytoplankton species (Thalersee). Diversity was also found to be high with high grazing pressure, which reduces or prevents the dominance of individual small species (as in the Einbessee) and with the dominance of small to middle-sized phytoplankton classes. This agrees with the results of Harris (1986): temporal variability in the mixing regime increases diversity and this diversity is reflected in the coexistence of a number of peaks in the size spectrum of algae.

Low diversity or decrease in diversity were found to be associated with the following factors: periods of structural stability (stagnation) (Pelhamer See, Thalersee) and frequently disturbed periods (turnover) (see also the introduction of 'constancy' for this circumstances in Reynolds, this volume). The decline of diversity with increasingly strong and frequent disturbances, as predicted by the Intermediate Disturbance Hypothesis (Connell, 1978) is explored in experiments from Gaedeke & Sommer (1986); however, field evidence is weak. Conditions which are extremely favourable for particular species (for example the equilibrium period in Thalersee, with dominance of *Planktothrix agardhii*) were also found to be correlated with low diversity as well as grazing pressure, leaving behind only those species that are resistant to grazing (Thalersee) and the dominance of large, K-selected size classes (Reynolds, 1987).

The examples investigated here support the hypothesis that phytoplankton communities of deeper, stably stratified lakes follow gradual climatic changes; phytoplankton communities of shallow, wind exposed lakes with weak stratification follow gradual climatic changes but are likely to be set back by sudden or repeated disturbance events. In this investigation of the scales of compositional change of phytoplankton community and the simultaneous changes of diversity it became apparent that repeated disturbance events prevent ecological equilibrium and so maintain species diversity.

Acknowledgements

I wish to thank all members of the working group of the Limnologische Station Seeon, especially J. Platzek for his helpful comments and W. Brosch and A. Hayn for technical assistance. My particular thanks go to Dr. J. Padisák and Dr. C.S. Reynolds for critical comments on an earlier version of the manuscript.

References

Clifford, H. T. & W. Stephenson, 1975. An introduction to numerical classification, Academic Press, London, 229 pp.
Connell, J., 1978. Diversity in tropical rain forests and coral reefs. Science 199: 1304–1310.
Elber, F. & F. Schanz, 1989. The causes of change in the diversity and stability of phytoplankton communities in small lakes. Freshwat. Biol. 21: 237–251.
Gaedeke, A. & U. Sommer, 1986. The influence of the frequency of periodic disturbances. Oecologia 71: 98–102.
Gigon, A., 1983. Typology and principles of ecological stability and instability. Mount. Res. Devel. 3: 95–102.
Harris, G. P., 1986. Phytoplankton ecology, 1. Chapman & Hall, London, 384 pp.
Harris, G. P. & B. B. Piccinin, 1980. Physical variability and phytoplankton communities; IV. Temporal changes in the phytoplankton community of a physically variable lake. Arch. Hydrobiol. 89: 447–473.
Kaneta, P. J., M. Levandowsky & W. Esaias, 1985. Multivariate analysis of the phytoplankton community in the New York bight. Mar. Ecol. Prog. Ser. 23: 231–239.
Marker A. E., E. A. Nusch, H. Rai & B. Riemann, 1980. The measurement of photosynthetic pigments in freshwaters and standardization of methods: Conclusions and recommendations. Arch. Hydrobiol. Beih. Ergebn. Limnol. 14: 91–106.
Padisák, J., L. G. Tóth & M. Rajczy, 1988. The role of storms in the summer succession of the phytoplankton community in a shallow lake (Lake Balaton, Hungary). J. Plankton Res. 10: 249–265.
Pielou, E. C., 1966. Species diversity and pattern diversity in the study of ecological succession. J. theor. Biol. 10: 370–383.
Reynolds, C. S., 1984. Phytoplankton periodicity; the inter-

action of form, function and environmental variability. Freshwat. Biol.
Reynolds, C. S., 1987. The response of phytoplankton communities to changing lake environments. Schweiz. Z. Hydrol. 49: 220–235.
Reynolds, C. S., 1988. The concept of ecological succession applied to seasonal periodicity of freshwater phytoplankton. Verh. int. Ver. Limnol. 23: 683–691.
Reynolds, C. S., 1993. Scales of disturbance and their role in plankton ecology. In J. Padisák, C. S. Reynolds & U. Sommer (eds), Intermediate Disturbance Hypothesis in Phytoplankton Ecology. Developments in Hydrobiology 81. Kluwer Academic Publishers, Dordrecht: 157–171. Reprinted from Hydrobiologia 249.
Reynolds C. S., S. W. Wiseman & M. J. O. Clarke, 1984. Growth- and loss-rate responses of phytoplankton to intermittent artificial mixing and their potential application to the control of planctonic algal biomass. J. appl. Ecol. 21: 11–35.
Rott, E., 1981. Some results from phytoplankton counting intercalibrations. Schweiz. Z. Hydrol. 43: 34–62.
Shannon, C. E. & W. Weaver, 1949. The mathematical theory of communication. University & Illinois Press, Urbana, 117 pp.
Sommer, U., 1989. The role of competition for resources in phytoplankton succession. In: U. Sommer (ed.), Plankton ecology: succession in plankton communities. Springer, Berlin, 57–106.
Sterner, R. W., 1989. The role of grazers in phytoplankton succession. In: U. Sommer (ed.), Plankton ecology: succession in plankton communities. Springer, Berlin, 107–170.
Utermöhl, H., 1958. Zur Vervollkommnung der quantitativen Phytoplanktonmethodik. Mitt. int. Ver. Limnol. 9: 1–38.
Wildi, O. & L. Orloci, 1990. Numerical exploration of community patterns. SPB Academic Publishing bv, The Hague, 124 pp.
Wrobel, J. P., 1989. Zur Hydrologie des Gebietes nördlich und nordwestlich des Chiemsees. In: O. Siebeck: 20 Jahre Limnol. Station Seeon, Heller, München, 147–149.

Hydrobiologia **249**: 111–116, 1993.
J. Padisák, C.S. Reynolds & U. Sommer (eds), Intermediate Disturbance Hypothesis in Phytoplankton Ecology.

Ecology of the phytoplankton of the River Moselle: effects of disturbances on community structure and diversity

J.-P. Descy
Unit of Freshwater Ecology, Department of Biology, rue de Bruxelles, 61, B-5000 Namur, Belgium

Key words: phytoplankton, river, community composition, disturbance, diversity

Abstract

A data set on community composition of the phytoplankton of the River Moselle (France) has been used for testing the 'intermediate disturbance hypothesis' (IDH). After a short presentation of the ecology of the river and of its phytoplankton, the main changes in composition and diversity of the suspended algal assemblage are described. It is emphasized that discharge fluctuations, related to weather changes, play a key role, but that biotic factors such as grazing and parasitism may also influence diversity in 'stable' summer conditions.

Introduction

Lake phytoplankton has been worked extensively, whereas few studies have been devoted to the ecology of river phytoplankton (see Descy, 1987, for a brief review). Large rivers, where suspended algae can develop dense and diverse communities, can be viewed as water bodies characteried by:

- short residence times, which vary with the flow rate, i.e. the river discharge through a given cross-section;
- turbulent mixing all over the year; the water column is typically not stratified, save perhaps at very low discharges;
- an abundant nutrient supply from the watershed and the sediment, so that nutrient depletion is seldom observed;
- a high suspended matter concentration, which reduces the transparency.

There is, however, a large difference of scale of disturbance by turbulent mixing between rivers and lakes, which must be emphasized: the river environment is always more or less disturbed, and never experiences periods of stability as they occur in lakes. According to Reynolds (1984 a, b, 1988 a), the river phytoplankton should be dominated theoretically by 'w' (or 'R') species, able to grow in a strongly disturbed and light-limited environment: it should therefore be similar to the phytoplankton of eutrophic (or mesotrophic) lakes during the major mixing events (spring and autumn circulation), with high availability of nutrients and a steep gradient of light related to the Zm:Zeu ratio.

The theory is consistent with the observations: the phytoplankton of large rivers is generally dominated by diatoms (*Asterionella* and *Stephanodiscus* spp.) during spring and autumn, while green algae (mostly Chlorococcales) and cryptomonads may develop best in summer conditions. Eventually, at very low discharges, a typical succession leading to Cyanobacteria dominance (*Microcystis aeruginosa* Kütz., *Oscillatoria agardhii* Gom., etc.) may be observed.

The river discharge is a major integrating factor related to important processes affecting algal communities (rate of turbulent mixing, nutrient supply from the watershed and the sediment, light regime in the water column *via* the amount of suspended matter). Furthermore, discharge seems to control phytoplankton growth through a 'dilution' process (Descy *et al.*, 1987): in a given river stretch of length x, the 'dilution rate' depends on the reciprocal of the cross-section area (1/A) and the linear lateral inflow (dQ/dx). Suspended algae can build up significant populations only when their net growth rate (taking into account the loss factors) exceeds the dilution rate. This control of algal development by discharge favors further fast-growing and disturbance-tolerant algae (i.e. the 'colonizers' or 'pioneers'), at least in fluctuating conditions.

In the present paper, particular attention is given to the effect of discharge and its variations on the diversity of the phytoplankton of the river Moselle (France). It must be emphasized that the data were not initially collected in order to test the intermediate disturbance hypothesis: the frequency of observations may be too low, so that the assessment of the relationship between diversity and disturbance is hardly based on a quantitative approach, but rather on statements which have not been verified on a quantitative basis. So, the choice of the data set was mainly influenced by the interest of testing IDH on river phytoplankton.

Description of the site studied

The River Moselle, a tributary of the Rhine, is 313 km long and has a catchment area of 15 200 km^2. Its mean discharge at Uckange, near the border between France and Germany, is 130 m^3 s^{-1} (extreme values: 35–2700 m^3 s^{-1}). The study site is located in the lower French part of the river, at Koenigsmacher, upstream of the nuclear power plant of Cattenom. With regard to the main physical and chemical variables, this part of the river Moselle can be characterized as follows:

- the concentration of major ions is high (mean conductivity greater than 1300 μS cm^{-1} at 25 °C); in particular, the Cl^- concentration may reach 400 mg l^{-1};
- the nutrient supply is important: the lowest measured concentrations are 0.7 mg l^{-1} NO_3^--N, 0.145 mg l^{-1} PO_4^{3-}-P and 0.6 mg l^{-1} Si;
- the vertical attenuation coefficient is higher than 2.5 m^{-1} during the growth period of the phytoplankton, with a maximum above 3 m^{-1}, so that algal growth is light-limited (maximum GP = 14 g O_2 m^{-2} d^{-1} was in 1986; maximum chlorophyll *a* was 57 μg l^{-1} during the same period); mean depth is about 4 m.

Materials and methods

The data set comprises phytoplankton counts from samples collected between October 1985 and October 1986. The techniques for collecting and treating the samples have been described in Descy & Willems (1991): briefly, unfiltered water samples were fixed with Lugol's iodine and concentrated through a sedimentation procedure, so that the final preparations contained the whole plankton, without any size selection.

The Shannon function, H′ (Shannon & Weaver, 1949 in Pielou, 1975), was calculated from the densities of the phytoplankton components (units ml^{-1}): a unit was a cell, even for colonial and filamentous forms, except for the very small-celled Cyanobacteria. The identification level of the algae was the species or an infraspecific taxon, except in a few cases, such as the small *Stephanodiscus* of the '*hantzschii-group*' (see Descy & Willems, 1991 for more details) or the small Thalassiosiraceae (*Cyclotella pseudostelligera* Hust., *Thalassiosira pseudonana* Hasle & Heimdal), which cannot be identified reliably at the light microscope level, even after proper mounting. Not all the taxa recorded in the taxonomic lists were used for the calculations of diversity, but a selection of algae which can be considered as true plankton (Chlorococcales, centric diatoms and some pennates, Cyanobacteria,

Euglenophyceae, Cryptophyceae). Most pennate diatoms were encountered as single specimens or empty frustules and were not taken into account. So, the number of taxa per sample which were actually used for determining the Shannon function was typically between 43 and 79. However, this selection does not exclude changes in diversity due to the input of algal cells from some tributaries where true phytoplankton does develop.

In the plankton samples, presence of zooplankton (mainly rotifers) and fungal parasites (chytrids) was simply noted. The water analytical data were obtained from the 'Laboratoire d'Ecologie', University of Metz, France, and the daily discharge measurements at Uckange in 1986 were kindly supplied by J.-L. Salleron, Agence de Bassin Rhin-Meuse, France.

Results

The number of taxa in the phytoplankton of the river Moselle totals 239; the main components are diatoms, Chlorophyceae (Chlorococcales), Cryptophyceae and Cyanobacteria. With respect to the abundances (Fig. 1), centric diatoms dominated over the whole growth period (mid-March to mid-October in 1986). The maximal density of phytoplankton units is close to 40000 ml^{-1}, but very few taxa developed large populations, e.g. 2 or 3 green algae, 7 centric diatoms (Fig. 2). In particular, *Skeletonema potamos* (Weber) Hasle accounts for as much as 50% of the phytoplankton cells over the whole period of observation. This filamentous diatom has a summer optimum, whereas, in the colder seasons, small *Stephanodiscus* dominate.

The mid-summer phytoplankton consists mainly of large unicells and filamentous or colonial forms: their relative abundance may result from an intense grazing pressure removing smaller algal cells, as many rotifers are present at that time of the year. At the same time, many chytrid vesicles can be seen on the filamentous diatoms: this strongly suggests that parasitism could be partly responsible of the phytoplankton decline in late summer. Further details on the phytoplank-

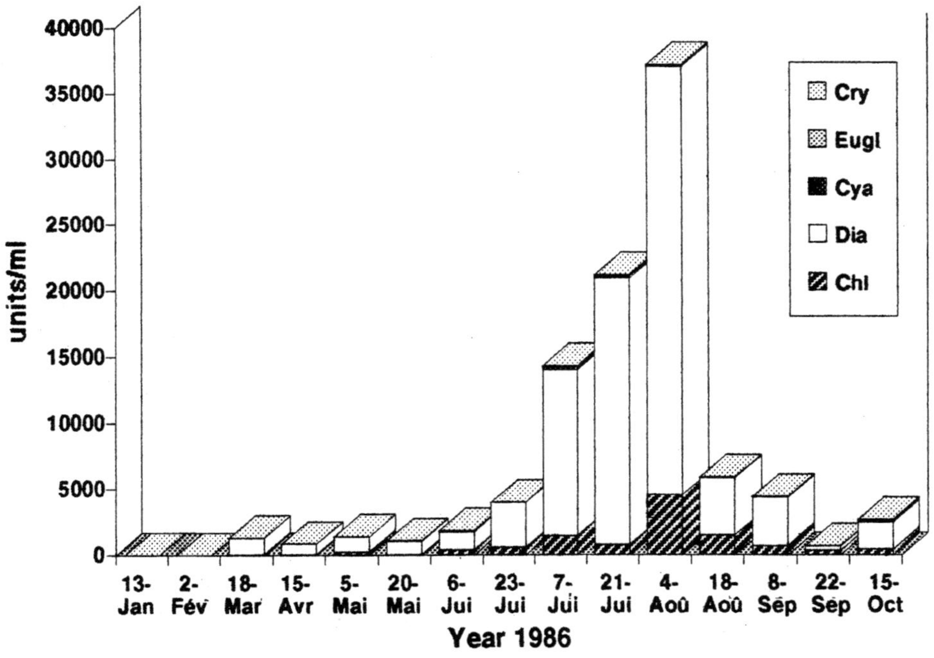

Fig. 1. Changes of the composition of the phytoplankton in the river Moselle at Koenigsmacher in 1986; CHL: Chlorophyceae; DIA: diatoms; CYA: Cyanobacteria; EUGL: Euglenophyceae; CRY: Cryptophyceae.

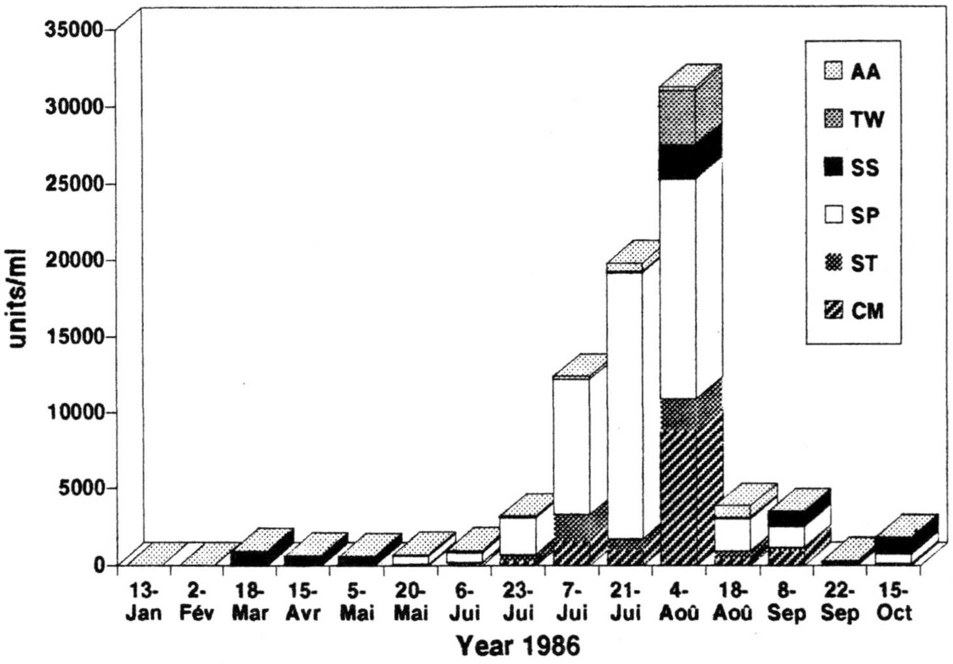

Fig. 2. Seasonal occurrence of some important diatoms in the River Moselle at Koenigsmacher in 1986; CM: *Cyclotella meneghiniana*; ST: small Thalassiosiraceae; SP: *Skeletonema potamos*; SS: small *Stephanodiscus*; TW: *Thalassiosira weissflogii*; AA: *Aulacoseira ambigua.*

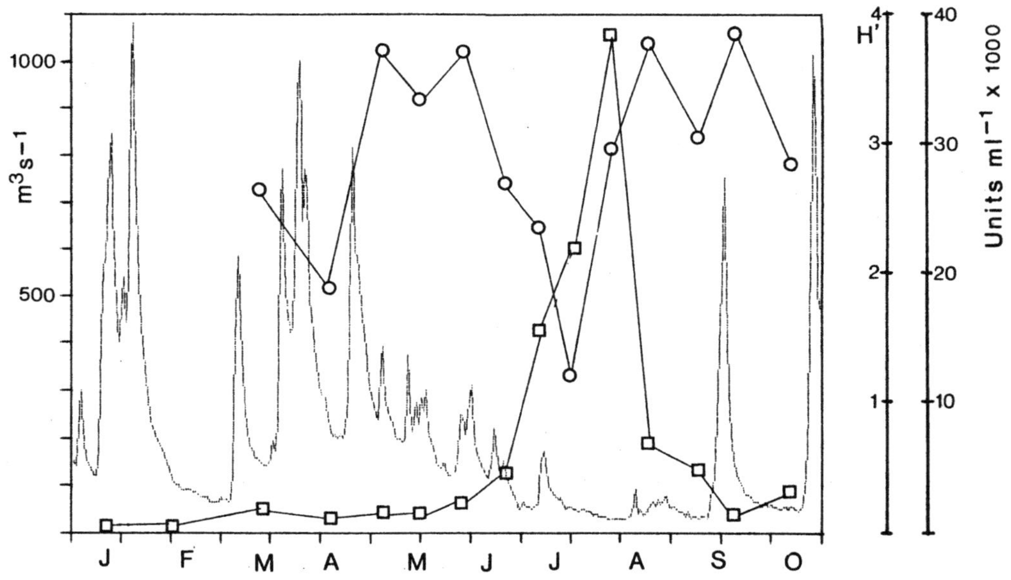

Fig. 3. Variations of the Shannon diversity index (circles) vs. total phytoplankton units (squares) and river discharge (thin line); data from the river Moselle at Koenigsmacher in 1986.

ton of the river Moselle can be found in Descy & Willems (1991).

The seasonal development of the Shannon-diversity is shown in Fig. 3, overlaid on the discharge and the total density of algal units. Four phases can be delimited:

- low diversity in March-April, corresponding to the 'small *Stephanodiscus*' dominance;
- high diversity in May-beginning of June;
- sharp decrease till mid-July, when the *Skeletonema* bloom builds up;
- high diversity values again in August-September, with some fluctuations.

Discussion

The main events in the phytoplankton community changes in the River Moselle can be related to the fluctuations in flow rate and to the biotic interactions. They can be summarized as follows:

- in March–April, which was a period of very large discharge fluctuations, diversity was rather low: the assemblage was dominated by 'small *Stephanodiscus*', able to stand the changing conditions, as well as low temperature and low light;
- in May–beginning of June, diversity increased as discharge decreased and as temperature and light conditions improved; this period of maximal diversity was also characterized by changeable weather, which is reflected by the fluctuations of discharge in the range of 100–400 m^3 s^{-1}; the frequency of these changes occured over the time-scale postulated to influence phytoplankton community structure;
- in June–July, discharge decreased and more 'stable' conditions favored a bloom dominated by *Skeletonema*; as a consequence, diversity declined and reached a minimum value;
- in the beginning of August, diversity increased again, as the best growth conditions for most species were met; grazing seemingly contributed to maintain a high diversity, and large unicells and filamentous diatoms developed important populations;
- the late summer and autumn phytoplankton was a low-biomass, but rather diverse community, associated with the *Skeletonema* decline due to parasitism by chytrids; afterward, the weather changes reflected in the discharge peak of September induced some diversity fluctuations: with the return of unstable conditions, a 'small *Stephanodiscus*' assemblage developed again, similar to that of the spring.

Conclusions

As presented above, the changes in the community structure and diversity of the Moselle phytoplankton are clearly connected with the amplitude of weather and discharge fluctuations: in fact, this type of data set is particularly convenient to show changes driven by physical factors, as they operate, apparently, in the absence of nutrient limitation – hence, in the absence of nutrient competition.

The observations tend to verify the three statements of the IDH (Connell, 1978):

- in steady conditions (e.g. the low discharge period in the summer), diversity reaches minimum levels (*Skeletonema potamos* is dominant);
- strong disturbances (high and variable discharge in spring) also induces minimum diversity ('small *Stephanodiscus*' dominance after the discharge peaks);
- a maximum diversity occurs when the disturbances (discharge and weather fluctuations) are in an intermediate range of intensity and frequency (phytoplankton assemblage of May–June).

In other words, large discharge variations clearly result in low diversity and biomass of the potamoplankton, whereas frequent and low-amplitude fluctuations of flow rate favour higher diversity, while biomass remains low, as the algal growth rate barely exceeds the dilution rate (Fig. 3, late spring situation). At this time, the periodicity of the discharge variations is in the range of several days: this is, in agreement with Reynolds (1988 b), a typical situation where allogenic changes interfere with the internal, autogenic processes leading to a 'stable' community dominated by one or few taxa.

However, these conclusions are merely statements, which should be verified by establishing quantitative relations between diversity and measurements of rate and amplitude of change of the physical environment (flow rate, light, turbulence, etc.). Unfortunately, such a quantitative analysis was not feasible with the data available.

In addition, the effect of loss factors of biotic origin (grazing and parasitism) could also be regarded as disturbances able to generate or to maintain a rather high diversity. Indeed, when dealing with the disturbance-diversity concept, disturbances are defined as 'non-biotic and stochastic events that result in distinct and abrupt changes in composition and which interferes with internally-driven progress toward ecological equilibrium' (Reynolds *et al.*, 1993). However, a selective biotic loss process affecting one or more dominant species can result in a higher diversity value, as far as H′ is taken as a measurement of diversity. In the river Moselle, such biotic interactions seem to play a significant role when the system has reached some physical constancy.

References

Connell, J., 1978. Diversity in tropical rain forests and coral reefs. Science 199: 1304–1310.

Descy, J.-P., 1987. Phytoplankton composition and dynamics in the River Meuse (Belgium). Arch. Hydrobiol., suppl. 78, 2, Algol. Stud. 47: 225–245.

Descy, J.-P., P. Servais, J. S. Smitz, G. Billen & E. Everbecq, 1987. Phytoplankton biomass and production in the river Meuse (Belgium). Wat. Res. 21: 1557–1566.

Descy, J.-P. & C. Willems, 1991. Contribution à la connaissance du phytoplancton de la Moselle (France). Cryptogamie, Algologie, 12: 87–100.

Pielou, E. C., 1975. Ecological diversity. Wiley Interscience, New York, 165 pp.

Reynolds, C. S., 1984a. The ecology of freshwater phytoplankton. Cambridge University Press, Cambridge, 384 pp.

Reynolds, C. S., 1984b. Phytoplankton periodicity: the interaction of form, function and environmental variability. Freshwater Biol. 14: 111–142.

Reynolds, C. S., 1988a. Functional morphology and the adaptive strategies of freshwater phytoplankton. In C. D. Sandgren (ed.), Growth and Reproductive Strategies of Freshwater Phytoplankton, Cambridge University Press, Cambridge, 388–433.

Reynolds, C. S., 1988b. The concept of ecological succession applied to seasonal periodicity of freshwater phytoplankton. Verh. int. Ver. Limnol. 23: 683–691.

Reynolds, C. S., J. Padisák & U. Sommer, 1993. Intermediate disturbance in the ecology of phytoplankton and the maintenance of species diversity: a synthesis. In J. Padisák, C. S. Reynolds & U. Sommer (eds), Intermediate Disturbance Hypothesis in Phytoplankton Ecology. Developments in Hydrobiology 81. Kluwer Academic Publishers, Dordrecht: 183–188. Reprinted from Hydrobiologia 249.

Hydrobiologia **249**: 117–124, 1993.
J. Padisák, C.S. Reynolds & U. Sommer (eds), Intermediate Disturbance Hypothesis in Phytoplankton Ecology.

Some notes about the Intermediate Disturbance Hypothesis and its effects on the phytoplankton of the middle Orinoco river

Humberto I. Carvajal-Chitty
Universidad Simón Bolívar, Instituto de Recursos Naturales Renovables, Apdo. 89.000, Z.P. 1080-A, Caracas, Venezuela

Key words: Orinoco river, disturbance effects, phytoplankton

Abstract

Phytoplankton behaviour in the middle reaches of the Orinoco river (middle basin) is discussed in relation to the Intermediate Disturbance Hypothesis. Seven sets of samples were taken from two channels formed below the confluence of Meta and Orinoco rivers between September 1990 and August 1991. The Orinoco river channel (ORISMA-1) had many more species than the Meta river channel (ORISMA-2). Chlorophyceae dominated both, except during the rising water period when Bacillariophyceae were dominant. The quantitative dominancy alternates among Chlorophyceae, Bacillariophyceae and Cryptophyceae in both river channels, with minima averaging 14 329 and maxima of 1 038 392 cells per liter. The phytoplankton concentration of both river channels is negatively correlated with variability in suspended sediments load. The Shannon-Weaver diversity for ORISMA-1 gives higher values than those from ORISMA-2 and oscillates between 1.37 to 3.52. Disturbances due to increases of water level and high concentration of suspended sediment can be the driving force governing phytoplankton behaviour in large rivers.

Introduction

Phytoplankton communities in large rivers play a fundamental role in the energetics of the ecosystems (Margalef,1983). As in lakes, they are tenuously organized around the availability of the light energy that is required to sustain photosysthesis (Reynolds, 1988).

In rivers the gradient of physical factors formed by the drainage network (Vannote *et al.*, 1980), flow-related turbulence and the pronounced seasonal cycles in the water chemistry (Hamilton & Lewis, 1987), result in seasonal variation in biotic communities and the establishment of species (Patrick, 1988). All these characteristics can be defined as disturbance factors (Reynolds, 1988) that may have a major selective influence over the phytoplankton population and, indeed, over ecosystem-level processes, including primary and secondary production, biomass accumulation, energetics and nutrient cycling (Sousa, 1984).

Connell (1978) introduced the Intermediate Disturbance Hypothesis to explain high species diversity in tropical rain forest and coral reefs. According to this hypothesis, diversity in communities increases with disturbances that are intermediate in frequency and intensity. They create a balance between colonizers and equilibrium species, with neither group finally excluded (Reice, 1985).

Most of the phytoplankton studies in the Orinoco have been made in the lower basin, especially floodplain lagoons (Lewis, 1988; Lewis & Weibezahn, 1985; Blanco & Sánchez, 1986; Sánchez & Vásquez, 1989; Vásquez & Sánchez, 1984). Recently, the hydro-ecological study of the Orinoco river (Proyecto Ecosistema Orinoco, PECOR) has contributed to a greater knowledge of the composition and structure of the plankton from the upper and middle Orinoco. Here are presented some phytoplankton data for the middle Orinoco river and its relationship with disturbance factors proposed by Connell (1978).

Description of site studied

The Orinoco river drains a watershed of 1,000,000 km^2 in Venezuela and Colombia (Fig. 1), receiving waters from geologically and ecologically diverse tributary watersheds (Saunders & Lewis, 1989; Hamilton *et al.*, 1990, Weibezahn, 1985). In the southern part of Venezuela, the Orinoco river flows along the northern edge of the Precambrian Guayana Shield and receives tributaries that drain this formation. These right margin tributaries carry low concentrations of chemical components and suspended sediments (Weibezahn *et al.*, 1989) and are typically clear- or black-water rivers, according to Sioli's (1968) classification. The tributaries that drain to the left side of the Orinoco river mostly have their headwaters in the Andes. These flow across alluvial plains supporting tropical savannas (Saunders & Lewis, 1989), whence they transport high concentrations of electrolytes and suspended sediments (Lewis *et al.*, 1990; Weibezahn, 1990) i.e. they are regarded as typically whitewater rivers according to Sioli's classification.

The Orinoco river has been subdivided in three reaches: The High Basin that begins in its headwaters at Sierra Parima, and ends at the confluence of the rivers Atabapo-Guaviare- Orinoco; the Middle Basin extends thence to the confluence with the Apure river, and the Lower Basin which extends from the Apure to the Orinoco Delta on the Atlantic Ocean (Weibezahn, 1985).

The Meta river rises in the Andes and flows

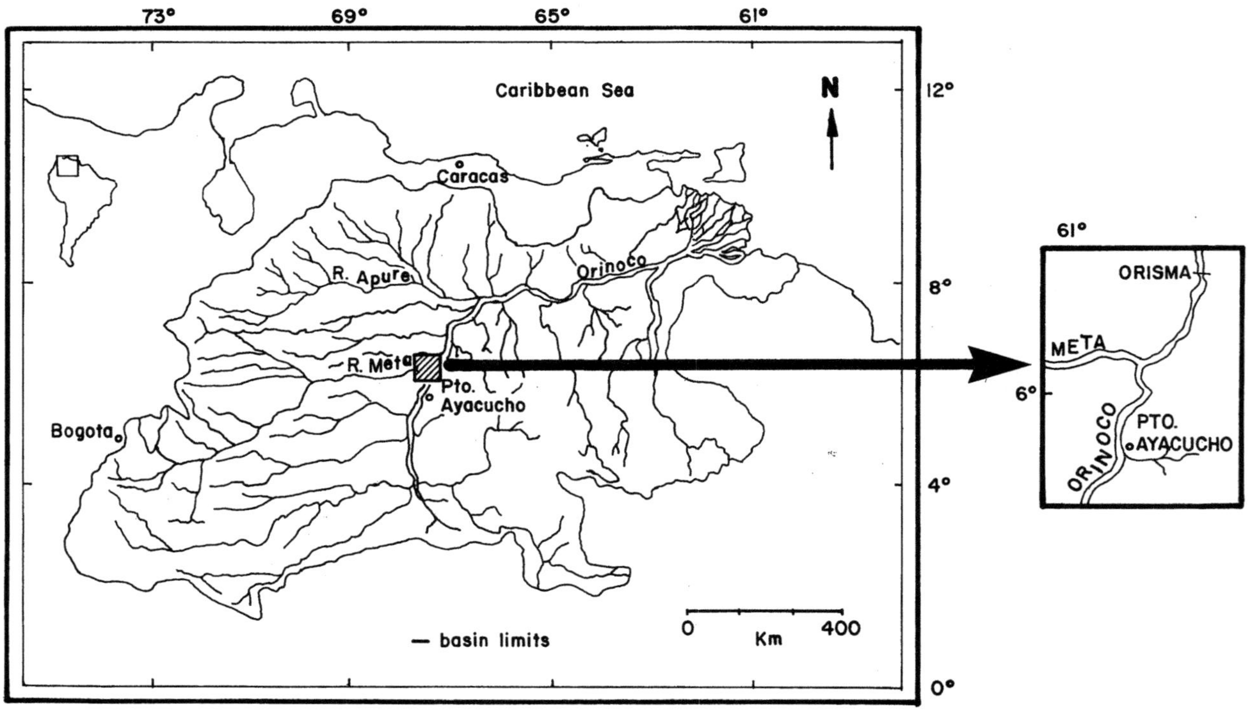

Fig. 1. Map of the Orinoco river basin showing the location of the sampling station (ORISMA).

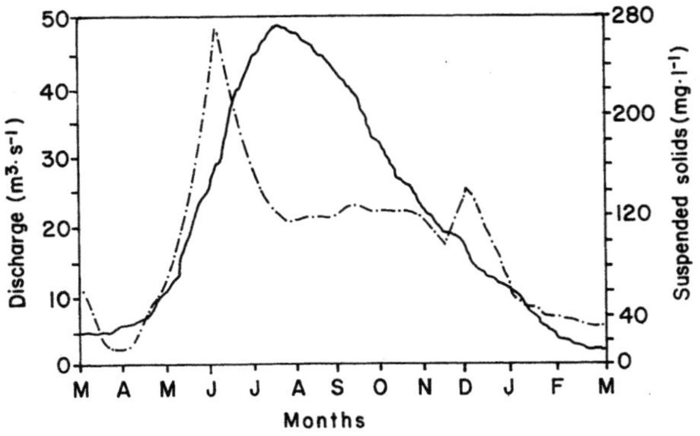

Fig. 2. Seasonal variation of the discharge (× 1000, solid line) and suspended sediments (dashed line) for ORISMA (Data from Weibezahn 1985).

across alluvial grassland, bringing high amounts of suspended sediment and electrolytes, so contrasting with the Orinoco river at their confluence.

Materials and methods

In this study, the PECOR Project's sampling station was located on the Orinoco below the Meta and Orinoco river's confluence, at Santa María (ORISMA, Fig. 1). Seven samples collections were made between September 1990 and August 1991 (Fig. 2), covering periods of falling water (Sept-90, Oct-90 and Dec-90); low water (Apr-91); rising water (May-91; Jun-91) and high water (Aug-91).

At this station, we established two sampling points, one for each river channel because the Orinoco and Meta rivers due not become fully mixed for more that 50 km. below the confluence. At each point, qualitative samples were taken in vertical hauls through the water column, using a net of 53 μm mesh and quantitative samples were taken with an integrating sampler (Nordin *et al.*, 1983), that averages cross-sectional heterogeneity of river water (Lewis *et al.*, 1984; Weibezahn *et al.*, 1989). These procedures were used for the physico- chemical and phytoplankton samples.

All the samples from the net were kept in 250 ml polyethylene bottles and an aliquot of 1000 ml was taken from the integrating bottle, for phytoplankton quantitative analysis, and both samples were preserved with Lugol's solution. In the laboratory, the qualitative samples were allowed to settle for at least 24 hours before analysis were done. Ten aliquots were taken from the bottom of each bottle and examined exhaustively under the light microscope. A species list was made to obtain a total of species for each occasion. For the quantitative analysis, five chambers of 10 ml were settled and three cross-sections were counted under inverted microscope (Wild M40) at × 200 magnification. This method was adopted because the high amount of suspended sediments made

Table 1. Averages of physical-chemical variables for ORISMA-1 and ORISMA-2, for the sampling period. Minimum and maximum values are between bracket. Minimum values are for the high water period with except of phosphorus and suspended sediments that have the minimum in the low water period. * Data from Weibezahn (1985).

Variables	Orisma-1		Orisma-2
pH	5.89 (5.6–6.05)		5.97 (5.2–6.2)
Conductivity (μS cm^{-1}) at 25	11.6 (11.0–30.2)		19.0 (24.1–30.9)
Alcalinity (meq l^{-1})	5.87 (3.2–6.4)		7.86 (5.3–11.2)
Transparency (m)	0.32 (0.20–0.60)		0.20 (0.10–0.50)
Current speed (m s^{-1})	0.89 (0.16–1.34)		1.14 (0.37–1.51)
Discharge (m^3 s^{-1})		22.630 (3.499–44.244)*	
$N\text{-}NO_3$ = (μg l^{-1})	73.1 (32.9–131)		89.4 (46.8–164)
$N\text{-}NO_2$ = (μg l^{-1})	< 1		< 1
$N\text{-}NH_4^+$ (μg l^{-1})	17.1 (0.25–29.7)		20.8 (0.30–33.6)
$P\text{-}PO_4$ = (μg l^{-1})		39.2 (10.2–83.4)*	
Suspended Solids (mg l^{-1})	97.0 (19.0–155)		192 (21.2–308)

counting difficult at larger magnifications. A Mann-Whitney test (Conover, 1980) was used to compare phytoplankton composition between ORISMA-1 and ORISMA-2, and Spearman's rho for linear regression analysis. All the statistics were calculated to within 95% levels of significance. The Shannon-Weaver diversity was calculated using the following expression:

$H' = p_i \log_2 p_i$ and p_i = biomass of i species/total biomass in the sample.

Results

The seasonal variations in discharge and suspended sediments for the Orinoco river at ORISMA are shown in Fig. 2 (data from Weibezahn, 1985). As can be seen, the discharge maximum is in August and the minimum in April. The suspended sediments have a different pattern with a maximum before the peak discharge and a second late during the falling water period. Lower

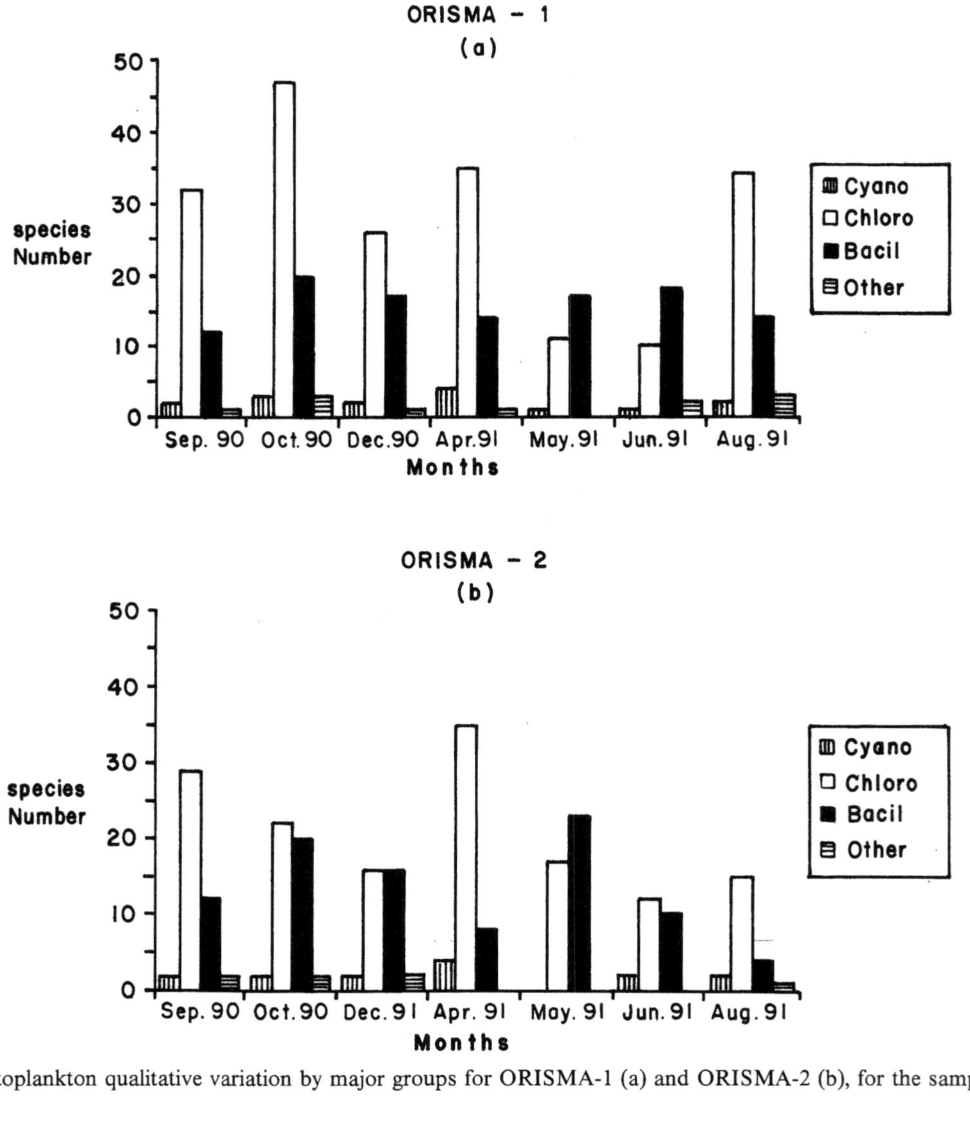

Fig. 3. Phytoplankton qualitative variation by major groups for ORISMA-1 (a) and ORISMA-2 (b), for the sampling period.

concentrations of suspended sediments were found during the low and high discharge peaks. These peculiar pattern was described by Meade *et al.* (1990). A summary of physico-chemical measures is shown in Table 1.

A total of 177 species (8 Cyanophyceae; 121 Chlorophyceae; 43 Bacillariophyceae; 2 Euglenophyceae; 2 Dinophyceae and 1 Chrysophyceae) were found for ORISMA-1 and 135 species (8 Cyanophyceae; 80 Chlorophyceae; 40 Bacillariophyceae; 2 Euglenophyceae; 2 Dinophyceae and 1 Chrysophyceae) were found for ORISMA-2, in net samples. Bacillariophyceae dominated during the rising water period at both sampling points (Fig. 3 a,b). In the rest of the year, Chlorophyta were dominant. It is important to note that throughout the year many more species were observed at ORISMA-1 that at ORISMA-2. In general, the qualitative composition in both river channels was very similar.

At ORISMA-1, the cells per liter ranged from 22211 (at maximum discharge, June) to 1003821 (minimum discharge, April) with an alternation of species dominance through the period. Bacillariophyceae cells averaged 31326 per liter; with a maximum in the low water period (April-91) of 565679 cell per liter (Fig. 4a). The minimum during the high water period was 3224 cell per liter. *Aulacoseira granulata* Ralfs; *Aulacoseira granulata var. angustissima* Müll. and *Cyclotella spp.* contributed the major portion of Bacillariophyceae mass throughout the year. Two maximum diver-

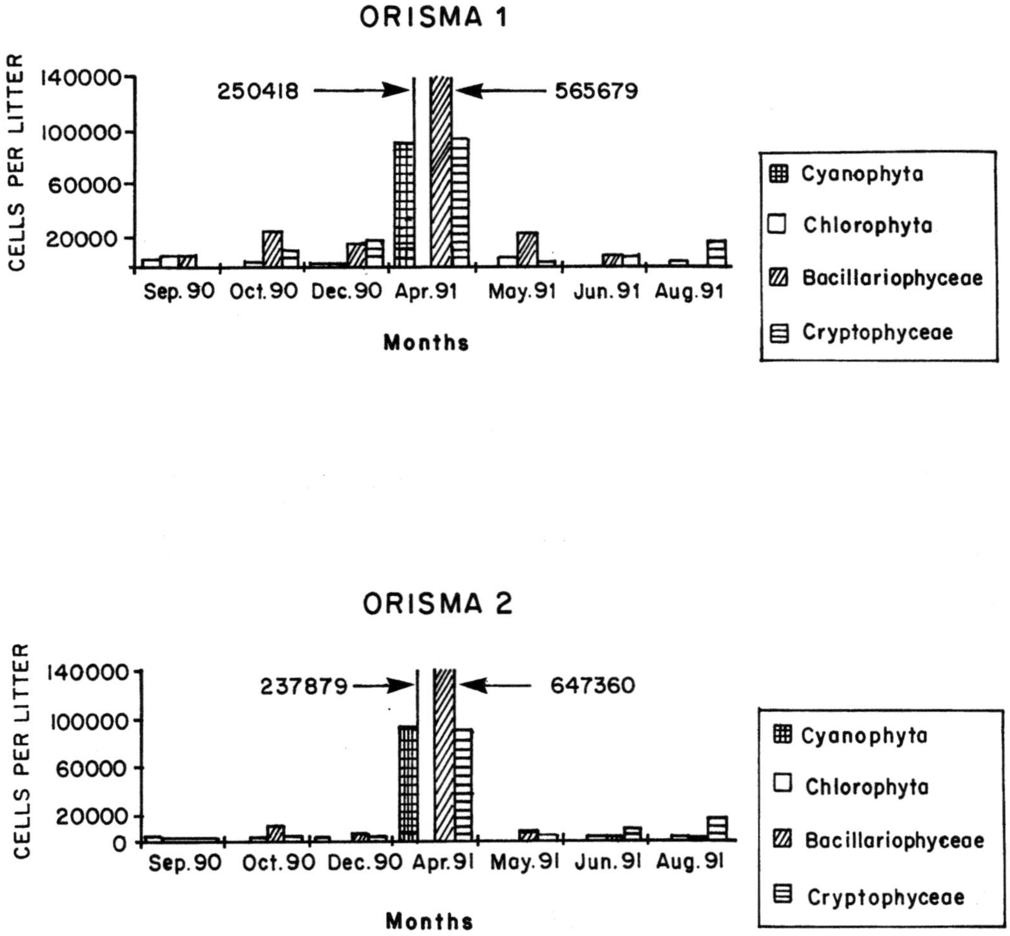

Fig. 4. Phytoplankton quantitative variation by major groups for ORISMA-1 and ORISMA-2 for the sampling period.

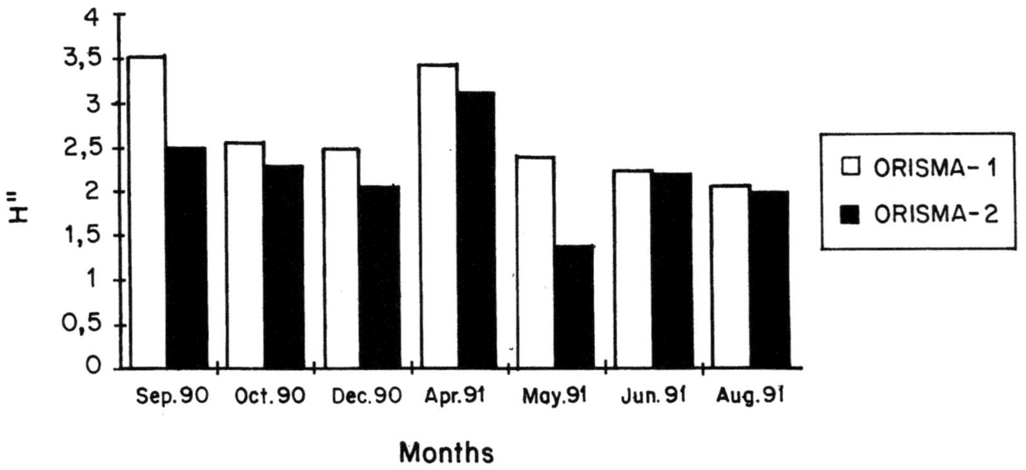

Fig. 5. Diversity index (H') variation for ORISMA-1 (a) and ORISMA-2 (b) during the sampling period.

sity occurred, the first in falling water period (Sep-90, $H' = 3.52$) and the second in low water (Apr-91, $H' = 3.44$), (see Fig. 5). Chlorophyceae averaged 40 124 cells per liter with a maximum during the low water level (April 91 = 250418 Figs. 4a) and the lower values at the maximum discharge (June-91: 1791 cells per liter). Chlorophyceae dominate the plankton during the high water period. Species that made the major contributions included: *Dictyosphaerium pulchellum* Wood; *Scenedesmus spp.*; *Schroederia setigera* (Schröd.) Lemm. and *Gonatozygon* sp. The Cryptophyceae constitute an important group with an average of 22 928 cells per liter, a maximum peak at low water level (92 787 see Fig. 4a) and a minimum at the maximum of water level (716). Sometimes, Cryptophyceae make an important contribution to the plankton. Cyanophyceae, Euglenophyceae and Dinophyceae were very sparse throughout the year. During the low water period, the Cyanophyceae produced a maximum of 92 070 cells per liter but in the rest of the time their density was below 10 000 cells per liter (Fig. 4a). The total algal cells at ORISMA-1 were negatively correlated with total suspended sediments ($r^2 = 0.68$; significant; $p = 0.05$); a significative positive correlation with transparency ($r^2 = 0.72$: $p = 0.05$).

At ORISMA-2, the number of cells per liter ranged from 4657 to 1 072 964. The maximum correspond to the minimum water level and the minimum cell density was recorded after the maximum water level (Fig. 4b). Phytoplankton dominance alterned conspicuously among algal representatives through the year. First, Bacillariophyceae dominated with an average of 97 393 cells per liter and a maximum of 647 360 at the low water period (Fig. 4b) and a minimum at high water (2866). Next, the Cryptophyceae with and average 21 077 cells per liter and the maximum density (90 617) at the minimum discharge (April-91, Fig. 4b). Chlorophyceae and Cyanophyceae were important only during the low water period with densities of 241 819 and 92 070 cells per liter, respectively. Total algae at ORISMA-2 were significantly and negatively correlated with suspended sediment load ($r^2 = 0.79$), but positively correlated with transparency ($r^2 = 0.91$, positive). In general, there were more algae at ORISMA-1 than ORISMA-2. A Mann-Whitney test showed that the phytoplankton population carried at ORISMA-1 was more dense than that at ORISMA-2.

Figure 5 shows the variation in the Shannon-Weaver diversity function. In general, values ranged from 2.05 to 3.51 for ORISMA-1 and 1.37 to 3.12 for ORISMA-2. Maximal values were found during the high ($H' = 3.52$, Sept-90) and low ($H' = 3.44$, April-91) water periods for ORISMA-1 and the same was for ORISMA-2

($H' = 3.12$ for April-91 and $H' = 2.50$ for August-91). The Orinoco river channel (ORISMA-1) have higher diversity values than the Meta river channel (ORISMA-2).

Discussion

The abundance and composition of phytoplankton of the Orinoco river show strong seasonal variation in response to hydrological fluctuations. A large cell concentration ocurred during the low water period while a low density coincides with during the peak flow. (Lewis *et al.*, 1990). The phytoplankton of Meta and the Orinoco rivers are strikingly different in their quality and quantity. The Orinoco river (ORISMA-1) has many more species than the Meta river (ORISMA-2), perhaps because the tributaries of the upper Orinoco carry lower concentrations of suspended sediments (average of 48.14 mg l^{-1}, Weibezahn, 1985) with respect to the Meta (average of 301 mg l^{-1}, Weibezahn, 1985), while the higher transparency of the Orinoco is more favourable to the phytoplankton growth. These explanations can also be applied to the quantitative composition but we have to add effect of the dilution of the Orinoco and Meta rivers phytoplankton communities at the maximum flow. Discharge in the Orinoco river, at ORISMA, averages 22630 m^3 s^{-1} (minimum 3499.44 m^3 s^{-1}; maximum 44244.90 m^3 s^{-1}; coefficient of variation = 62.21%; Weibezahn, 1985) may give a dilution factor up to 12.64 times. High dilution and elevated concentration of suspended sediments together force a decline in the phytoplankton density parallel with low diversity values, typical of highly selective systems. During low water periods, the low concentration of suspended sediments and the high transparency offers conditions suitable for the growth of many more species of phytoplankton and without no nutrient limitation (see Table 1). These have special importance during the low water period. At this time, the river does not have contact with the floodpalin and the phytoplankton has to grow from sources in the main channel (Lewis *et al.*, 1990), often with high diversity values. In the low water period, weak current and sediment transport in addition to the relative stagnant water of the river margin, probably favour phytoplankton growth (Sabater, 1990). In these Orinoco river sections, irregular flushing of phytoplankton, as a consequence of the variation of discharge (Lewis *et al.*, 1990), will bring algae into to the main river flow, so increasing the diversity.

In general, the phytoplankton density and patterns observed in both river channels seem to be more related to physical factors than to chemical conditions of the water as Saad & Antoine (1978) found in the River Tigris (Irak); and Lewis (1988) did in the lower Orinoco river. In contrast, phytoplankton densities are rather lower than those from the Ebro (Spain; Sabater & Muñoz, 1990) and Ter (Spain; Sabater, 1990) rivers and higher than those from the Caroní river (Venezuela; Sánchez & Vásquez, 1989).

Conclusion

Middle Orinoco river phytoplankton seemed to respond to disturbance factors (i.e. increase of water level and high concentration of suspended sediments) that limited their growth. Although differences in quantitative and qualitative composition, (by major groups), between seasons and between river channels support the view that recovery from disturbance favours maximum diversity, more data are needed to confirm that the Intermediate Disturbance Hypothesis of Connell (1978) truly applied to the plankton of large rivers.

Acknowledgements

This work is part of the PECOR project which is sponsored by Ministerio del Ambiente y Recursos Naturales of Venezuela. I appreciate the valuable suggestions and criticism from Dr J. Padisák, Dr C. S. Reynolds, Dr U. Sommer, H. Alvarez and an unknown reviewer. Also my grateful thanks to FUNDAYACUCHO, CONICIT and Prof. Z. Arenas who made possible the author's attendance at the I.A.P. Workshop at Baja, Hungary.

References

Blanco, L. & L. Sánchez, 1986. Contribucion al estudio taxonómico de las Euglenophyta, Cyanophyta, Chlorophyta y Chromophyta del Medio Orinoco, bajo Caroní, Oracoa y algunas lagunas de inundacion (Venezuela). MEM. Soc. Cienc. Nat. La Salle XLV 125–126: 7–48.

Connell, J. H., 1978. Diversity in tropical rain forest and coral reefs. Science 199: 1302–1310.

Conover, W. J., 1980. Practical nonparametric statitics. 2nd Edition. J. Wiley & Sons, Inc. New York: xiv + 493 pp.

Hamilton, S. & W. M. Lewis Jr., 1987. Causes of seasonality in the chemistry of a lake on the Orinoco river floodplain, Venezuela. Limnol. Oceanogr. 32: 1277–1290.

Hamilton, S., S. J. Sippel, W. M. Lewis Jr. & J. F. Saunders, 1990. Zooplankton abundance and evidence for its reduction by macrophyte mats in two Orinoco floodplain lakes. J. Plankton Res. 12: 345–363.

Lewis, W. M. Jr., 1988. Primary production in the Orinoco river. Ecology 69: 679–692.

Lewis, W. M. Jr., J. F. Saunders & R. Dufford, 1990. Suspended organisms and biological carbon flux along the lower Orinoco river. In: Weibezahn, F. H.; H. Alvarez & W. M: Lewis (Eds) The Orinoco river as an Ecosystem; Caracas: 269–300.

Lewis, W. M. Jr., J. F. Saunders & F. H. Weibezahn, 1984. Cross-sectional variation in the chemistry and the suspended sediment load of the Orinoco river at Ciudad Bol'var. Acta Cient. Venezolana 35: 382–385.

Lewis, W. M. Jr. & F. H. Weibezahn, 1985. Progress report on studies of the Orinoco river as an ecosystem. Informe IRNR. USB. 42 p.

Margalef, R., 1983. Limnologia. Ed. Omega, Barcelona. 1010 p.

Meade, R. H., F. H. Weibezahn, W. M. Lewis Jr. & D. Pérez Hernández, 1990. Suspended sediments budget for the Orinoco river. In: Weibezahn, F. H.; H. Alvarez & W. M. Lewis (Eds). The Orinoco river as an Ecosystem, Caracas: 55–80.

Nordin, C. F., C. C. Cranston & A. Mejia, 1983. New Technology for measurement water and suspended sediment discharge of large rivers. Proc. 2nd. Int. Symp. on River Sedimentation, Oct. 1983, Naujing. Publ. Water Resources and Electric Power Press, Beijing China. 1145.

Patrick, R., 1988. Importance of diversity in the functioning and structure of riverine communities. Limnol. Oceanogr. 33: 1304–1307.

Reice, S. R., 1985. Experimental disturbance and maintenance of species diversity in a stream community. Oecologia 67: 90–97.

Reynolds, C. S., 1988. Functional morphology and adaptive stratigies of freshwater phytoplankton. In: C. D. Sandgren (ed.), Growth and Reproductive Strategies of Freshwater phytoplankton. Cambrigde University Press, Cambrigde: 388–433.

Saad, M. A. H. & S. E. Antoine, 1978. Limnologycal studies on the river Tigris, Iraq. III. Phytoplankton. Int. Revue ges. Hydrob. 63: 801–814.

Sabater, S., 1990. Phytoplankton composition in a medium-sized Mediterranian river: The Ter (Sapin). Limnetica 6: 47–56.

Sabater, S. & I. Muñoz, 1990. Succesional dynamics of the phytoplankton in the lower part of the river Ebro. J. Plankton Res. 12: 573–592.

Sánchez, L. & E. Vásquez, 1989. Hydrochemistry and phytoplankton of a major blackwater river (Caroní and a hydroelectric reservoir (Macagua), Venezuela. Arch. Hydrobiol. 33: 303–313.

Saunders, J. F. & W. M. Lewis Jr., 1989. Zooplankton abundance in the lower Orinoco river, Venezuela. Limnol. Oceanogr. 34: 397–409.

Sioli. H., 1968. Hydrochemistry and geology in the Brazilian Amazon region. Amazoniana 1: 267–277.

Sousa, P. S., 1984. Ann. Rev. Ecol. Syst. The role of disturbance in natural communities. 15: 353–391.

Vannote, M. G., Wayne Minshall, K. W. Cummings, J. R. Sedell & C. E. Cushing, 1980. The river continuum concept. Can. J. Fish. aquat. Sci. 37: 130–137.

Vásquez, E. & L. Sánchez, 1984. Variacion estacional del plancton en dos sectores del río Orinoco y una laguna de inundación adyacente. MEM. Soc. Cienc. Nat. La Salle. XLIV 121: 11–34.

Weibezahn, F. H., 1985. Concentraciones de especies químicas disueltas y transporte de sólidos suspendidos en el alto y medio Orinoco y sus variaciones estacionales (Febrero 1984–Febrero-1985). Informe de resultados. Convenio MARNR-PDVSA-USB. xxv + 229.

Weibezahn. F. H., 1990. Hidroquímica y transporte de sedimentos en el alto y medio Orinoco. In: Weibezahn, F. H.; H. Alvarez & W. M. Lewis (Eds.) The Orinoco river as an Ecosystem, Caracas: 150–210.

Weibezahn, F. H., A. Heyvaert & M. Lasi, 1989. Lateral mixing of the waters of the Orinoco, Atabapo and Guaviare rivers after its confluence in the Southern Venezuela. Acta Cient. Venezolana 40: 263–270.

Hydrobiologia **249**: 125–133, 1993.
J. Padisák, C.S. Reynolds & U. Sommer (eds), Intermediate Disturbance Hypothesis in Phytoplankton Ecology.

Effects of the water discharge on periphyton abundance and diversity in a large river (River Danube, Hungary)

Éva Ács & Keve T. Kiss [1]
Botanical Department, Hungarian Natural History Museum, Budapest, Pf. 222, H-1476, Hungary;
[1] *Hungarian Danube Research Station, Hungarian Academy of Sciences, Göd, Jávorka S. 14, H-2131, Hungary*

Key words: periphyton, river, diversity, disturbance, abundance, artificial substratum

Abstract

This paper examines the relevance of intermediate disturbance theory in the context of in a large river the algal flora attaching to artificial substrata. It was observed that the theory is in good agreement with most of the communities analysed, however, with different frequency intervals compared to phytoplankton. Floods provided the most significant disturbances. As a rough generalization, we can say that in case of floods following each other in 8–14 days, respectively, the disturbance is of medium frequency. In case of 8 days, high frequency is more typical while in case of 14 days, low frequency is more characteristic. Owing to the fact that riverine periphytic algal organisms are attached in different ways and with differing efficiencies, the changes in the total mass occurring as a result of changes in water discharge lead to changes in diversity. The analysis of individual numbers can help in investigating the intensity of the disturbing effect on the algal communities. Negative correlation was found between water discharge and numbers of individuals, that is, the numbers of algae increase with low water discharge and decrease with high water level in the river. This second observation is valid only within a certain range of discharges (in this case up to 100 m^3 s^{-1}), which will be specific to particular rivers and individual reaches. Below the upper limit of the range, the increased water flows carry more nutrients to the organisms but with no severe wash-off effect. Beyond the upper limit (in this case in range 100–400 m^3 s^{-1}) the numbers of individuals is roughly in inverse proportion to the wash-off effect of the current, and the community is in equilibrium. In cases when discharge surpasses 400 m^3 s^{-1} decrease in individual numbers is general. We suggest that in the environments similar to the study area, the periphyton of the artificial substrata (e.g., pontoons) undergo similar processes. In respect of the periphyton formed on the surface of the riverbeds in large rivers with considerable changes in water level, a more complex system can be supposed where part of the periphyton is periodically dried and flooded alternately.

Introduction

The theory of intermediate disturbance was originally described for tropical rain forests and atolls (Connell, 1978), and later applied to phytoplanktonic successions by Reynolds (1988). Using this theory as a basis, Padisák (1993) concluded that the maximal diversity of a community is reached during medium frequency and intensity disturbances. More frequent disturbances lead to smaller species number; the scarcity of disturbance causing low evenness, thus decreasing diversity.

So far this theory has been mainly tested in the case of lacustrine phytoplankton (Padisák *et al.*, 1988; Reynolds, 1988). Sabater *et al.* (1991) comment on the possible use of diatoms in riverine periphyton communities in studying the effects of disturbance.

The aim of our paper is to argue the relevance of intermediate disturbance hypothesis in the context of a large river with respect to periphyton formed on artificial substrata.

Materials and methods

Periphyton was collected from a 10 cm^2 matt-glass plate in 1984 and from matt-glass tubes in 1985 and 1986. The substrata were placed in the River Danube at Göd (1669 rkm), to the North of Budapest (upstream) in the drift line in a container frame, 10 cm beneath the surface of the water. Sampling was performed weekly in 1985 and 1986. In 1984, samples were taken more frequently from the end of May till late October. Algae were identified under light and scanning electron microscope, counted by Utermöhl's (1958) method taking the statistical errors of Lund *et al.* (1958) into consideration. Diatoms were identified under light microscope after digestion with H_2O_2. For SEM identification, we followed Kiss' method (Kiss, 1986). Details on the procedure of collecting and elaboration were published by Ács & Kiss (1991). Species diversity (H) values were calculated using Shannon-Weaver's (1948) formula $H = -\Sigma n_i N_i^{-1} \log_2(n_i N_i^{-1})$, where n = the density measure of the i-th species and $N = \Sigma n_i$. The evenness (J) is determined from $J = H H_{max}^{-1}$, where $H_{max} = \log_2 S$, S is the number of species in the sample.

Only a part of the sampling period was used in the analysis. The first three weeks following the deposition of the substratum when colonization took place was not considered. The periods of low water in autumn and low temperature in winter-spring were not considered. Thus, the data used in the analysis were collected only during the summer months.

Results and discussion

In running waters (as in the River Danube), the most critical influence on periphyton communities is flood. Stevenson & Hashim (1989) also regard this as an important disturbance factor. It seems justified to suppose that floods influence the algal communities because of changes in water velocity. At Göd, for example, 1060 $m^3 s^{-1}$ water discharge corresponds to 0.8 m s^{-1} water velocity, while 3050 $m^3 s^{-1}$ means 1.02 m s^{-1}. Therefore an increase in water discharge of 2.88 times resulted in an increase of velocity of only 1.28. The relation is more complex than we would suppose. Increase in suspended matter caused by the increase in water discharge has on one hand, a shading effect on algae, while on the other hand, the floating sediment causes lots of damages to the algal organisms. According to a study on shallow-lake plankton algae (Padisák *et al.*, 1990), the numbers of individuals of some algal species were significantly lower after storm. The storm had a disturbing effect on the sediments and the turbulent motion of the inorganic particles increased mortality of the algae. In the periphytic assemblages a further disturbing factor is the shading of the drifted sediments as well as the mechanical drifting effect of the large sediment particles. Stevenson & Hashim (1989) observed that abrasion sets a limit to the maximal number of individuals present on the stones.

The relationship among some environmental parameters, including the amount of dissolved oxygen, biological (BOD) and chemical (COD) oxygen demand, pH, water temperature, suspended matter content, water discharge and algal abundance, were studied by principal component analysis in 1984 and 1985 (Fig. 1. For the sake of better understanding, the values for water discharge were plotted symmetrically against the X and Y axis. Thus we can observe the extent to which water discharge and algal abundance interact with each other.)

It is apparent from the graph that abundance and water discharge are in negative correlation. Negative correlation was also found between abundance and suspended matter content in both

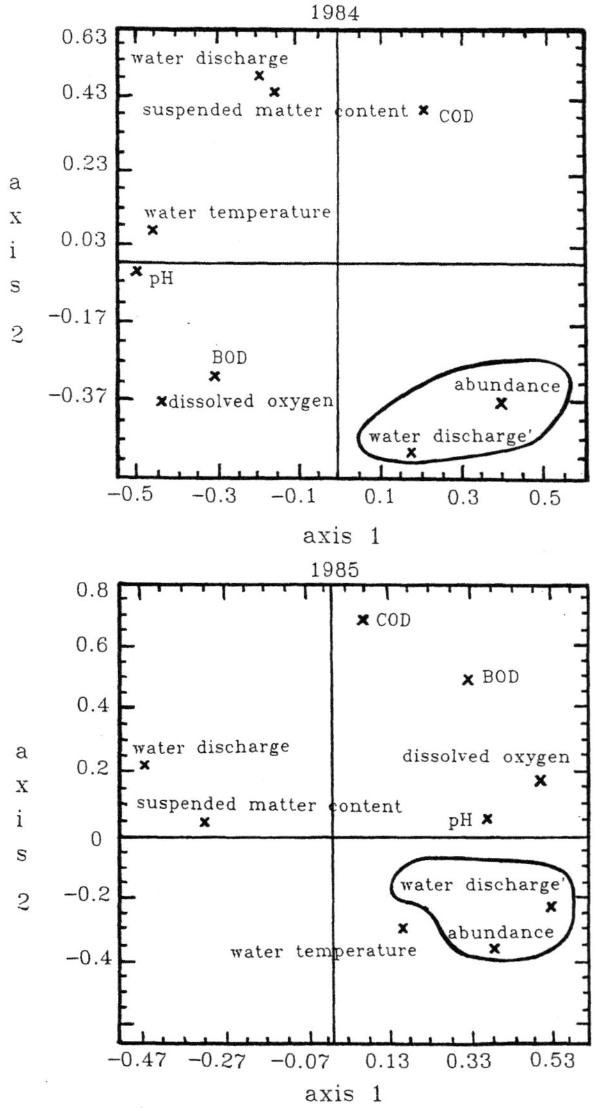

Fig. 1. Principal component analysis. Projection of the variety according to the first two axes.

years, but the coefficients of correlation were not less strong than that between water discharge and algal abundance. This is apparent from the results of 1985, featuring two distinct flood events. This can be explained by the fact that increase in water discharge means increase in suspended matter and related effects that were enumerated above, as well as some dilution of the nutrients. The negative correlation observed between the water discharge and the abundance is the complex outcome of all these factors, that is, as a result of the increase in water discharge, individual numbers decrease. This is, however, not so simple in case of periphytic algae. Some references (Stevenson, 1983, 1984, 1990) mention the increase in the numbers of individuals with increase in water discharge. It must be pointed out that these observations were made in context of 2.5–3 times slower current velocities than we have experienced in case of the Danube. We can put the following questions: up to what point will a rising discharge lead to an increase in periphyton abundance and above what limit will wash-off effects become significant (Fig. 2.)? Increase in water discharge can be, to a certain limit, favourable for algal abundance because it is advantageous for the penetration of nutrients to the cells inside the thick epilithic coating (Stevenson, 1990). The increase in number of individuals can be described with the help of a logistic graph for this phase, as the main process is reproduction within the assemblage. With further increases in water discharge, the wash-off effects of the higher water drift velocity become dominant. Calculating the correlation between water discharge and abundance, we observed that for an increment in water discharges $>=$ 400 m^3 s^{-1}, the algal increase between two samplings implied that significant wash-off had taken place, while an increment of only under 100 m^3 s^{-1} usually favoured a net increase in the population. Within the intermediate range 100–400 m s^{-1}, there was a substantial equilibrium between growth and wash-off in the coating, thus abundance could be represented by a constant level. It would be useful to give a general value for this, e.g. in percentages, but we would get different values in context of different velocities. For example, an increase of the water discharge by 400 m^3 s^{-1} would give different results at water discharge level 1000 m^3 s^{-1} than for 4000 m^3 s^{-1}.

During their phylogenetic evolution, periphytic algae have presumably adopted different adhesion strategies to survive in flowing waters. The diatoms *Achnanthes lanceolata* (Bréb.) Grun., *A. minutissima* Kütz., *Cocconeis placentula* Ehr. stick tightly to the substratum. *Achnanthes minutissima*, which is often found fixed onto the

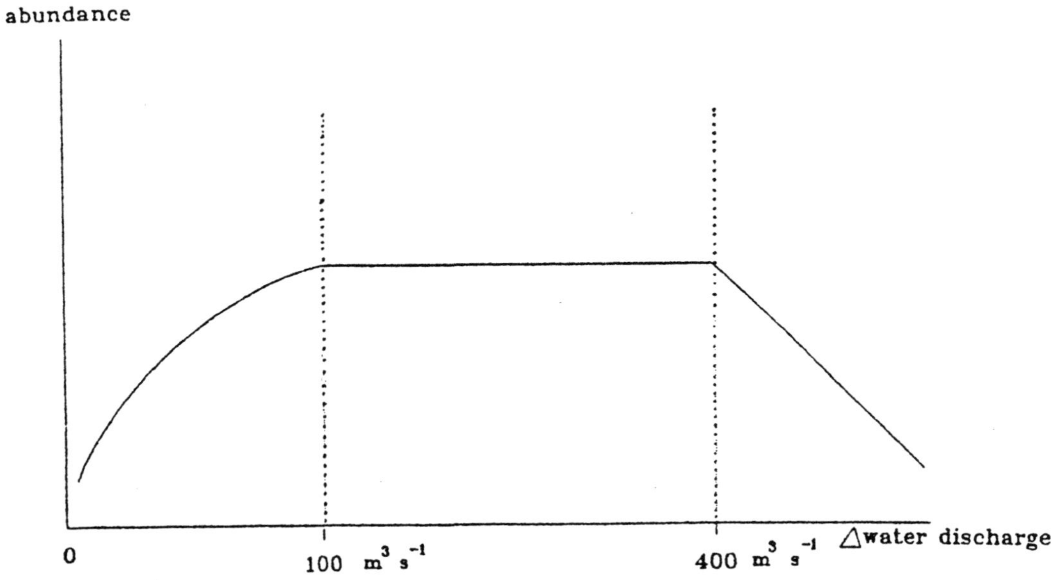

Fig. 2. A model of changes of abundance in function of water discharge.

substratum by a mucilage stalk in lakes, was not fixed in this way in the River Danube, probably because the heavy drift would tear away the algal cell from the thin and tiny (5 μm long) stalk. *Rhoicosphenia abbreviata* (Kütz.) Grun. is fixed by a strong, branching mucilage stalk. *Diatoma vulgare* Bory secretes a sticky substance from a pore field at one end of the cell, by which it can fix itself to the substratum (Stevenson, 1983). There is selection by velocity of flow of the different adhesion strategies. Change in water discharge affects not only the number of individual but also the diversity of species. For example, at first the euplanktonic algae will be washed out from the coating because they are unable to attach to the substratum, while in the case of slowing waters, more and more of these types will be sedimented on the substratum.

Just as in the case of changes in abundance, the amplitude and frequency of floods are important to the measurable diversity values. In 1984, at the beginning of the study period, we observed a definite decrease in diversity (Fig. 3.). Floods followed each other at intervals of 7, 4, 8 days respectively, leading to the decrease of species number. Diversity increased when the flood followed 10 days later and turned to decreasing tendency over the following 15 days. In such circumstances, a definite dominance of *Cocconeis placentula* and *Rhoicosphenia abbreviata* was formed, interrupted by a new flood maximum on the 1st of August. In the following 9 days period, diversity increased again. In the next 6 days, diversity decreased, and later (after 14 days) increased again. Flood maximum by the 30th of August caused a further decrease in diversity. In the following 12-day period, diversity was first increased; later, as the level of the water became more shallow, species number, individual number and evenness were all increasing.

Data for 1985 and 1986 must be interpreted with due caution because of the lesser frequency of sampling (weekly or less frequently, Padisák, 1993). In spite of this, some tendencies are worth noting here. In mid-August, 1985, a major flood took place on the Danube, causing an increase in water discharge of some 3000 $m^3 s^{-1}$(Fig. 4.) and resulted in a significant decrease in diversity. When the flood went down, the algal community remaining on the substratum reproduced quickly. This process was, however, interrupted dramatically by a new great flood and two smaller flow peaks kept diversity values fairly low. In 1986, there were no great floods during the period of

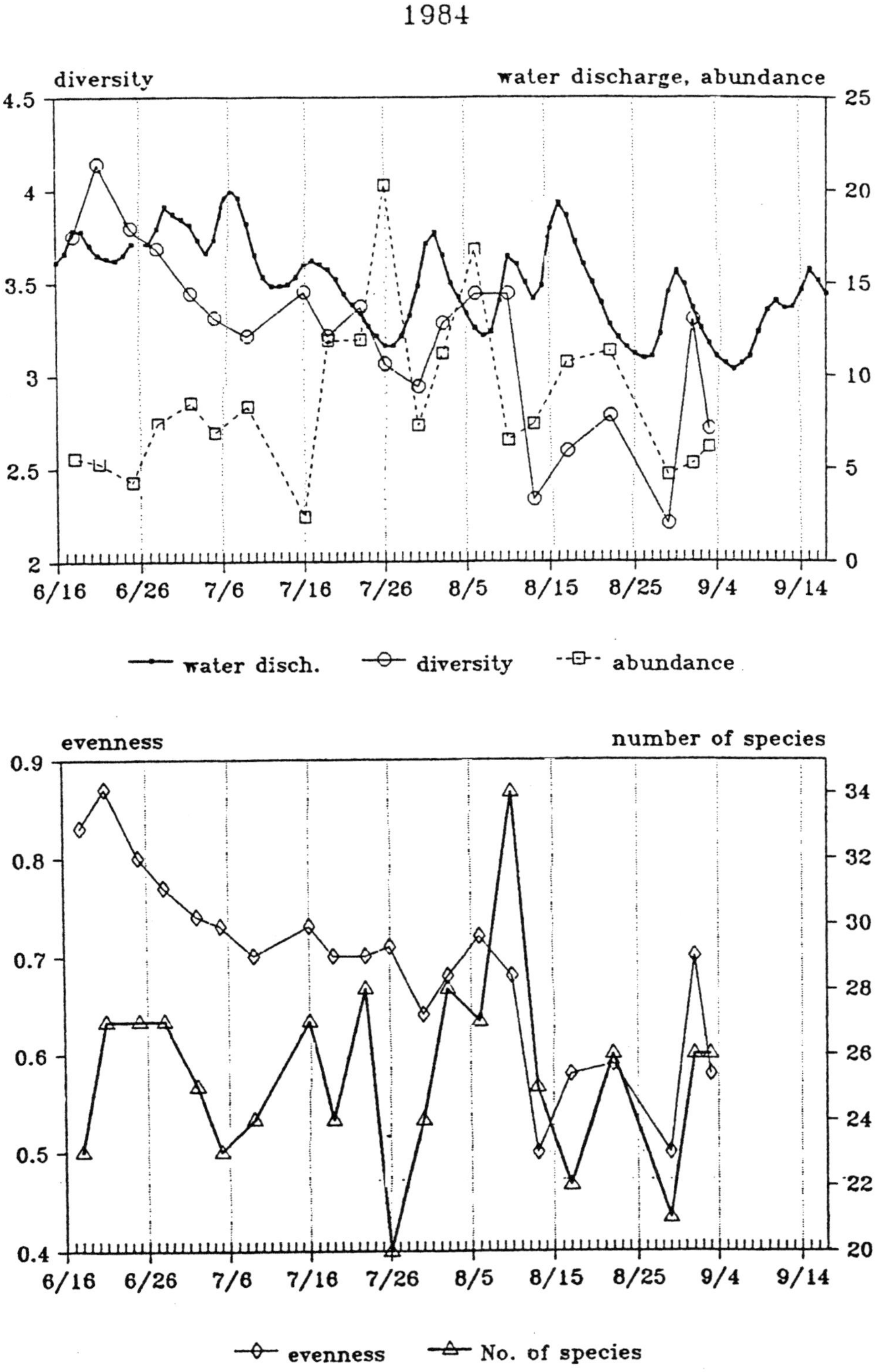

Fig. 3. Values of water discharge, diversity, evenness and individual number in the study period in 1984 (Water discharge: 10^2 m^3 s^{-1}, abundance: 10^5 ind. cm^{-2}.

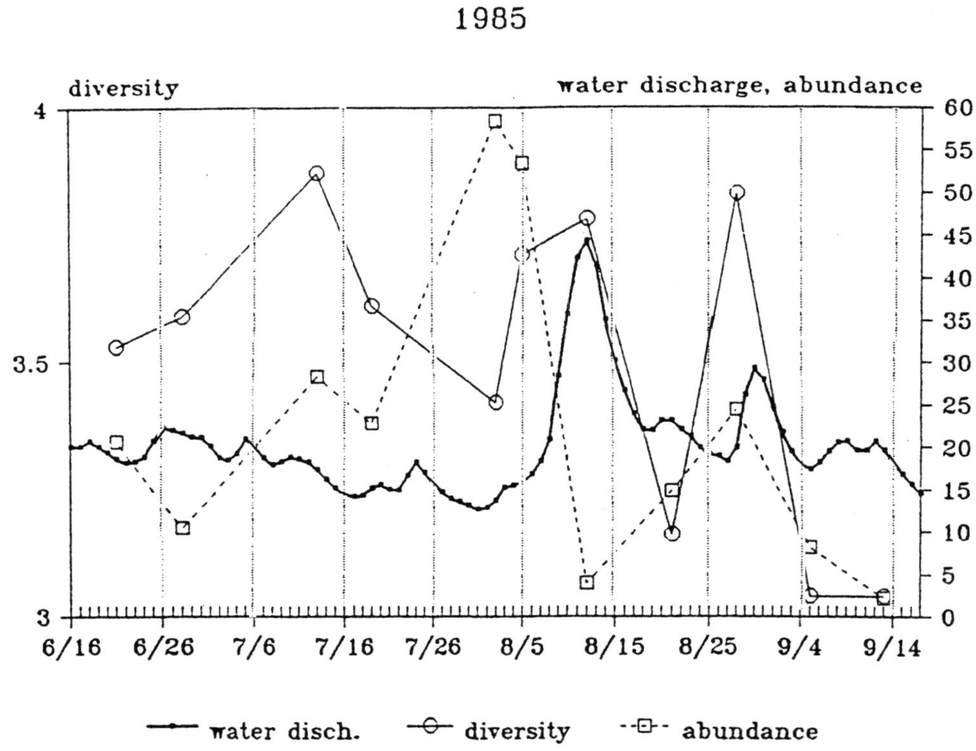

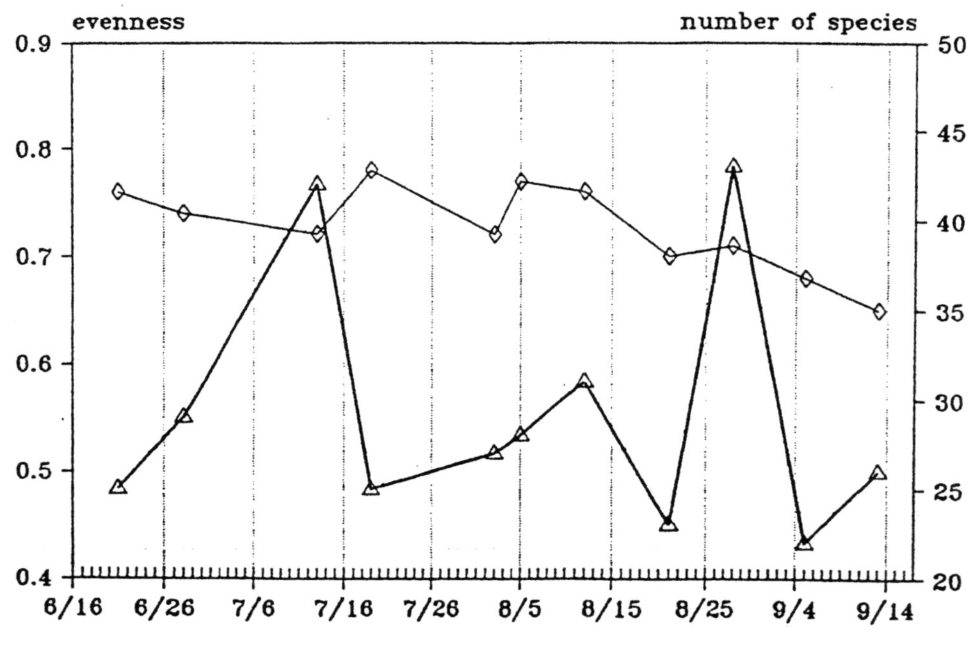

Fig. 4. Values of water discharge, diversity, evenness and individual number in the study period in 1985 (Water discharge: 10^2 m^3 s^{-1}, abundance: 10^5 ind. cm^{-1}.

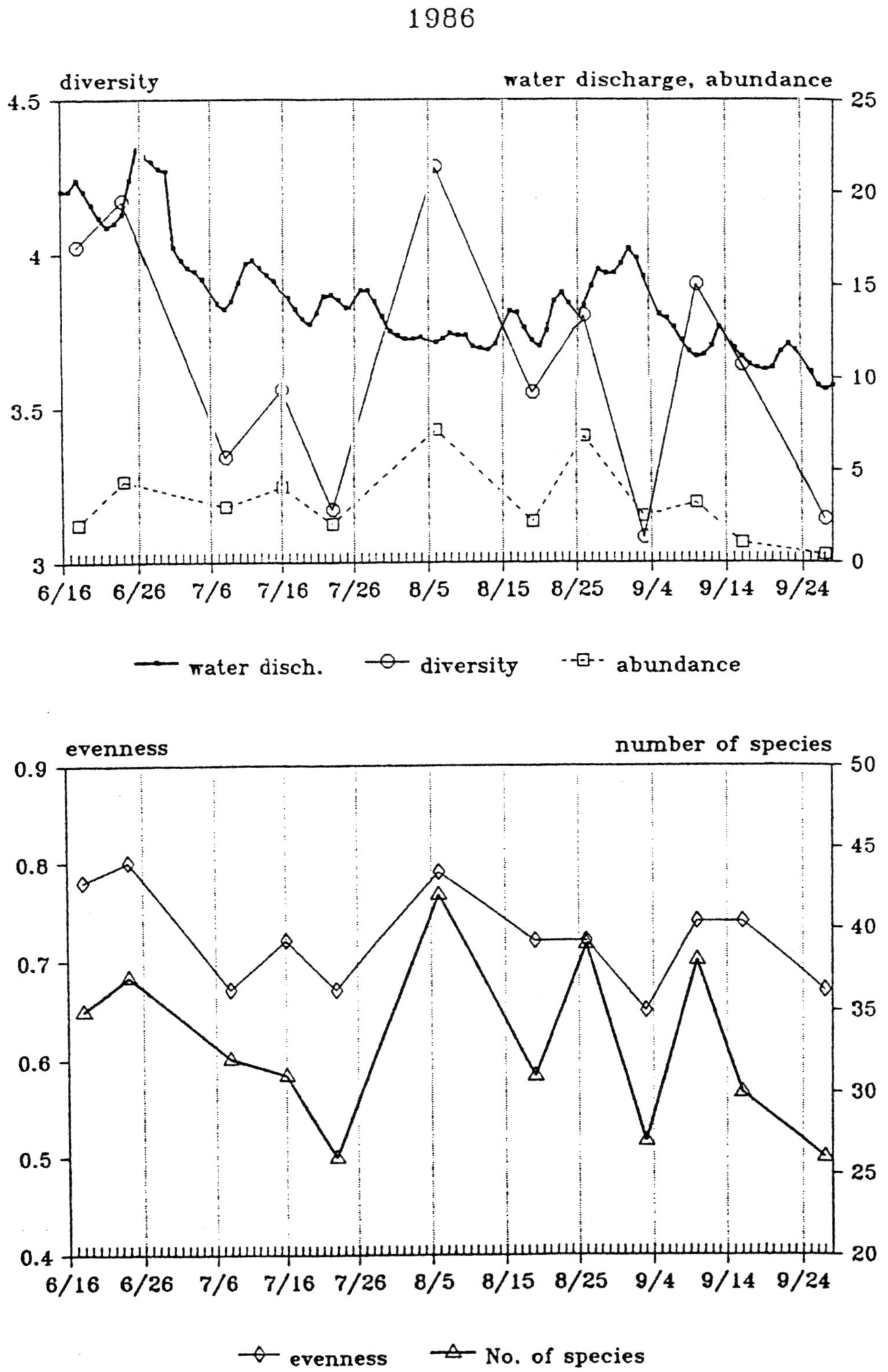

Fig. 5. Values of water discharge, diversity, evenness and individual number in the study period in 1986 (Water discharge: 10^2 m^3 s^{-1}, abundance: 10^5 ind. cm^{-1}.

observations. Diversity essentially decreased in a 15-day period from the end of June (Fig. 5.). At this moment, *Nitzschia romana* Grun. and *Amphora pediculus* (Kütz.) Grun. become dominant. Following a shorter increase in diversity (period of 11 days) a decrease followed (frequencies is 4 day), followed by increase again. Though we had a 4-day period between two 8-day ones, this short phase did not decrease diversity owing to the small amplitude of the flood. Following this period we observed a definite decrease (periods of 7, 5 and 4 days). Diversity started to increase again within 12 days of disturbance but decreased essentially within 6-day periods.

According to Padisák (1993), high frequency disturbance for phytoplankton of some central European shallow lakes, means interferences more frequent than 3 days, intermediate is 3–8 days, low frequency disturbance mean interferences separated by over 8–9 days. Disturbance frequency values are higher here than the ones ascertained by Padisák (1993) for shallow lacustrine environment. As a rough generalization, we can say that, in the case of periphytic communities, floods arriving every 8–14 days represent intermediate disturbance frequencies; more frequent events represent high-frequency disturbance; less frequent ones, low-frequency disturbance. It can be hypothesised, that the generation times are longer than those of lacustrine phytoplankton due to dimensional differences between the algae.

Relative to lacustrine circumstances, riverine circumstances represent for algae a constant disturbance factor with more turbulence, less light, stronger current and greater loads of suspended and transported matter. Consequently, the algal communities living in this environment are accommodated to some basic disturbance level (e.g., a more robust construction of the specimens, dominance of diatoms tolerating poorly lit environment). This is especially true for sessile periphyton communities where algae cannot 'fly away' from unfavourable effects.

On the basis of our analyses we can state that intermediate disturbance hypothesis proved to be valid for periphyton communities attached on artificial substratum in the river Danube. The frequency interval observed, however, are different from those of lacustrine phytoplankton assemblages because riverine periphyton communities have accommodated themselves to some 'background disturbance'.

Acknowledgements

The authors express their thanks to Dr. Judit Padisák, Dr. Colin Reynolds and an unknown referee for suggesting ideas concerning their work and for the revision of the manuscript. The Hungarian Credit Bank, through the 'Foundation for Hungarian Science' sponsored this work.

References

Ács, É. & K. T. Kiss, 1991. Investigation of periphytic algae in the Danube at Göd (1669) river km, Hungary. Arch. Hydrobiol. Suppl. Bd. 89., Algol. Studies 62: 47–67.

Connell, J. H., 1978. Diversity in tropical rain forests and coral reefs. Science 199: 1304–1310.

Kiss, K. T., 1986. Species if the Thalassiosiraceae in the Budapest section of the Danube. Comparison of samples collected in 1956–63 and 1979–83. In M. Ricard (ed.), Proc. 8th Internat. Diatom Symp. Publisher Paris: 23–31.

Lund, J. W. G., C. Kipling & E. D. Lecren, 1958. The inverted microscope method of estimating algal numbers and the statistical basis of estimations by counting. Hydrobiologia 11: 143–170.

Padisák, J., 1993. The influence of different disturbance frequencies on the species richness, diversity and equitability of phytoplankton in shallow lakes. Hydrobiologia 249: 135–156.

Padisák, J., L. G.-Tóth & M. Rajczy, 1988. The role of storms in the summer succession of phytoplankton in a shallow lake (Lake Balaton, Hungary). J. Plankton Res. 10: 249–265.

Padisák, J., L. G.-Tóth & M. Rajczy, 1990. Stir-up effect of wind on a more-or-less stratified shallow lake phytoplankton community, Lake Balaton, Hungary. In P. Biró & J. F. Talling (eds), Tropic Relationships in Inland Waters. Developments in Hydrobiology 53. Kluwer Academic Publishers, Dordrecht: 249–254. Reprinted from Hydrobiologia 191.

Reynolds, C. S., 1988. The concept of biological succession applied to seasonal periodicity of phytoplankton. Verh. int. Ver. Limnol. 23: 683–691.

Sabater, S., J. Armengol, E. Marti, F. Sabater & H. Guasch, 1991. Benthic diatom communities as descriptors of discontinuities in the River Ter, Spain. In B. A. Whitton, E. Rott & G. Friedrich (eds), Use of algae for monitoring rivers. Studia Studentenförderungs-Ges.m.b.H. Innsbruck: 157–163.

Shannon, C. E., 1948. A mathematical theory of communication. Bell Syst. tech. J. 27: 623–656.

Stevenson, R. J., 1983. Effects of current and conditions simulating autogenically changing microhabitats on benthic diatom immigration. J. Ecol. 64: 1514–1524.

Stevenson, R. J., 1984. How currents on different sides of substrates in streams affect mechanisms of benthic algal accumulation. Int. Revue ges. Hydrobiol. 69: 241–262.

Stevenson, R. J., 1990. Benthic algal community dynamics in stream during and after a spate. J. N. Am. Benthol. Soc. 9: 277–288.

Stevenson, R. J. & S. Hashim, 1989. Variation in diatom community structure among habitats in sandy streams. J. Phycol. 25: 678–686.

Utermöhl, H., 1958. Zur Vervollkommnung der quantitativen Phytoplankton-Methodik. Mitt. int. Ver. Limnol. 9: 1–38.

Hydrobiologia **249**: 135–156, 1993.
J. Padisák, C.S. Reynolds & U. Sommer (eds), Intermediate Disturbance Hypothesis in Phytoplankton Ecology.

The influence of different disturbance frequencies on the species richness, diversity and equitability of phytoplankton in shallow lakes

Judit Padisák
Botanical Department of the Hungarian Natural History Museum, Budapest/Balaton Limnological Institute of the Hungarian Academy of Sciences, H-8237 Tihany, Hungary

Key words: Intermediate Disturbance Hypothesis, community composition, seasonality, complexity, stress, characteristic disturbance pattern

Abstract

The relationships between the species richness, diversity and equitability of phytoplankton is discussed in the context of Connell's (1978, Science 199: 1304–1310) Intermediate Disturbance Hypothesis (IDH). The records of 759 vertical phytoplankton samples, which were obtained from four shallow central European lakes (Balaton, Neusiedlersee, and two small artificial ponds) at daily to weekly intervals were analysed.

1) The Shannon-Weaver function was used to measure diversity of the recorded species composition of the phytoplankton. It is shown on fictitious data that compositional diversity is sensitive to the number of coequilibrating species provided that the suspected interrelationship between diversity and 'complexity' is amenable to the application of this method.

2) The disturbance scale that was developed on the basis of the field records fits well to Reynolds' (1988, Verh. int. Ver. Limnol. 23: 683–691) derivation: <3 days qualifies as high frequency, approximately 3–8 days as intermediate frequency and >8–9 days as low frequency of disturbance for phytoplankton.

3) Arithmetical means of the compositional diversity of phytoplankton under different frequencies of disturbance support the hypothesis that maximal diversity appears at intermediate frequencies.

4) There are different reasons for decrease in diversity at higher and lower frequencies. Inequitability diminishes diversity at low disturbance; while species number decreases at high frequencies.

5) The case of Neusiedlersee calls attention to the fact that it is difficult, if at all possible, to differentiate between the indices under continuous stress and high frequency of disturbance in lakes in temperate regions. Similar species number-equitability pattern are induced by both and it is also presumable that high frequency disturbance can itself effect a serious stress.

6) The striking effects that regular major periodic events (e.g. significant changes in the grazing pressure at the onset of the clear-water phase, autumnal cooling) in the plankton have on its species diversity are evident. Thus, the relative importance of intermediate frequency disturbances has its own seasonality: it is increasingly important in periods (partly in the spring, but mostly in the summer-autumn equilibrium phases), in which competition among phytoplankton species is increasing. This observation suggests a way by which the stochasticity-based IDH can be incorporated into rather more deterministic explanations (e.g. PEG-model; Sommer *et al.*, 1986. Archiv für Hydrobiologie 106: 433–471) of plankton succession.

7) The most controversial issue and, therefore, the main difficulty, with IDH is that it not only maintains species richness in an ecosystem but it also supposes its presence. The lack of either early or late successional species in a given community can inactivate the mechanism. From the point of view of the diversity-species richness relationship, the persistence of disturbance at given frequencies is of greater importance than the temporal alterations themselves in the evolutionary ecology of the phytoplankton.

8) For characteristically unperturbed phytoplankton communities (no case was studied here), equilibrium concepts (niche diversification, etc.) should be more strongly applicable to their diversity and species richness.

Introduction

It has been widely recognized that natural lakes can be very rich in phytoplankton species. Many of the species are rather erratic in their appearance and abundance; others are more predictable but still very rare, when compared to the numbers of the few dominant species. Competitive Exclusion Theory (Hardin, 1960; Tilman, 1982) predicts that only as many species can coexist as there are limiting factors. Hutchinson (1961) termed the apparent contradiction between the number of species and the number of limiting resources the 'paradox of plankton'.

Although the paradox was dedicated to phytoplankton, it has been one of the central issues in general ecology. Wilson (1990) recently reviewed twelve explanations of the paradox based on evidence from New Zealand terrestrial plant communities. Most of these explanations (e.g. niche diversification, stabilizing coevolution, etc.) are equilibrium concepts, since the niche theory (two species cannot continuously share the same niche; Gause, 1934) has been widely acknowledged in terrestrial ecology. The basic assumption of the nonequilibrium solutions (Intermediate Disturbance Hypothesis [IDH]; [Connell, 1978], gradual climate change [Holloway, 1948, 1954]) is that there is never enough time for competitive exclusion to became completely effective because sudden or gradual disturbances always arrest the process and shift back the community to an earlier successional phase.

It has been satisfactorily proved in chemostat experiments that stable nutritional conditions lead to competitive exclusion (Sommer, 1983) and that intermittent nutrient pulses enable the coexistence of more species than predicted by the principle (Sommer, 1984, 1985; Gaedeke & Sommer, 1986). The PEG-model (Sommer *et al.*, 1986) successfully provides causal explanations for many of the commonly observed changes during the seasonal succession of natural phytoplankton. However, as stated in the description of the PEG-model, the shallowness of the lakes can lead to unpredictability, and the direction of phytoplankton compositional changes is strongly dependent upon the stochasticity of weather. In some cases, the 'noise' is such to disguise the main trends.

The aim of this paper is to test the Intermediate Disturbance Hypothesis based on summarized results of short-term studies (daily to weekly sampling frequency) in four shallow lakes in central Europe carried out during the last fifteen years. The theoretical background is given in Sommer *et al.* (1993).

Description of sites studied

For this study, records of vertical phytoplankton samples obtained from two central European shallow lakes (Balaton, Neusiedlersee) and two small artificial ponds were compared (Fig. 1).

Balaton

Lake Balaton is the largest shallow lake in central Europe. The lake has a surface area of 593 km^2, is 77.9 km long, 7.2 km wide on average

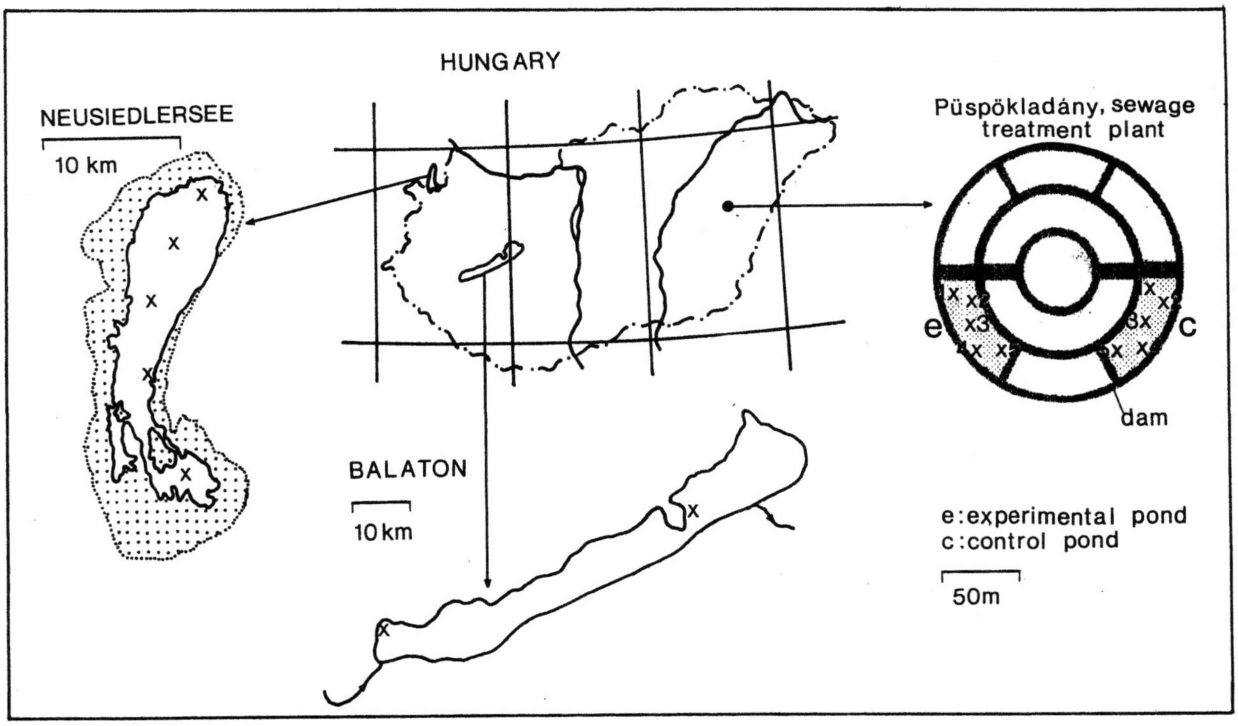

Fig. 1. Location of studied lakes. X indicates sampling stations.

(maximum width 15 km) and has a mean depth of 3.14 m (maximum 11 m). The theoretical retention time is 3–8 years. The lake was originally mesotrophic but underwent rapid eutrophication during the 1960s–1970s (Herodek, 1984). Because the majority of the nutrient load is received by the western basin, there is a sharp trophic gradient in the elongated lake: the western part is hypertrophic, while the eastern has been still meso-eutrophic. Its phytoplankton is limited primarily by nutrients, namely N or P or both. Silicon has never been found to be limiting for diatoms (Istvánovics & Herodek, 1985; Istvánovics *et al.*, 1986). When the algal biomass is high, or after storms the plankton may be subject to light limitation through self-shading or resuspended turbidity.

Phytoplankton was sampled daily in open water in the meso-eutrophic Tihany region of the lake during one-month periods in each of the summers of 1976, 1977, 1978 and 1982. In 1982, the sampling was continued at about weekly intervals until late autumn. In the eu-hypertrophic Keszthely Bay, offshore samples were taken daily between 1 April and 28 November 1980.

Neusiedlersee

The Neusiedlersee (Fertő) is another large shallow lake. Its surface area is 300 km^2; it is 35 km long, 8.6 km wide on average, and has a mean depth of 1.3 m (maximum 1.8 m). The theoretical retention time is 3 years. Approximately one third of the lake area is covered by reed swamp.

The mesotrophic lake has a high salt-content, is alkaline and very turbid. Conductivity ranges in 1000–2300 $\mu S\ cm^{-1}$, alkalinity is 8.0–10.5 mval l^{-1}, pH is 7.5–10. Secchi transparency in the open water is characteristically about 0.2 m (range: 0.06–0.8 m; higher values occur only under ice). The sediment is rich in silica, and the dissolved N and P concentrations are usually above the level that is considered limiting in other lakes. Because of the inherently high turbidity of the open water of the lake, growth of phytoplankton is presumably subject to frequent or continuous light limitation, with only brief exceptions.

Phytoplankton samples were taken at approximately weekly intervals by the staff of the Illmitz Biological Station from various sampling stations in the open water. Owing to the extremely high stohasticity of almost all biological variables (Herzig, 1990) and the fact that no consequent horizontal differences appear in phytoplankton distribution, samples taken at different open water sampling sites were considered to belong to the same time series.

Experimental ponds

In 1977, a detailed limnological project was organized to study two of the six small artificial hypertrophic ponds connected to a bio-mechanical combined oxidation-type-sewage-treatment plant (Dévai, 1977; Dévai & Woynarovich, 1981). The ponds are located in Püspökladány, eastern Hungary, and have surface areas of about 1920 m^2, respectively. Their water depth is 1.5 m. The calculated retention time is 9.2 days. On average, each receives 312 m^3 day^{-1} of household sewage with some industrial waste water. The sewage overflow is continuous through the system depending upon the level of household discharge. This has a morning and an evening peak, and which provides an approximately 12-h cycle of disturbance. The ponds did not receive sewage during the study period; they were disconnected from the system. For the purposes of food-web manipulation, crustaceans were killed in one of the two ponds (the experimental pond, hereafter) with a short half-life pesticide on the first day of the experiment. The other was considered as an untreated control (control pond, hereafter). The effect of disconnecting and/or pesticide treatment on soluble and particulate N/P ratios and on the phytoplankton/zooplankton ratio is shown in Fig. 2 as background information.

There was plenty of orthophosphate in both ponds during the study period. Decrease in both dissolved and particulate N/P ratios and the in-

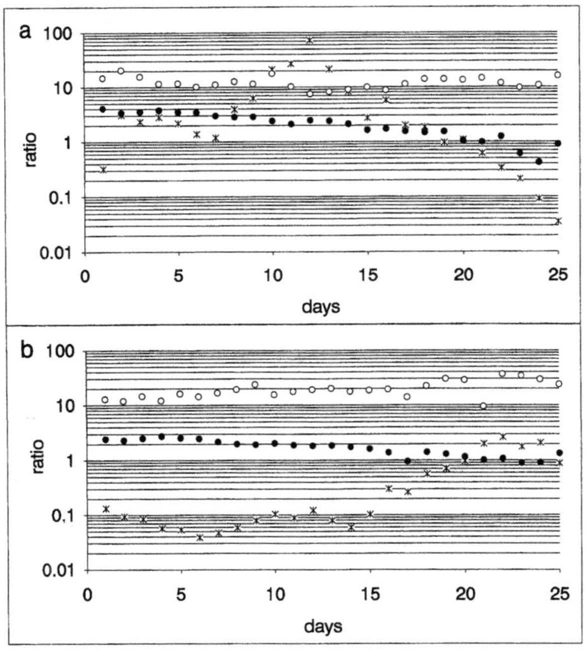

Fig. 2. Particulate (○) and dissolved (●) atomic N/P ratios, phytoplankton/zooplankton ratio (*) in the experimental (a) and control (b) pond between 16 August and 9 September, 1977. Days are numbered consecutively; 1 corresponds to 16 August.

creasing difference between the particulate and dissolved ratios may indicate a relative nitrogen deficiency. However, at the lowest dissolved N/P ratios the concentration of soluble inorganic nitrogen compounds were 370 $\mu gN\ l^{-1}$ (experimental) and 1890 $\mu gN\ l^{-1}$ (control). Secchi transparency changed between 0.25– 0.42 m (experimental) and 0.38–0.93 m (control). There was some indication of light limitation based on chlorophyll-biomass ratios (Vörös & Padisák, 1991). pH varied between 8.54–8.83 (experimental) and 8.34–8.53 (control). Conductivity (with only minor deviations) was 1570 $\mu S\ cm^{-1}$ in the experimental pond and 1630 $\mu S\ cm^{-1}$ in the control.

Phytoplankton was sampled daily between 16 August and 9 September 1977 at five points in either pond. Except when noted, records of samples deriving from the central sampling station (point 3 on Fig. 1) of the ponds are analyzed and discussed in this paper.

Material and methods

Phytoplankton countings and estimates

Samples collected from the above three locations were fixed with 3–4% formaldehyde (Balaton, Keszthely, 1980) or with Lugol's Iodine (all other cases) immediately after the sampling. Algae were counted on an inverted microscope. A minimum of 400 units (cells, colonies or filaments) were counted in each sample, giving a counting accuracy, expressed in terms of 95% confidence limits, of < 10% for the whole phytoplankton (Lund *et al.*, 1958). A similar counting method was applied in each of the studies described here; total chamber surface counting for large algae and transect counting for smaller forms. Volumetric biomass was calculated as the product of the cell or coenobial-unit numbers and their specific volumes, obtained by geometrical approximations. Biomass data were used in all of the calculations described in this paper.

Further details on all three studies are given in Bondár *et al.* (1981), Padisák (1980, 1991, 1992), G.-Tóth & Padisák (1986), Padisák *et al.* (1988, 1990), Padisák & G.-Tóth (1991), Kiss & Padisák (1991), Dokulil & Padisák (1993).

Diversity

Although there are numerous weighted and unweighted methods to estimate biotic diversity (Pielou, 1975), the specialized form of the Shannon-Weaver function (Shannon, 1948) is the most widely applied:

$$H'' = -\sum_{i=1}^{s} \frac{n_i}{N} \log_2 \frac{n_i}{N}, \tag{1}$$

where n_i is the density measure (in this case the biomass) of the *i*-th species; s is the number of species in the sample; and

$$N = -\sum_{i=1}^{s} n_i \tag{2}$$

Throughout this study, equation (1) was used to calculate diversity.

The mathematical properties of the Shannon function have been subjected to detailed analyses (e.g. Khinchin, 1957; Kullback, 1957). However, when testing Connell's IDH based on the principles given in Sommer *et al.* (1993), it is worth stressing those mathematical features of the function that have special relevance to the testing of IDH.

i) The more species that are present in a sample, the higher the diversity. The theoretical maximum (H''_{max}) of diversity in any sample can be expressed as

$$H''_{max} = \log_2 s \tag{3}$$

ii) The more equally distributed the total density among the species, the higher the diversity. Equitability (or evenness, E) can be expressed as

$$E = \frac{H''}{H''_{max}} \tag{4}$$

or as a percentage

$$E\% = 100 \frac{H''}{H''_{max}} \tag{5}$$

Thus, the diversity of any community can decrease in two ways: either by a decrease in species number or by a decrease in equitability.

iii) It can be shown that the share in diversity of any given species increases until its contribution approximates the about 1/4–1/3; afterwards it begins to decrease. This characteristic of the function was used to estimate species importance in diversity changes (Rajczy & Padisák, 1983). It follows that if diversity is plotted against the percentage contribution of an increasing species, the result is a maximum type curve. Figure 14 demonstrates a field example.

iv) From the point of view of Connell's IDH the number of coequilibrating species is irrelevant, although it has a large impact on the observed diversity. Let us imagine a fictitious 10-species assemblage in which either one or five species begin to increase at time t1 (when the equitability is maximal: $H'' = H''_{max}$) and reach a contribution of 90% by time t9. Contingency tables are given for the cases when a single species increases (Table 1a) and when five do so (Table 1b), simultaneously and at the same rate. When only one species is progressing to equilibrium dominance diversity reduces steeply. When five do so, the decrease is slight (Fig. 3). This example provides a way in which diversity may be related to 'complexity', if it is assumed that the number of coequilibrating species can be used as

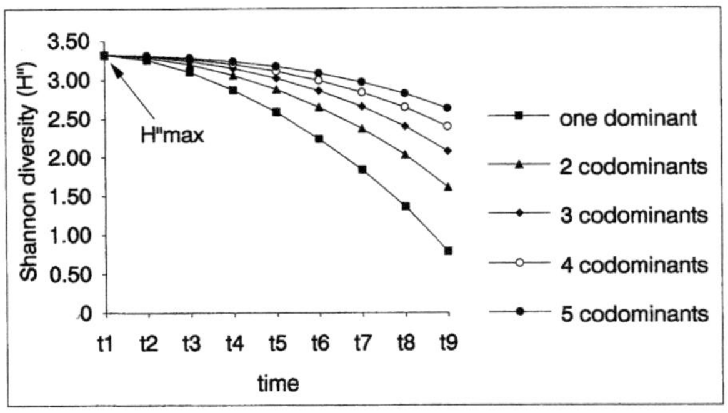

Fig. 3. Changes in Shannon diversity in cases in which one to five species reach an equilibrium by t9 time. Data are fictitious.

Table 1. Contingency tables for cases in which one (a) and five (b) species reach anequilibrium (altogether >90% dominance) between t1 and t9 time. Records are fictitious and can be understood as contributions to some total density measure.

a	T1	T2	T3	T4	T5	T6	T7	T8	T9
Sp1	10.00	20.0	30.00	40.00	50.00	60.00	70.00	80.00	90.00
Sp2	10.00	8.89	7.78	6.67	5.56	4.44	3.33	2.22	1.11
Sp3	10.00	8.89	7.78	6.67	5.56	4.44	3.33	2.22	1.11
Sp4	10.00	8.89	7.78	6.67	5.56	4.44	3.33	2.22	1.11
Sp5	10.00	8.89	7.78	6.67	5.56	4.44	3.33	2.22	1.11
Sp6	10.00	8.89	7.78	6.67	5.56	4.44	3.33	2.22	1.11
Sp7	10.00	8.89	7.78	6.67	5.56	4.44	3.33	2.22	1.11
Sp8	10.00	8.89	7.78	6.67	5.56	4.44	3.33	2.22	1.11
Sp9	10.00	8.89	7.78	6.67	5.56	4.44	3.33	2.22	1.11
Sp10	10.00	8.89	7.78	6.67	5.56	4.44	3.33	2.22	1.11
Species number	10	10	10	10	10	10	10	10	10
Total count	100	100	100	100	100	100	100	100	100
H''_{max}	3.32	3.32	3.32	3.32	3.32	3.32	3.32	3.32	3.32
Equitability	1.00	0.98	0.93	0.86	0.78	0.67	0.55	0.41	0.24
H''	3.32	3.26	3.10	2.87	2.58	2.24	1.83	1.36	0.79

b	T1	T2	T3	T4	T5	T6	T7	T8	T9
Sp1	10.00	11.11	12.22	13.33	14.44	15.56	16.67	17.78	18.89
Sp2	10.00	11.11	12.22	13.33	14.44	15.56	16.67	17.78	18.89
Sp3	10.00	11.11	12.22	13.33	14.44	15.56	16.67	17.78	18.89
Sp4	10.00	11.11	12.22	13.33	14.44	15.56	16.67	17.78	18.89
Sp5	10.00	11.11	12.22	13.33	14.44	15.56	16.67	17.78	18.89
Sp6	10.00	8.89	7.78	6.67	5.56	4.44	3.33	2.22	1.11
Sp7	10.00	8.89	7.78	6.67	5.56	4.44	3.33	2.22	1.11
Sp8	10.00	8.89	7.78	6.67	5.56	4.44	3.33	2.22	1.11
Sp9	10.00	8.89	7.78	6.67	5.56	4.44	3.33	2.22	1.11
Sp10	10.00	8.89	7.78	6.67	5.56	4.44	3.33	2.22	1.11
Species number	10	10	10	10	10	10	10	10	10
Total count	100	100	100	100	100	100	100	100	100
H''_{max}	3.32	3.32	3.32	3.32	3.32	3.32	3.32	3.32	3.32
Equitability	1	1.00	0.99	0.98	0.96	0.93	0.89	0.85	0.79
H"	3.322	3.31	3.29	3.24	3.17	3.09	2.97	2.83	2.63

a complexity measure. Field examples are given in Figs. 5, 8.

Results and discussion

Disturbances of intermediate frequency

When Reynolds (1988) quantified the different frequencies of disturbance relevant to successional responses in the phytoplankton, he postulated that the order of a few hours (< 1 generation time) qualifies as high frequency disturbance; 10 days or more as low frequency disturbance; between the extremes (200–20 h) disturbances might be considered to be intermediate.

Sampling frequency should always suit the problem in question. If around 1–8 days is the period designated as the range of intermediate disturbance, studies based on daily sampling are particularly pertinent to its description.

It has been shown (Padisák *et al.*, 1988, 1990) that in the short-term, storms have a fundamental effect on the summer succession of phytoplankton in Lake Balaton. Existing hydrographic structures break down and the abundance of one or more of the main species (in the cited case all were filamentous algae) immediately decreases. After storms, algae with small cell sizes (including picoalgae; G.-Tóth, 1980, described as colonial bacteria) exhibit synchronous changes, while large algae begin to synchronize only after several days. If there are 10–15 days between the storms (disturbance), the above pattern occurs repeatedly. Diversity appears wave-like with maxima and minima of this frequency (Fig. 4).

There are several notable periods on the diversity plot for the 1980 study (Fig. 5a), when sampling was done every day during a 212-day period.

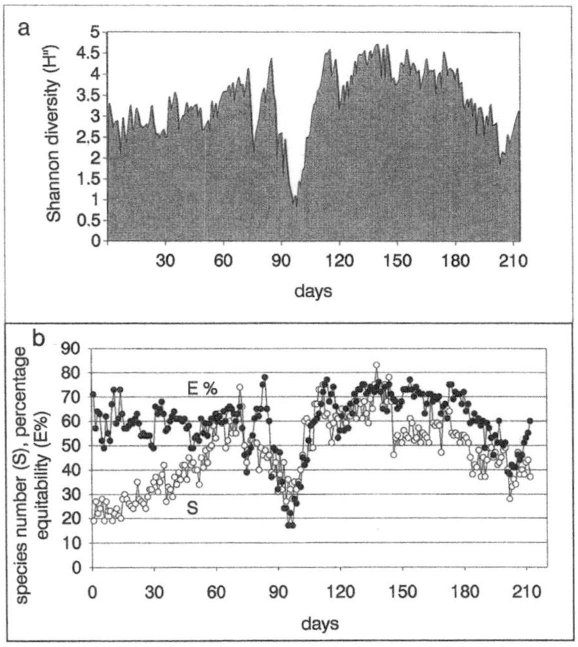

Fig. 5. Changes in phytoplankton diversity (a), species number and equitability (b) on 212 consecutive days in Lake Balaton between 1 April and 28 October 1980. Days are numbered consecutively; 1 corresponds to 1 April.

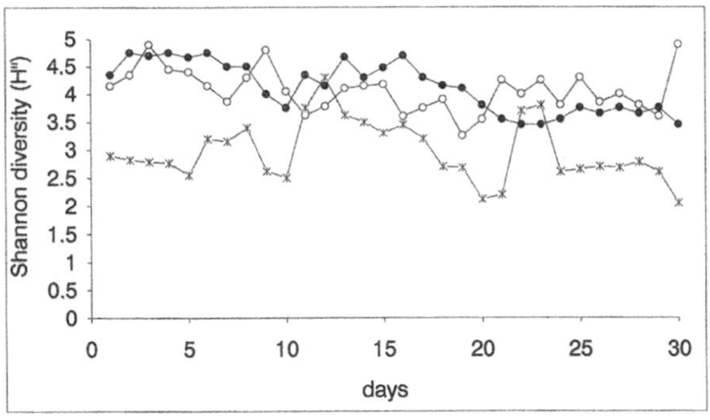

Fig. 4. Changes in diversity on 30 consecutive days in Lake Balaton during the summers of 1976 (●), 1977 (○) and 1978 (*). Days are numbered consecutively; 1 corresponds to 20 July in 1976 and 1977 and to 24 July in 1978.

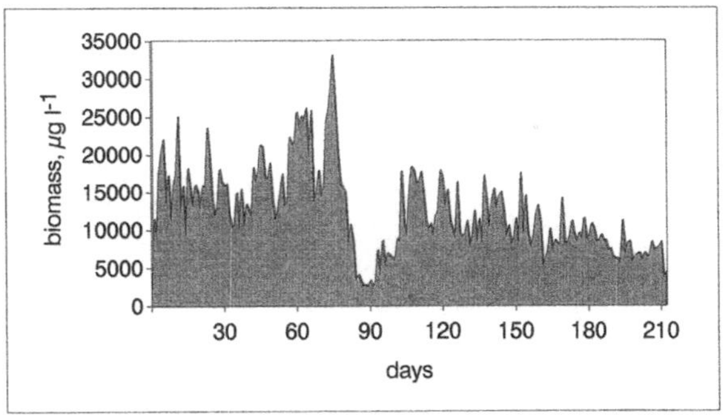

Fig. 6. Changes in phytoplankton biomass on 212 consecutive days in Lake Balaton between 1 April and 28 October 1980. Days are numbered consecutively; 1 corresponds to 1 April.

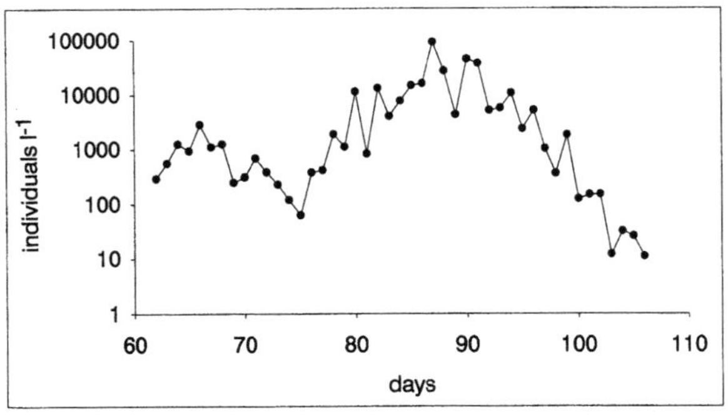

Fig. 7. Changes in numbers of *Bosmina longirostris* (logarithmic scale; data were made available by Dr. Pál Gulyás) in Lake Balaton between 30 May and 15 July 1980. Days are numbered consecutively; 60 corresponds to 30 May.

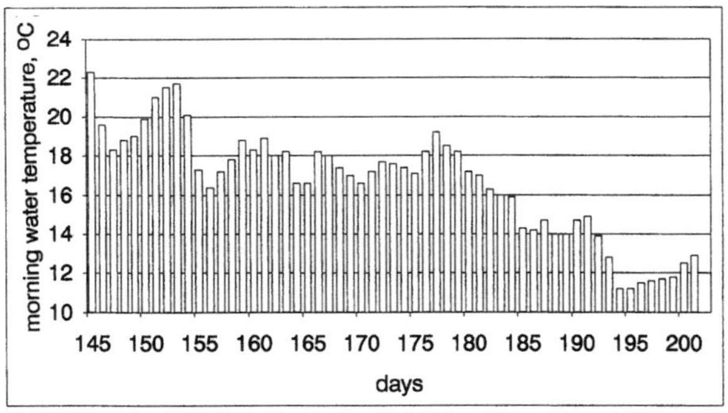

Fig. 9. Morning water temperature in Lake Balaton between 23 August and 15 October 1980 (data were made available by Dr. György Dévai). Days are numbered consecutively; 145 corresponds to 23 August.

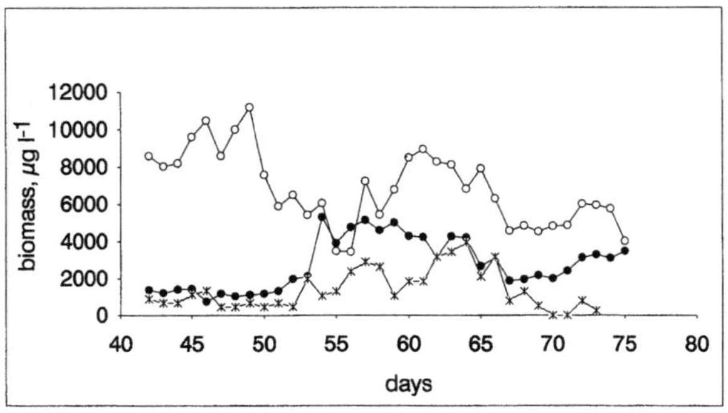

Fig. 8. Biomass of *Lyngbya limnetica* (●), *Synedra acus var. radians* (○) and *Aphanizomenon flos-aquae* (*) between 10 May and 14 June 1980 in Lake Balaton. Days are numbered consecutively; 40 corresponds to 10 May.

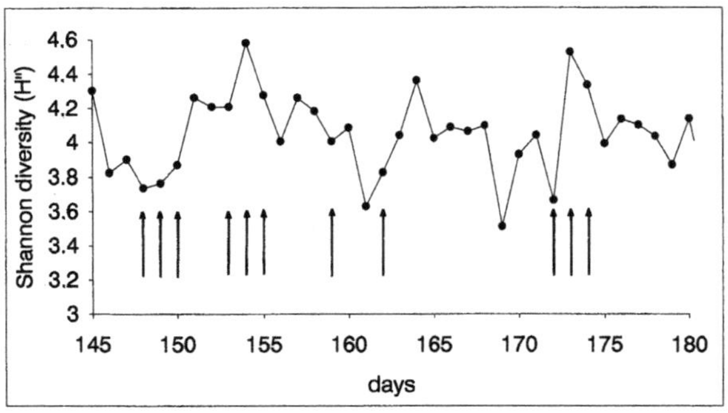

Fig. 10. Shannon diversity of phytoplankton in Lake Balaton between 23 August and 15 October 1980. Days are numbered consecutively; 145 corresponds to 23 August. Arrows indicate days when average wind velocity exceeded 1.5 m sec^{-1} (data were made available by Dr. György Dévai).

(1) The radical change in early July coincided with the clear-water phase (see also Fig. 6; population dynamics of dominant phytoplankton species are given in G.-Tóth & Padisák, 1982; Padisák, 1992), when *Bosmina longirostris* 0. F. Müller peaked. Both peaks in the *Bosmina* population density (Fig. 7) were reflected as drops in diversity with a time lag of only several days.

(2) There is a considerable increase in diversity (Fig. 5a) towards the end of the spring equilibrium phase (40–75th days). At the beginning of this period, *Synedra acus var. radians* Kütz. was overwhelmingly dominant. By about the 53rd day, *Aphanizomenon flos-aquae* Elenk. and *Lyngbya limnetica* Lemm. began to increase, and by the 60–65th days a three-species co-equilibrium had developed (Fig. 8). The effect on diversity of mono versus multi-species equilibria has been shown above (cf. Fig. 3).

(3) The dramatic decrease in diversity coincided with autumnal cooling (cf. Fig. 9), which reflects the climatic impact on phytoplankton succession (cf. Sommer, 1991).

(4) In periods when there was no clear trend (as above) in diversity it changed in rather distinct 5–13 day 'waves' (days 20–45, 145–180; Fig. 10). In the latter period (early autumn), the weather was changeable and without either persistent warming or cooling (Fig. 9). The diversity always increased after windy days; it decreased in calmer periods.

(5) Whenever dramatic decrease in diversity

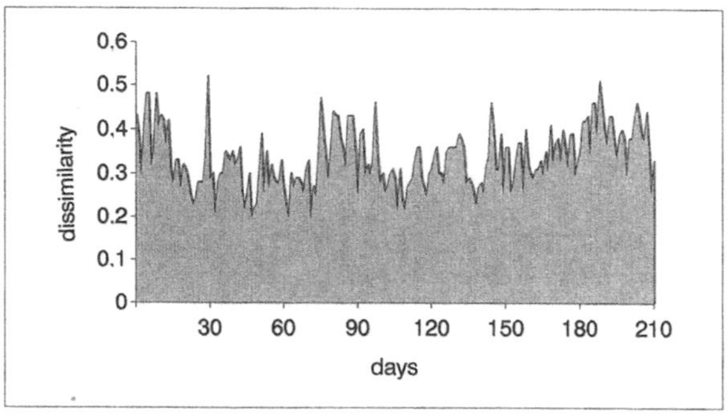

Fig. 11. Sørensen's dissimilarity index of phytoplankton calculated among every two consecutive samples taken in Lake Balaton between 1 April and 28 October 1980. Days are numbered consecutively; 1st record corresponds to the dissimilarity among samples taken on 1 and 2 April.

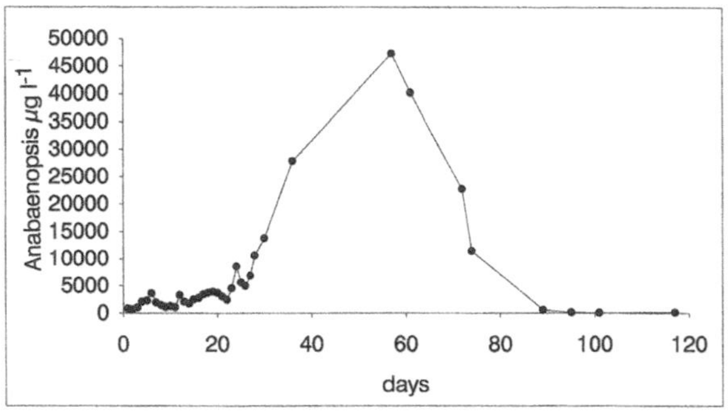

Fig. 12. Biomass of *Anabaenopsis raciborskii* in Lake Balaton between 15 July and 11 November 1982. Days are numbered consecutively; 1 corresponds to 15 July.

occurred (clear-water phase, autumnal cooling), both equitability and species number decreased (Fig. 5b). In other periods changes in diversity followed those in equitability, which means that in periods when IDH is supposed to play an important role, it has compositional causes. However, it should be stressed that considerable community changes can occur without being reflected in compositional diversity. Therefore

(6) another measure, a quantitative dissimilarity index (Sørensen, 1948) has also been used to describe community changes (Fig. 11). High dissimilarity values characterize the transition periods (*Cyclotella comta* [Ehr.] Kütz.→*Synedra* in April, clear-water phase). The wave-like behavior of community change in non-transition periods is also evident in dissimilarity. This variable clearly

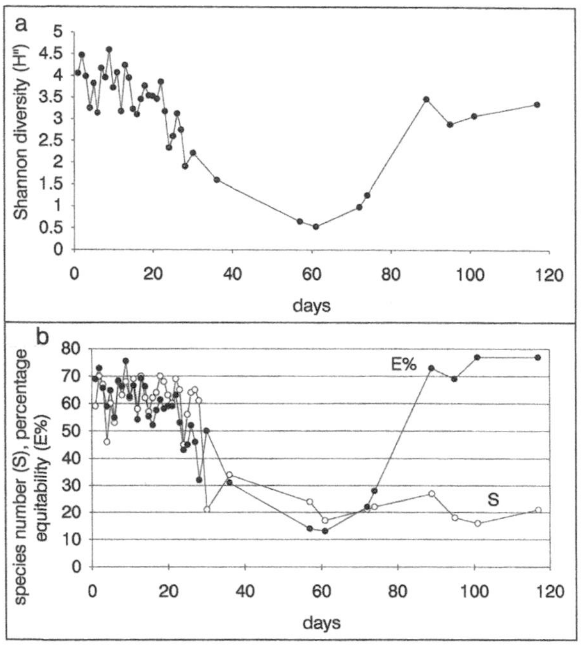

Fig. 13. Shannon diversity (a), species number and equitability (b) of phytoplankton in Lake Balaton between 15 July and 11 November 1982. Days are numbered consecutively; 1 corresponds to 15 July.

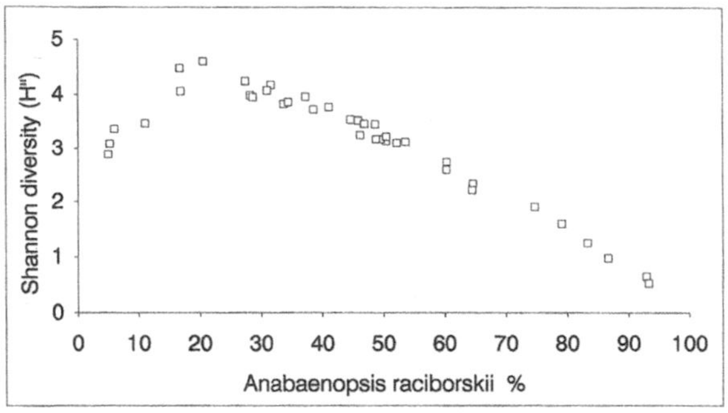

Fig. 14. Shannon diversity of phytoplankton plotted against the percentage contribution of *Anabaenopsis raciborskii* to total biomass in Balaton between 15 July and 11 November 1982.

indicates, that beside the 5–10 day disturbance frequency that generally obtained in Lake Balaton that year, highly variable periods with high dissimilarity values were separated by periods of 'quiescence' of several days during which low dissimilarities were measured. This agrees with the earlier observation (Padisák *et al.*, 1988) that 10–15 day calm periods are needed for the de-

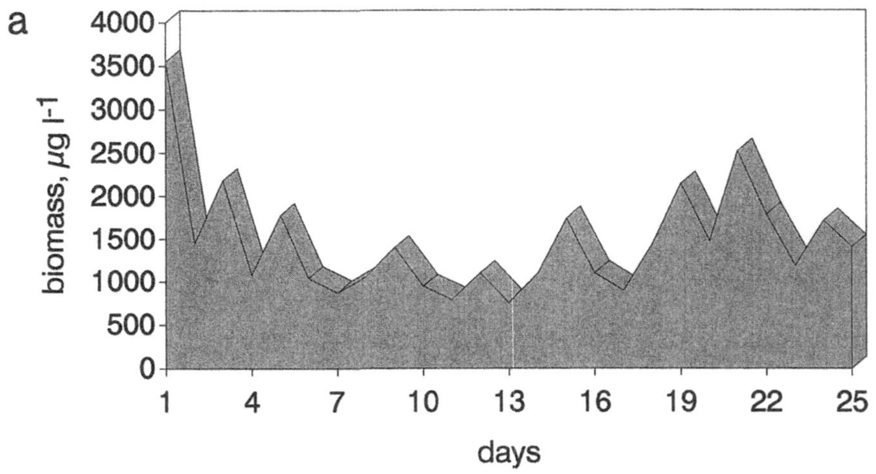

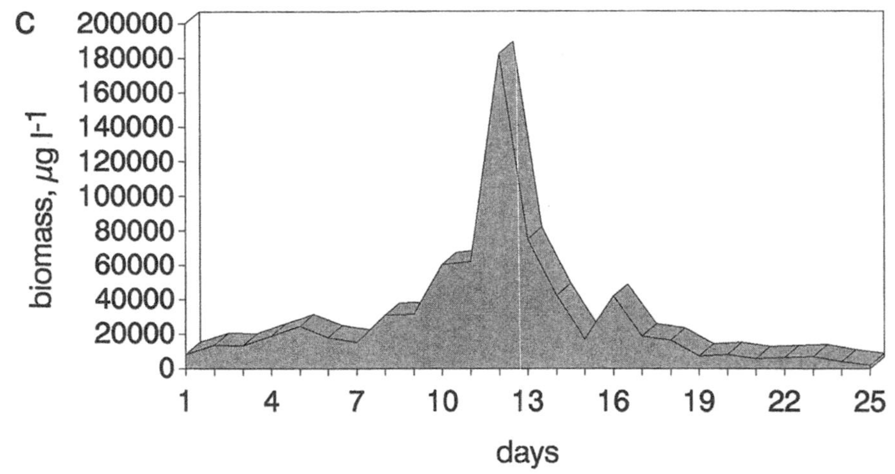

velopment of a phase of dominance of K-strategist species.

Disturbances of low frequency

The year 1982 was meteorologically exceptional in central Europe. After an extremely rainy July and early August (130–140 mm precipitation instead of the 60–70 mm average of the previous 10 years; in consequence there was increased nutrient loading from the tributaries), late August and September were very calm. For example, the water temperature in Lake Balaton in September exceeded the average of the previous 10 years by 3.84 °C.

Anabaenopsis (Cylindrospermopsis) raciborskii Wolosz., a heterocytic blue-green alga, began to increase early in the calm period (Fig. 12). The temporal stop at the beginning of the exponential phase of growth coincided with a storm. It was also observed in several other cases (Rajczy & Padisák, 1983; Padisák *et al.*, 1990) that storms interrupt the increase of filamentous algae that are dominating in Lake Balaton in summers. In the eastern part of the lake, where the phenomenon was studied in detail, *Anabaenopsis* exceeded all previously recorded algal counts and biomass. By the time the bloom had developed the contribution of all the other phytoplankton species to the total density measures was less than 10% (G.-Tóth & Padisák, 1986). The Shannon diversity

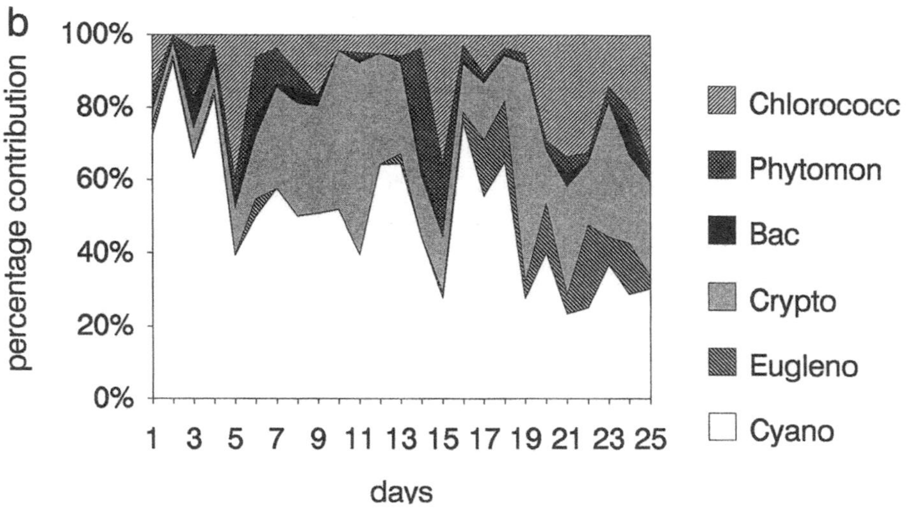

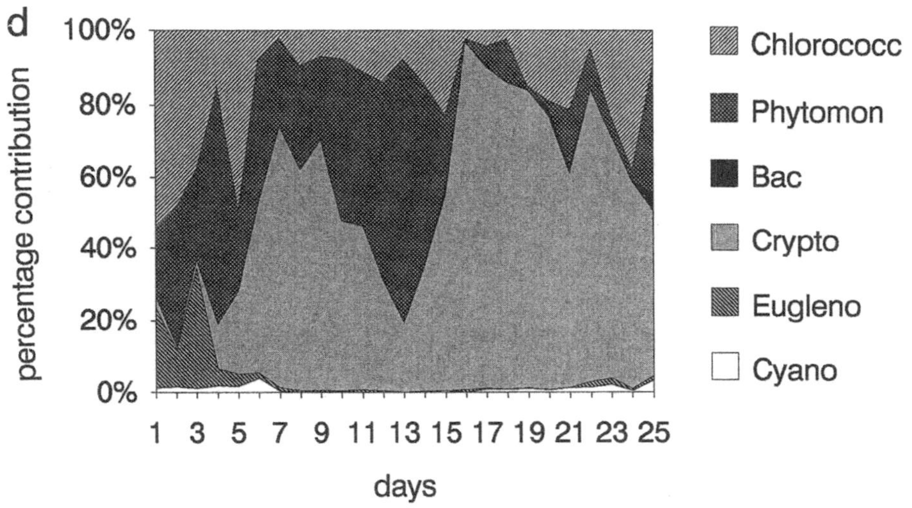

Fig. 15. Biomass (a,c) of phytoplankton and percentage contribution of main algal groups to total biomass (b,d) in the control (a,b) and experimental (c,d) ponds between 16 August and 9 September 1977. Days are numbered consecutively; 1 corresponds to 16 August. Records represent the average of the five sampling stations.

decreased continuously while the bloom developed and reached its minimum when it peaked (Fig. 13a). There is a clear relation between the percentage contribution of *Anabaenopsis* to total phytoplankton biomass and diversity (Fig. 14). The drop in diversity was mostly a consequence of a decrease in equitability (Fig. 13b). The observed decline in species number during the bloom comes from the applied counting technique (the total number of enumerated specimens was standardized as 400), and indicates only that the relative frequency of other species decreased. The fact that despite equitability species number did not rise after the bloom may be a consequence of autumnal cooling (cf. year 1980, Fig. 5b).

High frequency disturbances

In contrast to the previous two cases, it is rather difficult to find examples in which the effect of high-frequency disturbances on species diversity of phytoplankton can be demonstrated. The average frequencies of meteorological fluctuations

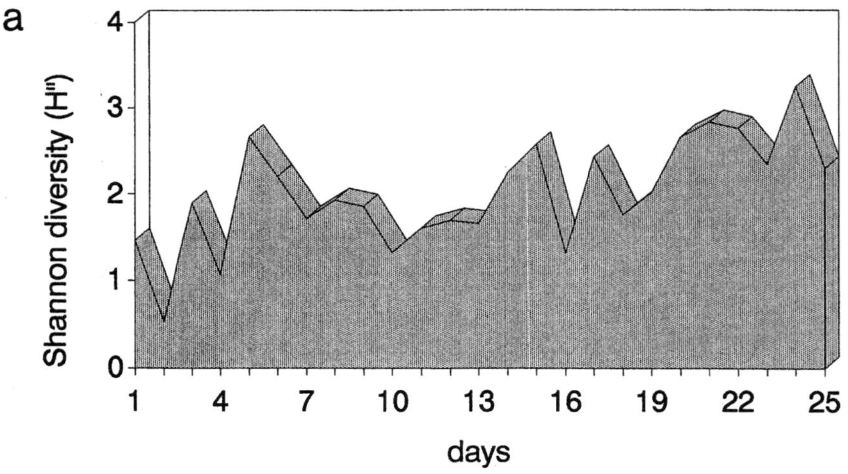

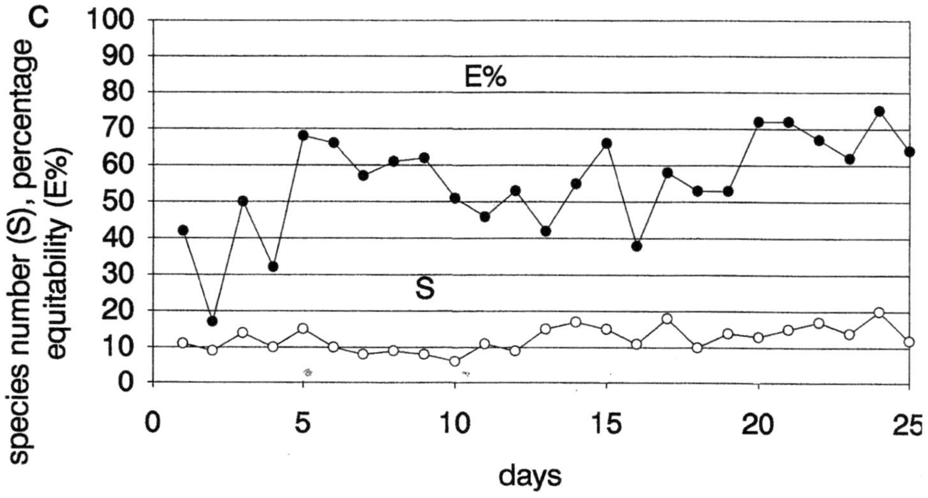

in temperate climates mostly fit the range that is here considered as the timescale of intermediate disturbances for phytoplankton (Trimbee & Harris, 1983) and the length of exceptional, (but in a long-term overview regularly occurring) calm periods can reach the time requirement of low frequency disturbance (> 20 d). Therefore, meteorological parameters alone are hardly enough to provide a high disturbance-frequency background: some additional disturbance factor is needed.

The Neusiedlersee is located in a wind-channel and its longitudinal axis more or less parallel with the prevailing wind directions. Its load of inorganic particles is dominated by small grains which sediment very slowly. For this reason, phytoplankton of Neusiedlersee was here considered to experience frequent disturbance.

The case of experimental ponds is much simpler. In general, they receive sewage in 12 h fluxes, which provide a high disturbance frequency. The act of disconnection allows us to observe the case in which frequently disturbed phytoplankton is released from frequent disturbances.

Experimental ponds

Because the ponds were disconnected from the sewage treatment plant on the first day of the experiment, only the initial values of parameters

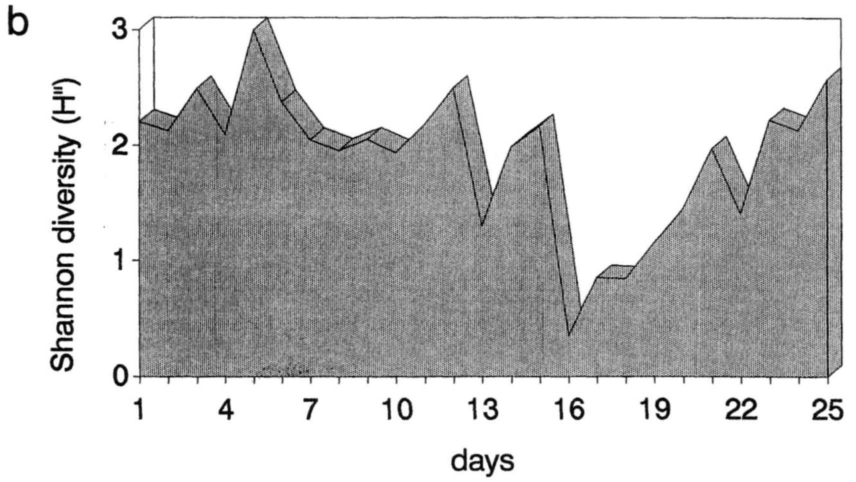

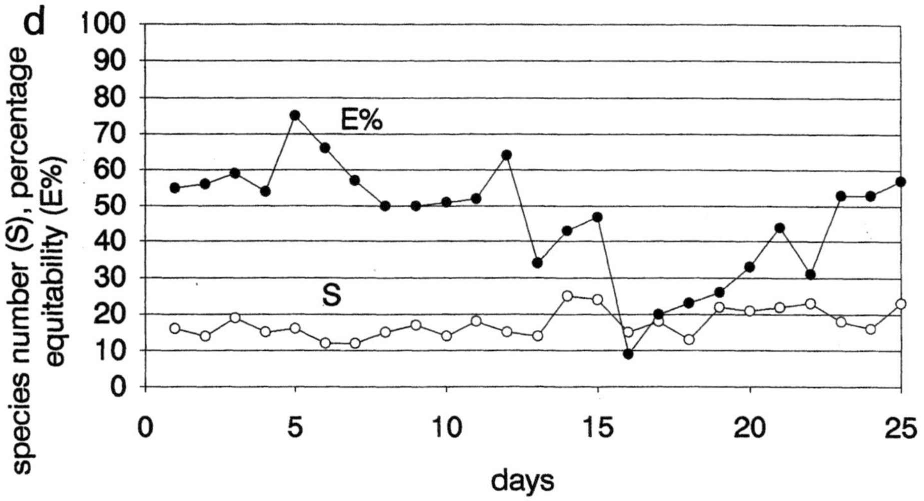

Fig. 16. Shannon diversity (a,b), species number and equitability (c,d) in the control (a,c) and experimental (b,d) ponds between 16 August and 9 September 1977. Days are numbered consecutively; 1 corresponds to 16 August.

(diversity, evenness, species number, composition of phytoplankton) can be considered as those that characterize highly perturbed conditions. The effect of the disconnection itself can be better studied in records of the control basin, where no other treatment was carried out.

On the basis of the IDH concept, a definite increase in diversity within several days can be predicted as a result of the release from high-frequency imposed disturbance. In the control pond, a small *Chroococcus sp.* was dominant at the beginning (Fig. 15 a,b); its share dropped to about 50% by the fifth day then remained fairly constant. In addition, *Rhodomonas sp.*, *Ankyra ancora* (G. M. Smith) Fott, *Ankyra ocellata* (Kors.) Fott and *Korshikoviella gracilipes* (Lamb.) Silva were the other quantitatively significant species. As a consequence of the compositional changes at the beginning, Shannon diversity (Fig. 16a) increased rapidly. The progressive diversity increase, which began at about the tenth day, was the result of a contemporaneous increase in species number (Fig. 16c).

In the experimental pond, rapid changes in phytoplankton biomass and dominance relations (Fig. 15b,d) followed food-web manipulation

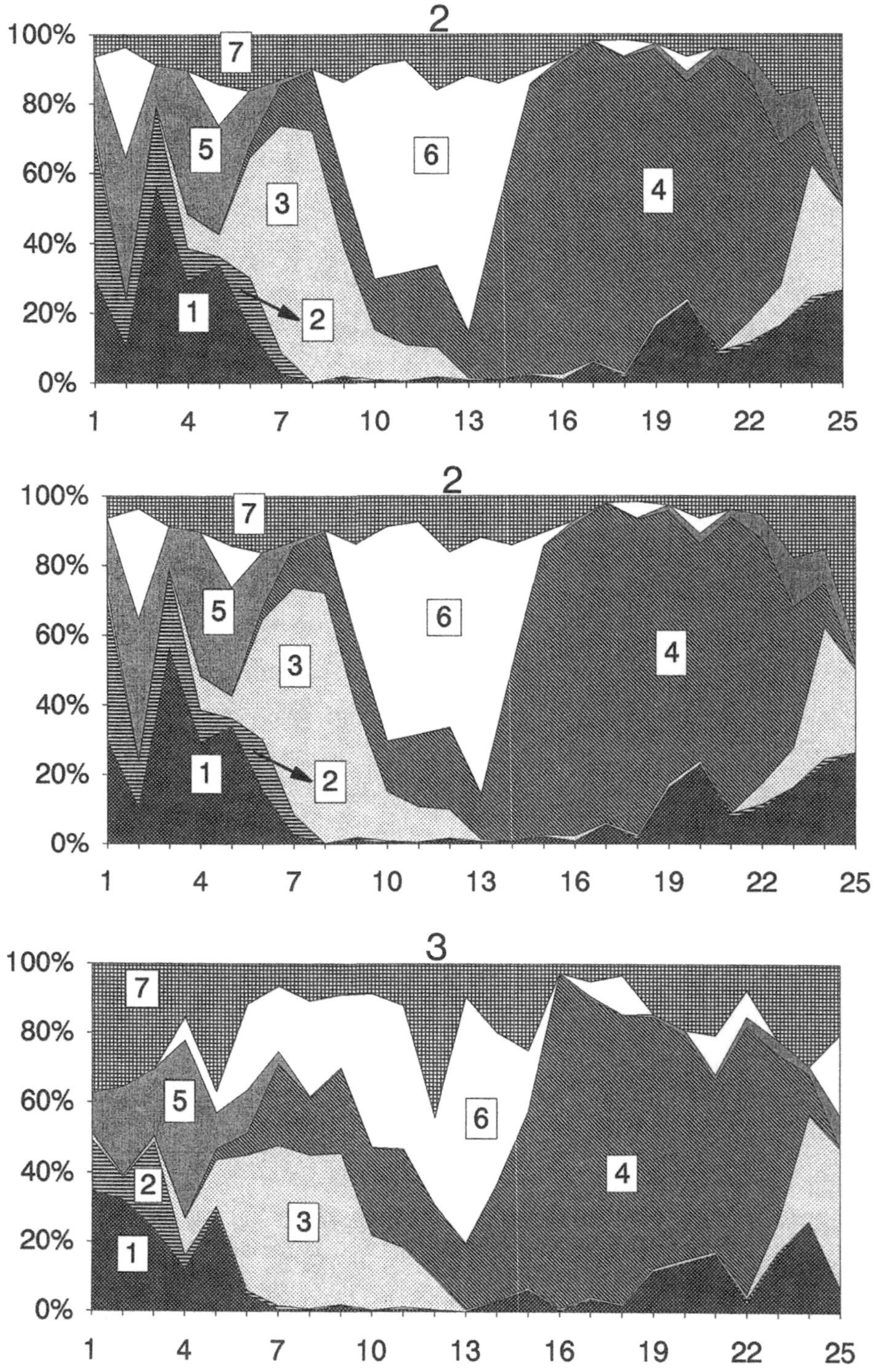

(crustacean zooplankton was killed on the first day, which, on one hand, radically reduced the grazing pressure and, on the other hand, increased the nutrient supply of phytoplankton). Dense populations of *Chlamydomonas reinhardtii* Dang., *Cryptomonas erosa* Ehr., *C. ovata* Ehr. and

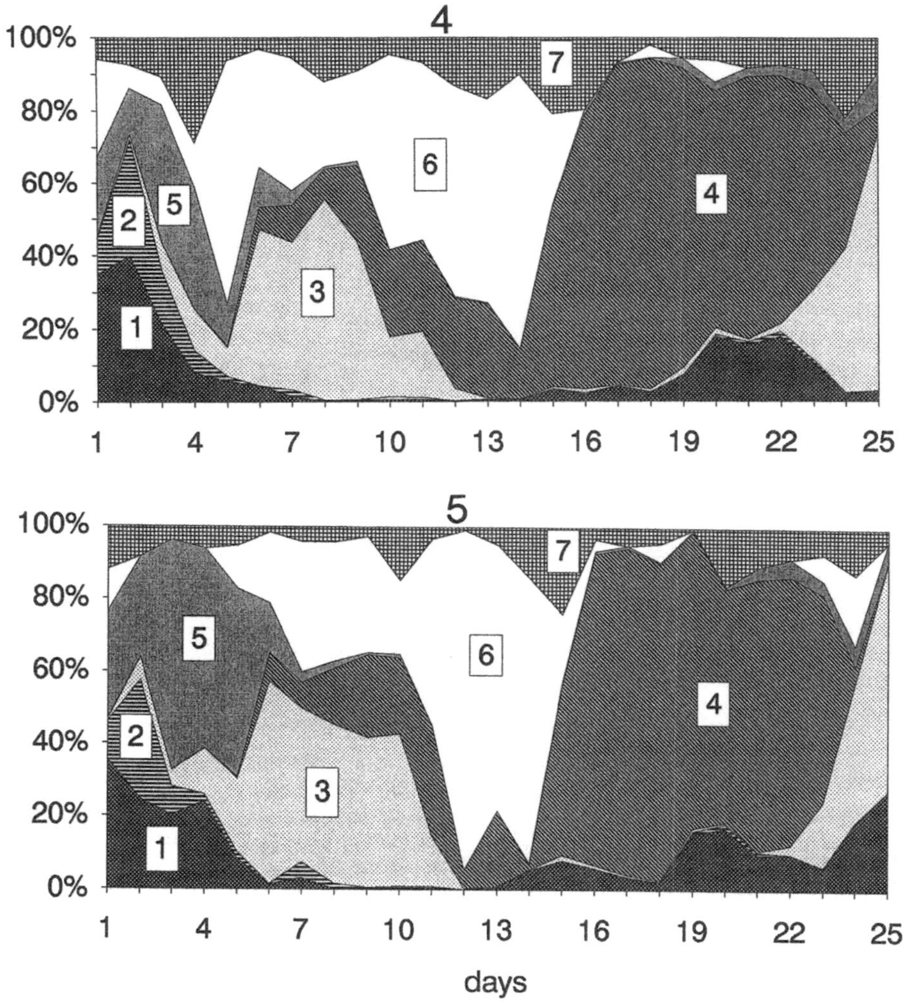

Fig. 17. Percentage contribution of *Chlorella spp.* (1), *Euglena polymorpha* Dang. (2), *Cryptomonas erosa* (3), *C. ovata* (4), *Chlamydomonas reinhardtii* (5), *Pandorina morum* (6) and all other species (7) to total biomass at the five sampling sites (compare Fig. 1) in the experimental pond between 16 August and 9 September 1977. Days are numbered consecutively; 1 corresponds to 16 August.

Pandorina morum (O. F. Müller). Bory followed each other, with the peak-biomass contributed by *Pandorina*. Despite the overall similarities, there were considerable differences between the five sampling stations in the small pond. For example, the *Pandorina* phase was much longer-lasting at sampling station 4 than on 2, etc (Fig. 17). The rapid and large community changes were not clearly reflected in diversity. Like equitability, Shannon diversity did not show clear, trend-like changes until the 12th day, when it decreased to a low level (Fig. 16b). Afterwards a continuous increase began. The changes in species number were the same as in the control pond: a continuous, slow increase during the whole period (Fig. 16d).

Neusiedlersee

The most characteristic feature of Neusiedlersee phytoplankton is a very weak seasonality involving both phytoplankton biomass and species composition; the several dozen main planktonic species can be found with almost equal chance at any time; few diatom species dominate all over

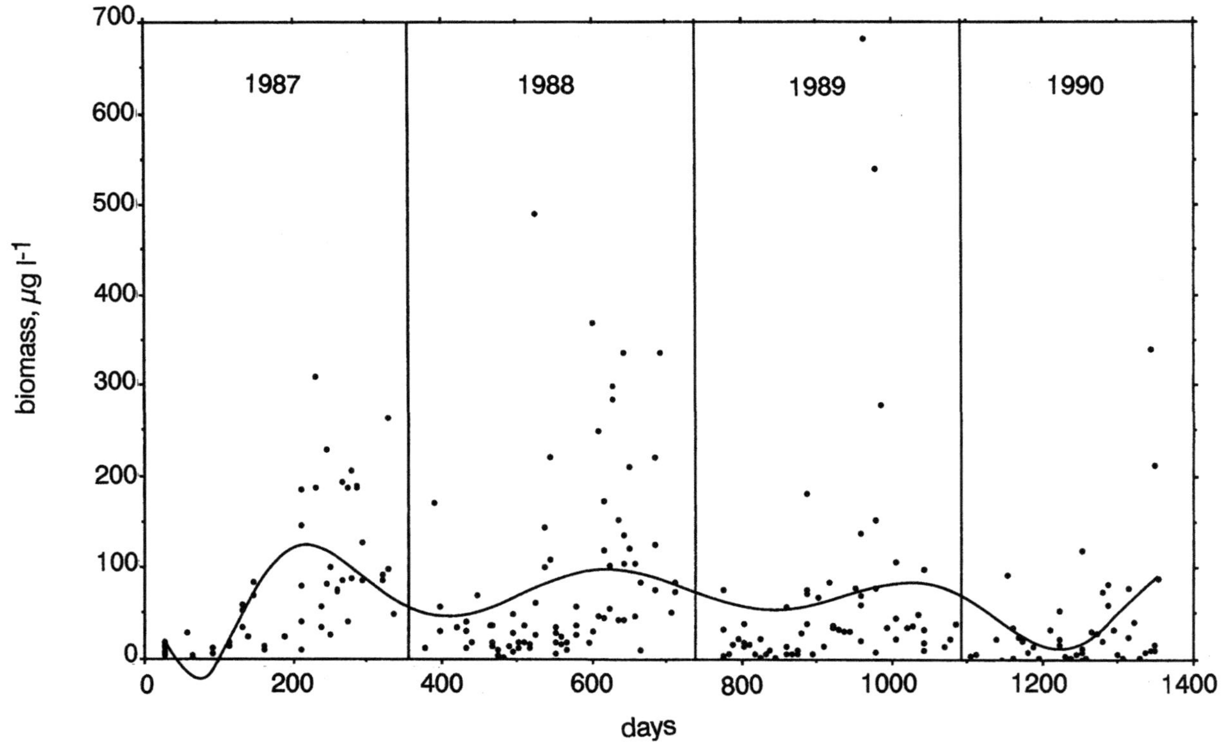

Fig. 18. Biomass of blue-green algae in Neusiedlersee between 1987 and 1990 (running numbers). An orthogonal polynom of the 6th order (continuous line) was fitted on the records to indicate the trend.

the year. The most expressed seasonality is exhibited by the quantitatively unimportant blue-green algae (their share is usually < 10%), which have a quite predictable autumnal increase. This is shown on Fig. 18; the graph is given only to demonstrate the level of stochasticity that characterizes this lake. Compared to that of other lakes, the variability of the time-series records of phytoplankton is very great. The above described characteristic also appears in diversity and equitability: highly variable records, no trend, and no seasonality (Fig. 19a,b). The species number (Fig. 19b) is rather low throughout the year. The winter minimum, although it is not sharp, seems to be rather consequent.

A comparison of disturbances of different frequency

The general applicability of Connell's (1978) IDH to phytoplankton is well justified by the case studies presented above. However, after its test on

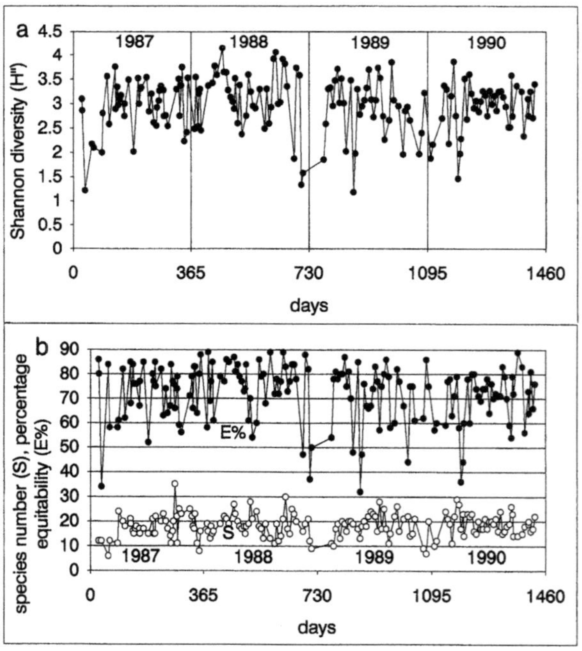

Fig. 19. Shannon diversity (a), species number and equitability (b) of phytoplankton in Neusiedlersee between 1987 and 1990 (running numbers).

short-term phytoplankton records trends, problems and uncertainties arise, and unanswered questions remain.

Quantifying the timescale

In Fig. 20 the arithmetical means of the diversity/equitability/species number data of the studies detailed in this paper are shown on a 'disturbance'-scale. Although many physical background data are known in each of the case studies, none of measured possible background factors correlated clearly with diversity indicating that disturbance is a 'compound variable'. Moreover, it was shown in daily samples that in characteristically intermediately disturbed ecosystems 'quiescence' is much easy to trace, and even quantify, than 'disturbance'. A similar and both theoretically and practically better detailed concept was elaborated by Chorus & Schlag (1993) analyzing the diversity-disturbance relation. In the case of the sewage ponds, the indicated frequency is a real disturbance frequency. Although the overall average did not differ significantly, only averages of the first three days were used, because the 12 hr disturbance frequency was switched off at the beginning of the studies. The number of

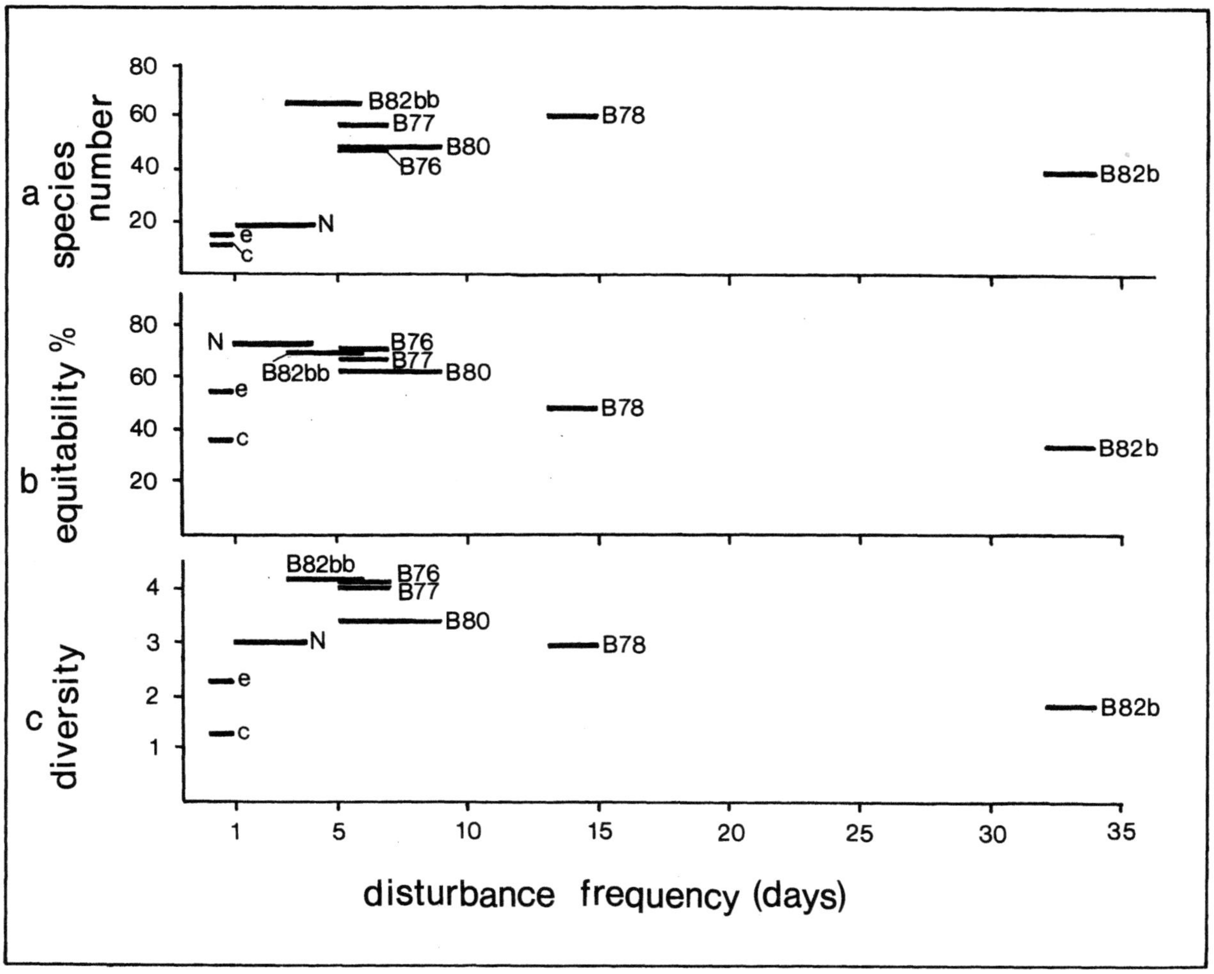

Fig. 20. Arithmetical means of species number (a), equitability (b) and Shannon diversity (c) in case studies analyzed or discussed in this study. Abbreviations: B76 – Balaton, 1976; B77 – Balaton, 1977; B78 – Balaton, 1978; B80 – Balaton, 1980; B82bb – Balaton, before *Anabaenopsis* bloom; B82b – Balaton, 1982 during *Anabaenopsis* bloom; N – Neusiedlersee; e – experimental pond; c: control pond. See text for more details.

initial days were established by cluster-analysis: the first three clustered consequently at each sampling stations and by all similarity measures that were applied. In cases of Balaton and Neusiedlersee disturbance frequency was established as the characteristic length of calmer periods between stormy ones. The 1982 records in Lake Balaton were divided into pre-bloom and bloom period. The records of the clear-water phase were omitted (1980 records in Balaton). In addition to what is described in this paper, reasoning behind this semi-quantitative scale is available elsewhere (G.-Tóth & Padisák, 1986; Padisák *et al.*, 1988).

Thus, disturbance ranges in this representation were established in an empirical way and reveal mainly the most characteristic disturbance-quiescence frequency in the given study.

The diversity – species number – equitability relation

Accepting that the above disturbance-quiescence scale is as correct as it can be, the diversity behaves in the predicted way: it has a maximum at intermediate frequencies (Fig. 20c). The most notable feature of this diversity plot is its asymmetry: a sharp increase until 4–6 days frequency then a slow decrease. On this basis, diversity is most sensitive to a disturbance frequency of 2–4 days in agreement with Reynolds' (1993) conclusion that the generation time is the key unit in plankton ecology (and perhaps in ecology in general).

As a result of the comparison of the diversity-disturbance relation with the way in which equitability and species number relate to disturbance-frequency, it is evident that diversity reduces at low and high frequencies of disturbance for different reasons. At high frequencies, basically, species number diminishes diversity, and low equitability only sharpens the decrease (Fig. 20a,b). For example, in the experimental ponds diversity rapidly increased in the first several days because equitability increased. Later and of greater importance low species number (control pond) or low species number combined with exclusion (experimental pond) provided the main limitation to diversity increase. At low frequencies, small diversity values come from small equitabilities. If one species is overwhelmingly dominant (as was

Table 2. Number of species that were found in case studies analyzed or discussed in this study. Since species number in quantitative samples is a function of the sample size, the number of samples to which the indicated 'species richness' refers to is also given.

	Exp. pond	Control pond	Neusiedlersee	Balaton				
				1976	1977	1978	1980	1982
Chroococcales	2	2	12	7	8	12	6	10
Hormogonales	5	4	2	5	10	9	15	8
Euglenophyta	3	2	8	6	7	6	15	14
Cryptophyxeae	4	4	2	4	3	1	3	3
Dinoflagellata			3	2	3	2	5	3
Xanthophyceae			1				11	1
Chrysophyceae		1	2	3	2	4	2	
Centrales	2	2	5	3	3	3	5	4
Pennales	6	5	38	28	33	31	36	33
Phytomonadina	6	2	3		2	2	5	3
Chlorococcales	26	20	41	30	55	54	153	67
Desmidiales		1	4	4	9	8	17	10
Filamentous green			2	1	1	3	2	1
Total species number	54	42	122	92	137	133	277	159
Number of samples	125	125	168	30	30	30	212	39

Anabaenopsis in Balaton during 1982), relative frequency (therefore the observed species number in quantitative samples) of other species necessarily decreases, but it does not mean that the community necessarily less 'species rich', cf. Table 2.

Stress and disturbance

The case of the Neusiedlersee points out the difficulties that arise in differentiating between stress and disturbance. Although in this study this phytoplankton was considered to be under the influence of high frequency disturbances, it does not follow that the disturbance rather than stress-factors, led to the observed values of diversity and species number. Each phytoplankton species in this lake has to tolerate the high salinity and an insufficiency of light. These factors may provide such a highly extreme environmental background that a species poor, rather 'Gleasonian' (each species has its own independent reasons to exist or not to exist in the given environment; Gleason, 1926) phytoplankton may became established. Balaton phytoplankton in 1980 also provides some indication that the dominance of physical stress factors leads to reduced species number, parallel with a high equitability. Early April records (Fig. 5b) of species number and equitability would provide the same ratio as that in the Neusiedlersee throughout the year or that in control pond (where the Secchi transparency was also very small) during the whole study period. On the basis of the results of these case-studies, it is impossible to differentiate between stress and disturbance. However, it is an observed fact that a similar species number-equitability pattern appears in both explanations, and it is also possible that high frequency disturbance in itself can be a major stress factor.

Effect of zooplankton on phytoplankton diversity, intensity of disturbance, seasonality

Although it was not the aim of these studies to observe the effect of zooplankton on phytoplankton diversity, it is clear that changes in grazing pressure, independent of the characteristic disturbance frequency, led to a decrease in phytoplankton diversity by reducing both equitability and species number. In Lake Balaton, 1980, both subsequent *Bosmina* peaks were followed by reduced diversity, while in the experimental pond a decrease in grazing pressure had the same effect. It seems that not the direction (increase or decrease), but the change itself and perhaps its rate have the greatest importance in the maintenance of diversity. This observation also shows that the stochasticity based IDH can be incorporated into the mainly deterministic views on plankton succession. According to the PEG-model, zooplankton grazing before the clearwater phase and autumnal cooling have the most invariable effects on phytoplankton succession. In terms of IDH both can be considered as high intensity disturbance leading to a reduced level of diversity. In periods (spring diatom bloom, post clear-water phases) when competition among phytoplankton species is the main mechanism controlling planktonic events, the disturbance pattern can have an increased role.

Species number and species richness

The most controversial issue, therefore the main difficulty, with IDH is that it not only maintains species richness in an ecosystem, but, as a mechanism, it requires its *de facto* existence. A comparison between events in the experimental pond and Lake Balaton, 1982, clearly demonstrates the above statement. The initial conditions were similar in both cases: increased nutrient availability combined with lack of disturbance. In the otherwise intermediately disturbed, species-rich Balaton, one of the K-strategists *(Anabaenopsis)* was immediately selected and it eventually bloomed. In the otherwise frequently perturbed pond, none of the existing species could establish a persistent large population; one r-strategist replaced another with high frequency in a rather stochastic pattern at different sampling stations in the small pond (no similar sampling station to sampling station differences were observed in the control pond). Wilson's (1990) reasoning in connection with terrestrial vegetation and IDH is rather similar in an opposite way: in New Zealand IDH (if relevant at all) is more unimportant than elsewhere, because there is a lack of pioneers and several late successional species can act as pioneers. Despite IDH (and it is perhaps a weak point in the concept), other non-equilibrium concepts (e.g. gradual climate change; Holloway, 1948, 1954) of species co-existence postulate the necessity of a pool of potential recruits (remnants of the previous climate that can survive but cannot propagate because their seedlings are shade intolerant, etc.). The existence of a similarly defined 'community memory' is also described in the phytoplankton (Padisák, 1992).

In this paper, the two terms 'species number' and 'species richness' have been used with different meanings. 'Species number' refers to the number of species that were found in one quantitative sample, while 'species richness' pertains to the total number of species that can be found in a given lake (at least during the study period). A comparison of species richness in the studies analyzed or discussed here (Table 2) proves the overall species poverty of highly disturbed/ stressed environments (Neusiedlersee, ponds) compared to the intermediately disturbed one (Balaton). It has also been seen that when diversity decreased radically in a compositional way in the Balaton (1982, *Anabaenopsis*), the species richness itself (Table 2) did not decrease. It means that in discussing the diversity-species richness relation, the characteristic frequency, not the temporal alterations, should be considered. The above observation allows one to conclude that in phytoplankton, as in terrestrial plant communities, an initial floristic composition (Egler, 1954) has a significant role in successional changes.

It has been shown above that the species-number decrease under low frequency disturbance has statistical causes. Meanwhile, it can be hypothesized that characteristically undisturbed environments (no example is given in this study) are less species rich than the intermediately disturbed ones. In all these cases, equilibrium explanations for the maintenance of high diversity retain their validity.

Acknowledgements

I am extremely grateful to Pál Juhász-Nagy for his supervision of my early work and his encouragement for dealing with the diversity-disturbance relation ever since and to Colin Reynolds for valuable comments on the manuscript.

I thank Dr. György Dévai and Dr. Pál Gulyás for making their unpublished meteorological and zooplankton data (Balaton, 1980) available. For diversity and similarity calculations BP (Hungarian Natural History Museum) and SYN-TAX III (Podani, 1988) program packages were used. Dr. Éva Ács helped me in data processing.

Samples were taken by the staff of the i) Illmitz Biological Station, Austria, Burgenland (Neusiedlersee); ii) Balaton Limnological Institute of the Hungarian Academy of Sciences Tihany, Hungary (Balaton, mesotrophic part): iii) Hydrobiological team of Kossuth University, Debrecen, Hungary (Balaton, hypertrophic part, sewage ponds). This work was supported by the Hungarian National Science Found (OTKA N° 3172).

References

Bondár, E., Gy. Dévai, I. Dévai, L. G.-Tóth, Cs. Heim, A. Kovács, J. Moldován, J. Padisák & I. Wittner, 1981. A case-study on some hydrobiological interactions in a sewage treatment plant. Acta Biologica Debrecina Suppl. 18: 1–292.

Chorus, I. & G. Schlag, 1993. Importance of intermediate disturbances for the species composition and diversity of phytoplankton in two very different Berlin lakes. Hydrobiologia 249: 67–92.

Connell, J., 1978. Diversity in tropical rain forests and coral reefs. Science 199: 1304–1310.

Dévai, I., 1977. Eutrophication and oligotrophication as exampled by BMKO, a sewage treatment plant. Acta biologica Debrecina 14: 67–78.

Dévai, I. & E. Woynarovich, 1981. Eutrophication and oligotrophication process occurring in a BMKO sewage treatment plant. In: M. Sudzuki (ed), Some Approaches to Saprobiological Problems. Sanseido Co. Ltd, Tokyo: 37–47.

Dokulil, M. & J. Padisák, 1993. Langzeitveränderungen der Zusammensetzung und der Populationsdynamik des Phytoplanktons im Neusiedlersee (1958, 1968–1990). – BFB Bericht.

Egler, F. E., 1954. Vegetation science concepts. I. Initial floristic composition – a factor in old-field vegetation development. Vegetatio 4: 412–417.

Gaedeke, A. & U. Sommer, 1986. The influence of the frequency of periodic disturbances on the maintenance of phytoplankton diversity. Oecologia 71: 98–102.

Gause, G. F., 1934. The struggle for existence. Williams-Wilkins, Baltimore.

Gleason, H. A., 1926. The individualistic concept of the plant association. Torrey Bot. Club Bull. 53: 7–26.

G.-Tóth, L., 1980. Short term investigations on the bacterioplankton of Lake Balaton at Tihany. Acta Bot. acad. Sci. Hung. 26: 425–435.

G.-Tóth, L. & J. Padisák, 1986. Meteorological factors affecting the bloom of *Anabaenopsis raciborskii* Wolosz. (Cyanophyta: Hormogonales) in the shallow Lake Balaton, Hungary. J. Plankton Res. 8: 353–363.

Hardin, G., 1960. The competitive exclusion theory. Science 131: 1292–1297.

Herodek, S., 1984. The eutrophication of Lake Balaton: Measurements, modelling and management. Verh. int. Ver. Limnol. 22:1087–1091.

Herzig, A., 1990. Zur limnologischen Entwicklung des Neusiedler Sees. AGN, Internat. Symp. Schutz und Entwicklung *großer* mitteleuropäischer Binnenseenlandschaften, Bodensee – Neusiedlersee – Balaton, Tagungsband: 91–97.

Holloway, J. T., 1948. Ecological investigations in the *Nothofagus* forests in N. Z. N. Zeal. J. Forestry 5: 401–410.

Holloway, J. T., 1954. Forests and climates in the South Island of New Zealand. Trans. Royal Soc. New Zeal. 82: 329–410.

Hutchinson, G. E., 1961. The paradox of plankton. Am. Nat. 95: 137–147.

Istvánovics, V. & S. Herodek, 1985. Orthophosphate uptake of planktonic microorganisms in Lake Balaton. Hydrobiologia 122: 159–166.

Istvánovics, V., L. Vörös, S. Herodek, S., L. G.-Tóth & I. Tátrai, 1986. Changes of phosphorus and nitrogen limitation in enriched lake enclosures. Limnol. Oceanogr. 31: 798–811.

Khinchin, A. I., 1957. Mathematical Foundations of Information Theory. Dover, New York.

Kullbach, S., 1957. Information Theory and Statistics. Wiley, New York.

Kiss, K. T. & J. Padisák, 1990: Species succession of Thalassiosiraceae: Quantitative studies in a large, shallow lake (Lake Balaton, Hungary). In H. Simola (ed), Proceedings of the 10th Internat. Symp. on Living and Fossil Diatoms. Koeltz Scientific Books, Koenigstein: 481–490.

Lund, J. W. G., C. Kipling & E. D. Le Cren, 1958. The inverted microscope method of estimating algal numbers by counting and the statistical basis of enumeration by counting. Hydrobiologia 11: 143–170.

Padisák, J., 1980. Short-term studies on the phytoplankton of Lake Balaton in the summers of 1976, 1977 and 1978. Acta Bot. Hung. 26: 397–416.

Padisák J., 1991. Relative frequency, seasonal pattern and possible role of species rare in the phytoplankton (Lake Balaton, Hungary). Verh. int. Ver. Limnol. 24: 989–992.

Padisák, J., 1992. Seasonal succession of phytoplankton in a large shallow lake (Balaton, Hungary) – a dynamic approach to ecological memory, its possible role and mechanisms. J. Ecol. 80: 217–230.

Padisák, J. & L. G.-Tóth, 1991. Some aspects of the ecology of the subdominant green algae in a large nutrient limited shallow lake (Balaton, Hungary). Arch. Protistenkunde 139: 225–242.

Padisák, J., L. G.-Tóth & M. Rajczy, 1988. The role of storms in the summer succession of phytoplankton in a shallow lake (Lake Balaton, Hungary). J. Plankton Res. 10: 249–265.

Padisák, J., L. G.-Tóth & M. Rajczy, 1990. Stir-up effect of wind on a more-or-less stratified shallow lake phytoplankton community, Lake Balaton, Hungary. In P. Biró & J. F. Talling (eds), Trophic Relationships in Inland Waters. Developments in Hydrobiology 53. Kluwer Academic Publishers, Dordrecht: 249–254. Reprinted from Hydrobiologia 191.

Pielou, E. C., 1975. Ecological diversity. Wiley & Sons Inc., New York.

Podani, J., 1988. SYN-TAX III. Computer programs for data analysis in ecology and systematics. Abstr. Bot. 12: 1–183.

Rajczy M. & J. Padisák, 1983. DIVDROP analysis – a new method for the interpretation of species importance in diversity changes. – Ann. Hist.-nat. Mus. Nat. Hung. 75: 97–105.

Reynolds, C. S., 1988. The concept of biological succession applied to seasonal periodicity of phytoplankton. Verh. int. Ver. Limnol. 23: 683–691.

Reynolds, C. S., 1993. Scales of disturbance and their role in plankton ecology. In J. Padisák, C. S. Reynolds & U. Sommer (eds), Intermediate Disturbance Hypothesis in Phytoplankton Ecology. Developments in Hydrobiology 81. Kluwer Academic Publishers, Dordrecht: 157–171. Reprinted from Hydrobiologia 249.

Shannon, C. E., 1948. A mathematical theory of communication. Bell Syst. tech. J. 27: 623–656.

Sommer, U., 1983. Nutrient competition between phytoplankton species in multispecies chemostat experiments. Archiv für Hydrobiologie 96: 399–416.

Sommer, U., 1984. The paradox of plankton: fluctuations of phosphorus availability maintain diversity in flow-through cultures. Limnol. Oceanogr. 29: 633–636.

Sommer, U., 1985. Comparisons between steady state and non-steady state competitions: experiments with natural phytoplankton. Limnol. Oceanogr. 30: 335–346.

Sommer, U., 1991. Phytoplankton: directional succession and forced cycles. In H. Remmert (ed), The Mosaic-Cycle Concept of Ecosystems. Springer Verlag, Berlin: 132–146.

Sommer, U., Z. M. Gliwicz, W. Lampert & A. Duncan, 1986.

The PEG-model of seasonal succession of planktonic events in freshwaters. Archiv für Hydrobiologie 106: 433–471.
Sommer, U., J. Padisák, C. S. Reynolds & P. Juhász-Nagy, 1993. Hutchinson's heritage: the diversity-disturbance relationship in phytoplankton. In J. Padisák, C. S. Reynolds & U. Sommer (eds), Intermediate Disturbance Hypothesis in Phytoplankton Ecology. Developments in Hydrobiology 81. Kluwer Academic Publishers, Dordrecht: 1–7. Reprinted from Hydrobiologia 249.
Sørensen, T., 1948. A method for establishing groups of equal amplitude in plant sociology based on similarity of species content and its application to analyses of the vegetation on Danish commons. Biol. Skr. 5: 1–34.
Tilman, D., 1982. Resource competition and community structure. Princeton Univ. Press.
Trimbee, A. M. & G. P. Harris, 1983. Use of time series analysis to demonstrate advection rates of different variables in a small lake. J. Plankton Res. 5: 819–833.
Vörös, L. & J. Padisák, 1991. Phytoplankton biomass and chlorophyll-*a* in some shallow lakes in central Europe. Hydrobiologia 215: 111–119.
Wilson, J. B., 1990. Mechanisms of species coexistence: twelve explanations for Hutchinson's 'paradox of the plankton': evidence from New Zealand plant communities. New Zeal. J. Ecol. 13: 17–42.

Hydrobiologia **249**: 157–171, 1993.
J. Padisák, C.S. Reynolds & U. Sommer (eds), Intermediate Disturbance Hypothesis in Phytoplankton Ecology.

Scales of disturbance and their role in plankton ecology

Colin S. Reynolds
Freshwater Biological Association, NERC Institute of Freshwater Ecology, Windermere Laboratory, Ambleside, LA22 0LP, UK

Key words: succession, phytoplankton, equilibrium, intermediate disturbance, diversity

Abstract

The role of hydraulic and hydrographic disturbances in delaying, arresting or diverting successional sequences from achieving stable, climactic equilibria is discussed by reference to case studies. The critical time scale is expressed in terms of planktonic reproductive generation times. Environmental constancy persisting over some 12–16 generations may permit a climactic condition to be achieved. An Intermediate Disturbance, if sustained, can establish a new successional sequence or, if not, can lead to a reversion to a sequence similar to the predisturbance succession. At intervals of ~ 1 generation time, species are selected according to their ability to accommodate disturbances at the physiological level. Highly disturbed environments are considered to be likely to maintain 'plagioclimactic' associations.

Introduction

This paper addresses issues posed for discussion by the International Association of Phytoplankton Taxonomy and Ecology at its 8th Workshop, convened at Baja, Hungary, in July, 1991. The issues relate to the applicability of the Intermediate Disturbance Hypothesis, originally elaborated by Connell (1978), to the structure, organisation and dynamics of plankton communities.

In general terms, the hypothesis attempts to reconcile disparities between, on the one hand, compelling experimental evidence for the predictability of interspecific interactions determining a steady-state outcome wherein one or relatively few species achieve overwhelming dominance through competitive exclusion (Hardin, 1960) and, on the other, a wealth of observational data pointing to the apparent co-existence of several species, ostensibly competing for resources in supposedly isotropic environments (Hutchinson, 1961). Connell's (1978) attention was focused principally on the high diversity of species comprising the well-established communities of coral reefs and tropical rain forests. Hutchinson (1961) termed the analogous dubiety with respect to pelagic communities 'the paradox of the plankton'. Explanations for the simultaneous coexistence of species have been offered, invoking environmental heterogeneity ('contemporaneous disequilibrium': Richerson *et al.* 1970) and the differing demands made by individual species for limiting resources (or physiological niche packing: Petersen, 1975; Tilman *et al.*, 1982). Both can be said to envisage a steady state outcome of competitive interaction, only now applying merely to a smaller physical or conceptual subdivision of the environment. Reynolds (1987a), however, considered the rates of response of populations to altered environmental conditions, recognising that a finite passage of time was necessary for the achievement of an equilibrated steady state. Stochastic external forcing applied to the evolving equilibrium might set it back to a less mature condition. Recurrent forcing might render a permanently non-equilibrated state.

Ecologists have vehemently debated the equilibrium vs non-equilibrium view of living communities. The argument is not wholly academic, since it is closely enmeshed with the concepts of ecological succession and climax. It is also related to questions about the management and exploitation of ecosystems, including their productivity and species diversity and to concerns about the roles of interspecific competition for resources and of cascading 'top-down control' effected by consumers in regulating community structure.

This paper develops the idea that intermediate disturbance reconciles the equilibrium and non-equilibrium arguments within a single explanative theory. Using planktonic communities as the prime example, it is argued that their organisational state depends upon the extent of the progress towards equilibrium achieved since the last disturbance was sustained. Using ecological succession as a template, this article seeks to demonstrate that it is the scale of intermediate disturbances (i.e. their severity and frequency) which determines the state of community organisation and diversity.

The successional template

In many ways, the equilibrium – non-equilibrium debate is spurious. Far from being mutually incompatible, the opposing arguments are the different sides of the same coin (Reynolds, 1991). The unifying concept is that of ecological succession and its interruption by externally-imposed disturbances. The understanding of successional processes evolved gradually, owing to a long series of individual contributors. However, formalisation of a theory, with a set 'rules' describing the development of ecosystems, is properly attributed to E. P. Odum (1969). From the starting point of its initial creation, available ecological space (e.g. bare or freshly exposed land) is open first to colonisation by arriving pioneer species; it is progressively modified by subsequent arrivals and, eventually, approaches a climactic steady state wherein the competitively-determined best-fit species achieve unchallenged dominance. This outcome of succession, reached exclusively and unidirectionally through internal, mainly biotic mechanisms, has often been regarded as being at equilibrium. Strictly speaking, there can be no such thing as a precisely stabilised ecological equilibrium while the scales of signal varability (e.g. hour-to-hour differences in insolation and water-movement) and of response (e.g. rates of photosynthesis, growth, movement, cell-division and death) continue to be separated. Nevertheless, Odum (1969) has given a very clear anticipation of what its properties might be. Thus, although the transition from early to later developmental stages is accompanied by increases in the organic biomass (B) and in gross primary production per unit area (P), the proportion of the energy expended on its maintenance (R, representing community respiration) also increases until, at equilibrium, $R = P$, $P/R = 1$ and net productivity, *sensu* the ratio of net production to biomass, $(P\text{-}R)/B$, decreases to zero; at the envisaged steady state, B is the maximum biomass that can be supported by the available energy and resources. To avoid any semantic argument about the conceptual use of the word 'equilibrium' it might be preferable to regard successional climax as being the achievement of the succession's relatively most stationary and nearly-equilibrated state.

During the intermediate 'phase of accumulation' (Price, 1984), the organisms making up the community move away from the small, invasive species allocating resources predominantly towards rapid reproduction, growth and mobility (r-selected) and progressively towards more specialist (K-selected) species, having more complex adaptations to compete for depleted extrabiotic resources and to retain them intrabiotically. From the point of view of community metabolism, successions bring about more effective closure of nutrient cycles, slower biomass-to-environment interchanges and increased reliance upon regenerative detrital pathways. This is Price's (1984) 'phase of maturation'.

The control and direction of successional change thus becomes increasingly community-led and internally-determined; its acquired 'information' (Margalef, 1961) and stability (i.e. resilience to external forcing) are enhanced, while the entropic dissipation of energy is reduced. The eventual equilibrium condition, indeed, is the predictable outcome of the entirety of community processes. It may be reasoned therefore that such diversity as the equilibrium community might possess should be attributable only to the packing of available niches by narrow-specialist species.

Intermediate disturbance

Ecological successions, however, rarely progress smoothly from their pioneer to mature stages without disturbance from external sources. Connell (1978) cited the impact of storms, landslips, lightning strikes and insect plagues on tropical rain forests. The imposition of agriculture on former woodland, for example, represents a form of management by disturbance. Such impacts must, at least locally, arrest the succession and probably revert it to a more primitive successional stage: from closed canopy to clearing or to bared land. The affected area is reopened to colonising species, so renewed succession, analogous to the development of the damaged ecosystem, is invited.

The succession following a severe disturbance need not be identical but is likely to be initiated by those species with prolific seed-, spore- or juvenile-propagation and short generation/colonisation times and which would be expected to arrive and establish themselves first. It is but a short step in reasoning to deduce that the successional state of the community and its species composition will be related to the severity and frequency of major disturbances. If sufficiently frequent, the community may become dominated by species capable of surviving the disturbance or which are capable of quickly reaching reproductive maturity: the erstwhile woodland cleared for pastoral farming now carries a quite different permanent vegetation dominated by plagioclimactic grassland communities while tree growth is excluded by an appropriate application of grazing or cutting.

Compositional changes in planktonic communities

Though borrowed from terrestrial ecology, the term 'succession' has often been given a rather wide meaning in plankton ecology, where its use encompasses most sequential changes in species dominance (see, for instance, Smayda, 1980). Changes in species dominance of the phytoplankton can be mediated by a variety of mechanisms, including changes in ambient temperature, light income and nutrient supply; variations in the nature and intensity of differential rates of biomass loss through settling out, through removal by grazing zooplankton, and through attacks by pathogenic organisms (fungal parasites, bacteria and viruses); by the variations in pH and related changes in the availability of inorganic carbon; by selective toxicants, some of which may be produced by the algae themselves; and, most particularly, by abrupt changes in the physical environment due to alterations in the balance between buoyancy generation and the dissipation of kinetic energy and in the rate of hydrological exchanges. Dominance of the zooplankton may also be liable to alteration through analogous processes, as well as through changes in food availability and rates of consumption by planktivores. Nevertheless, year-to-year variation in the general sequence of dominance is often relatively modest and, hence, the outcome can usually be predicted with a high level of probability (Sommer *et al.*, 1986).

In previous considerations of plankton periodicity (Reynolds, 1976, 1980, 1984, 1988a), I have attempted to distinguish the effects of changes driven by density-dependent, external (*allogenic*) forcing, which generally weaken the organisation of the system, and internal, community-led (*autogenic*) processes, which strengthen it. I also proposed (Reynolds, 1980) that only the latter are likely to comply with Odum's (1969) criteria of successional development (see also Reynolds, 1988a) and, in the context of the plankton, the use of the term 'succession' should be applied to these instances alone.

Scaling in pelagic successions

Before passing to the scope and properties of planktonic successions and their vulnerability to allogenic disturbance, it is appropriate to emphasise the correct choice of scale to the interpretation of successional phenomena. Lakes are among the least permanent of geomorphological features; their tendency to accumulate materials, either washed in from their catchments or generated by the *in situ* organic production they support, in the form of bottom sediment, ultimately leads to their extinction. In many of the small lakes of the temperate regions, created at the end of the last glaciation ($\sim 10^4$ years B.P.), the disappearance of open waters has been assisted by relatively rapid encroachment from the margins by a macrophytic suspension (reedswamp→marsh→fen or bog→woodland) known as 'hydrosere' (Tansley, 1939; Tallis, 1973). By analogy, the seaward advance of coastal dune formation and saltmarsh development or the accumulation of biogenic deposits (e.g. chalk) could, in theory, bring about an analogous extinction of the ocean. Even assisted by tectonic movements and eustatic uplift of sedimentary deposits, the time scales are of the order of tens to thousands of million (10^7–10^9) years. To ecologists accustomed to successional time scales of a century or more, the concept of planktonic-species successions may seem quite fatuous: the annual wax and wane of the spring bloom of diatoms might be compared to the flowering season of a perennial shrub! However, it is pertinent to recall that individual planktonic organisms are characterised by small spatial and temporal scales: most planktonic plants fall within the size range 10^{-6} to 10^{-3} m, while planktonic animals span 10^{-5} to 10^{-2} m; their generation times may be of the order of 0.5 to 50 days (i.e., about 10^{-3} to 10^{-1} years). It may well require some 75–80 days to accommodate the 15 or 16 generations of the diatom, *Asterionella formosa* Hass., for the inoculum to attain its vernal maximum in Windermere, England (see Lund, 1949). Reynolds (1989) suggested that as few as six generations of *Microcystis*, recruited to the plankton from its benthic phase, are required to exclude its competing species; depending on the water temperature and light received, this can often be equivalent to 20–30 days. Sommer's (1985, 1989) experimental investigations into the outcome of interspecific competition in isolates of natural phytoplankton showed that some 35–60 days was required to achieve a steady-state equilibrium.

Assuming these figures to be representative, it is quite reasonable to deduce that 70 or more phytoplankton generations and, perhaps, three or four full successional sequences (or rather more truncated ones) could be accommodated within one calender year. These numbers are comparable to the likely generations of arboreal dominants in the European temperate zone since the end of Weichselian or Devensian Glaciation and constituting the well-established postglacial vegetation phases (see e.g. Pennington, 1969), featuring especially *Betula-Pinus*, *Corylus-Alnus-Ulmus* and *Quercus-Fagus*. The assembly of these phases does not itself necessarily conform to a successional sequence: each phase, however, would have represented the climatic or near-climatic outcome of the contemporaneous ecological successions, involving colonist herbs and shrubs. The changes in the dominant vegetation type occurred in response to the fluctuating postglacial climate, through which different equilibrium species have been selected in turn, according to their own specific adaptive abilities and tolerances.

To pursue the analogy one stage further, the broadly cyclical changes in the environmental characteristics of a water body through the calendar year – for instance, day length, heat fluxes and water temperature; stratification and mixing;

hydraulic and nutrient exchanges – are, in terms of generation times of the climactic dominants, comparable to interstadial climatic amelioration and deterioration on the land; yet through each phase the same principles of ecological succession apply. What changes is the *potential natural vegetation*, that is, the successional outcome selectively favoured by the contemporaneous blend of environmental factors. Thus, in the plankton as in postglacial terrestrial ecosystems, there is no single succession and no single recognisable climax condition.

Provided ecologists can accept this analogy, the assessment of succession in plankton offers important scaling advantages. For instance, the aspects of succession still hotly debated – how to recognise climax and whether species diversity increases or decreases through succession – are not only answered by the evidence from planktonic communities but are also amenable to searching experiment.

The course of planktonic successions

The pelagic analogues of ecologically-open land surface are furnished by water bodies, newly created through rainwater collecting in a depression in the earth's surface or, more likely, by existing small, river-fed lakes immediately following severe hydraulic washout (see, for instance, Reynolds & Lund, 1988) or by the reisolation of flood-pain, vàrzea lakes, when the river falls back within its channels (e.g. Tundisi, 1980; Hamilton & Lewis, 1990). The cyclical re-filling of artificial sewage-oxidation tanks (Uhlmann, 1971), and fish ponds (Grygierek & Wasilewska, 1979; Kořínek *et al.*, 1987), similarly provide open habitat for planktonic colonisation.

Assuming the 'new' habitat to be suitable to, and supportive of, species, the initial colonists will be those whose propagules arrive first and/or those which are able to grow fastest or, in the case of disturbance to an existing community, those with the greatest number of survivors. Reynolds (1984, 1988b) recognised that functionally-similar algae also shared common morphological features and that, typically, the colonist species were relatively small ($<10^3\ \mu m^3$), usually unicellular algae, spherical, ovoid or fusiform in shape giving a surface area/volume ratio of $>1\ \mu m^{-1}$ and a common ability to generate rapid specific rates of growth (up to $1.0–1.4\ d^{-1}$ at 22–25° on a 16 h temperate-region summer day). This group (X in Reynolds, 1984) was later subdivided among the non-motile, predominantly Chlorophyte association (X_1) of more eutrophic waters – including *Chlorella* and *Ankyra* – and the motile colonists (X_2) exemplified by *Rhodomonas* and *Chlamydomonas*; a further subdivision (X_3) might be proposed to accommodate Chrysophyte (s.l.) genera of rather more acid or oligotrophic lakes, such as *Ochromonas* and *Chrysolykos*.

The colonising populations may continue to expand, at a diminishing rate, either until they run out of nutrients (through corporate depletion) or light energy (self-shading) or until they come under the direct control of grazing animals. In each case, the habitat, now modified as a consequence of community activity, begins to offer more stringent selective stresses on species, the more successful of which will be better adapted and, thus, better 'competitors' for the limiting resource. Reynolds (1984, 1988b) gave several examples of species associations characteristic of intermediate successional stages, which included the *Sphaerocystis-Gemellicystis* grouping (F), widespread in clear, mesotrophic lakes, and the *Anabaena-Aphanizomenon-Cylindrospermopsis* group (H) of cyanobacteria of more eutrophic waters.

As the community becomes progressively more structured and the limiting gradients of (e.g.) nutrient become increasingly inverse to those of light, selection moves towards species better able to exploit the segregated resources and to conserve, rather than propagate, their biomass. There is considerable evidence (reviewed in Reynolds, 1987b; Pollingher, 1988) that (L,M) associations dominated by *Microcystis*, *Ceratium* and *Peridinium* are resistant to invaders and are able to accommodate some environmental variability. These algae of maturing successions are usually 'large' ($>10^4\ \mu m^3$) with low surface area/volume ratios ($<0.3\ \mu m^{-1}$), motile (potentially 10^{-4} m s^{-1}) and with relatively high nutrient-storage capacities (Reynolds, 1988b).

Considering 'model' successions in limnetic mesocosms (the Blelham Enclosures), Reynolds (1988a) argued that compositional changes not only coincided with the successional attributes listed by Odum (1969) but also confirmed the declining diversity and equatability of late successional stages, as he and Connell (1978) had hypothesised. These attributes are summarised in Fig. 1, which also shows the changes in community energetics and resource partitioning.

In highly stabilised, segregated environments, however, niche diversification and differential patterns, leading to the stable co-existence of several species, much as proposed by Petersen (1975) and according to hypotheses developed by Kilham & Kilham (1980) and, especially, Tilman (1982). Among the best examples are those observed in permanently ice-covered lakes in Antarctica, which support distinct horizons dominated by sparse flagellate populations (in the uppermost, nutrient depleted layer), by gas-vacuolate, photosynthetic bacteria (lowermost, light deficient layer) and intermediate, often plate-like layers of cryptomonads or filamentous cyanobacteria (Vincent, 1981; Vincent & Vincent, 1982). The net production is close to zero while the nitrogen cycle is virtually closed (Priscu *et al.*, 1989). What is of particular interest is that almost precisely analogous population structures have been described from physically-stable water columns in small, deep, karstic dolines (Vicente & Miracle, 1988), in wind-sheltered forest lakes in the tropics (Reynolds *et al.*, 1983) and in brackish temperate lakes where the stability is reinforced by salinity gradients (Lindholm, 1982). So far as planktonic successions go, these might represent the nearest approximations to a tangible, climactic equilibrium.

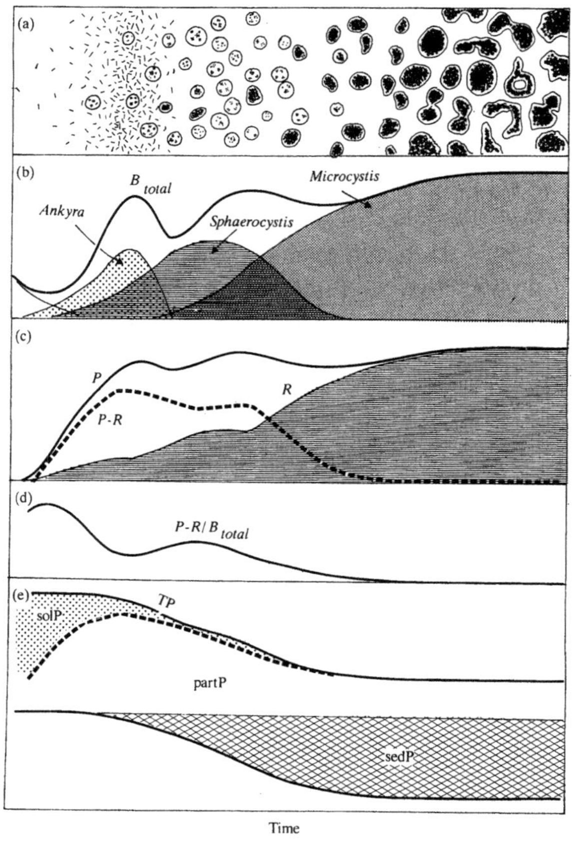

Fig. 1. Features of a simple, managed phytoplankton succession, described in full by Reynolds (1988a), involving (a,b) *Ankyra*, *Sphaerocystis* and *Microcystis* through a nine-week period of summer. Trends in (b) B_{total} (the aggregate algal biomass) are compared with those of (c) gross carbon fixation (*P*), of gross losses of mass, including respiration (R) and the net production (P-R), and (d) of net productivity, viz $(P\text{-}R)/B_{total}$. The distribution of phosphorus between the dissolved and particulate fraction is also shown (e) as well as the cumulative sediment flux (sed P) during the same period.

Disturbance of planktonic succession

It is hypothesised that in freshwater and marine environments, as on land, such fully equilibrated community structures are exceptional and, again as on land, that most of them exist far short of an equilibrium state. If the time-scale of the successional template is correctly applied, then it is also reasonable to deduce that other forces of sufficient vigour to set back, deflect or slow the progress towards equilibrium, intervene with frequencies greater (i.e. at intervals shorter) than the time required for community-determined stabilisation. These events can be termed 'disturbances', in the sense that they can reduce the 'minimal structure' of smaller scale components at which the present community is permitted to exist (Pickett *et al.*, 1989). In this way, a change in flushing rate or in the intensity of wind-induced turbulence would be perceived by a fish as a property of a homogeneously variable environment but may nevertheless represent an abrupt change in the environment of the actively-growing phytoplankton (Reynolds, 1991). It is therefore appropriate to consider the impact upon the idealised planktonic succession (Fig. 1) of disturbance interventions of various frequencies and intensities.

Community development with infrequent disturbances (2–8 Yr^{-1})

The origins of the concept can be traced back to Round's (1971) seminal consideration of 'environmental shock periods', wherein major seasonal alterations in the vertical structure of the lake (for instance the onset and breakdown of thermal stratification) triggered quite profound changes in the species composition of the phytoplankton (Fig. 2a). In each case, the currently dominant vegetation type becomes stressed, or otherwise adversely affected, by the altered environmental conditions. Other species, however, are positively selected but the new environment must be sufficiently persistent to permit several consecutive generations to establish a new dominant and initiate a new ('shifted'; Reynolds, 1980) successional sequence.

In attempting to explain the phytoplankton periodicity of small, eutrophic lakes of the English north-west Midlands (Shropshire and Cheshire meres), Reynolds (1976) proposed an assembly of short successional episodes, each preceded by structural alteration. The onset of thermal stratification in the spring was followed by the abrupt decline of the dominant *Asterionella*-*Stephanodiscus* diatom population (Reynolds Group C) and the inception of what was adjudged to be a *Eudorina* (Group G)→*Anabaena* (Group H)→*Microcystis*/*Ceratium* (Group M) succession. However, before this sequence culminated in a low-diversity, near-climactic community, interventions of increased wind mixing tended to restore diatoms (not always the same ones as in spring and often with proportionally more Group-P *Fragilaria* and *Aulacoseira*, sometimes together with desmids of the genera *Staurastrum* and *Closterium*) which, indeed, were more persistently reestablished after the autumnal return to full column mixing. In the summer period, however, the mixing subsided and the column re-stratified. The brief episode of diatom growth was accordingly followed by a recapitulation, in full or in part, of the G→H→M succession.

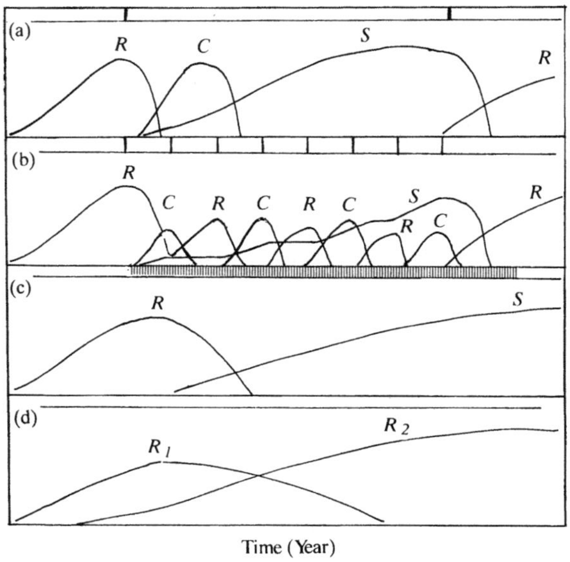

Fig. 2. Stylised seasonal progressions in a temperate, stratifying lake system involving three types of species, C (fast-growing colonists such as *Chlorella* or *Ankyra*), *S* (slow-growing summer dominants such as *Microcystis* or *Ceratium*) and *R* (e.g. diatoms and *Oscillatoria* spp.), and in which stratification and destratification represent (a) the only annual disturbances, (b) alternate at intervals of $\# \geq 20$ d, or (c) alternate daily; (d) is continuously mixed.

Reynolds (1980) advocated the terms 'succession' be used for the autogenic 'main sequence', exemplified by the G→H→M pattern. The allogenic establishment of an alternative succession could, if sustained, be referred to as 'shift' (here, C→G or M→P) or, if followed by renewed thermal stability, lead to a 'reversion' to something close to the main (G→H→M) sequence.

The extent of the reversion would depend, in part, upon the severity and persistence of the disturbance and their influence on the survival of the pre-disturbance populations. The greater is the carry-over through the adverse conditions, then the greater is the inoculum available to re-establishment of the post-disturbance reversion. In Margalef's (1961) terms, the more extensive is the community memory, then the more information can be carried forward. As an illustration, Reynolds *et al.* (1986) described the responses of *Melosira-* (or *Aulacoseira-*) dominated plankton through a 7-d period of reduced mixing intensity: although the population declined, it nevertheless remained dominant, until a phase of more aggressive mixing again selected in its favour. In contrast, Reynolds & Lund (1988) showed that high rainfall episodes flushed Grasmere, England, so efficiently that the next phase was initiated by recolonising populations.

Effects of more frequent disturbances (8–50 Yr^{-1})

The experiments in the Blelham enclosures reproduced analogous effects artificially and on a cycle alternating episodes of mechanical mixing of 10–14 days' duration with similar periods of 'quiescence' (Reynolds *et al.*, 1984). The anticipated coincidence of diatom increase and *Eudorina-Sphaerocystis* increase with the imposed cycle was, indeed, attained with population decline in the intervening period, while the population of *Anabaena*, though persisting through the mixing periods, increased only in the quiescent phases (Fig. 2).

The fidelity of these responses can be verified from the numerous studies of responses to more frequent stimulation. Lewis' (1978) interpretations of the responses of phytoplankton to profound alternations between (15–60 d) episodes of mixing and restratification in the large, ateliomictic Lake Lanao (Mindanao, Philippines: lat 7° 52'N, long 124° 13'E) came, quite independently, to similar conclusions. Sommer's (1986) review of sequences in a selection of central European lakes detected analogous alternation between phytoplankton favoured by episodes of mixing beyond the photic zone and of increased vertical stability. Time-series analysis of data for Guelph Lake, Ontario, Canada revealed analogous dynamic responses of comparable populations driven by weather fluctuations associated with frontal passage at a frequency of 7–11 days (Trimbee & Harris, 1983).

In larger or more exposed lakes, the stability may be relatively more fragile (with greater exposure to wind fetch and mechanical dissipation of energy); in western Lake Ontario, for instance, the dominance of the plankton community often hovers between diatoms and chlorophyte colonies of *Oocystis* (Haffner *et al.*, 1980). Ferguson & Harper (1982) described fluctuations in the total population and in the species dominance of phytoplankton in the highly eutrophic reservoir, Rutland Water, England; the remarkable feature here is that the 'helixors' which were operated in this reservoir to prevent thermal stratification were only adequate to weaken it, nevertheless enhancing its susceptibility to even quite mild wind action. Dominance of the phytoplankton oscillated between diatoms and simple chlorophytes, on cycles of change averaging 10 days or so. It was as if neither potential successional series was given adequate time to establish a large population. In consequence, aggregate biomass remained at acceptably low levels; certainly, they were well below the range of means predictable from Vollenweider's (1976) equation. That this case attests to control by frequent and intermediate disturbance of natural successional development is reinforced by the fact that, in 1984 (which, in the UK, was a particularly warm, dry summer), the circulating failed to overcome stratification. There was, that year and for the first time, a serious problem with the abundance of phytoplankton, dominated by *Microcystis* (Anglian Water, personal communication). Disturbances in hydrographic and hydrological stability, set around 2–3 generation times for algae ($<$ 10 days) clearly make an important impact on algal abundance and community structure in the plankton (Fig. 2(b)).

Effects of high-frequency disturbances (50–365 Yr^{-1})

With episodes of a few days or less, the frequency begins to coincide with the time-scales of cell generation. The inoculum effect now overwhelms the population response but the selectively-favoured species will be those that can accommodate to the full variability. Certainly, in the several examples of circadian cycles of near-surface stratification (in response to the day-time heat income) alternating with afternoon or early-evening periods of convectional cooling or wind-driven mixing (e.g. Ganf, 1974; Vincent *et al.*, 1984; Zohary & Robarts, 1989; Ibelings *et al.*, 1991), individually entrained algae might be subject, within periods of several hours, to extremes ranging between almost continuous exposure to full-sunlight near to the surface or relative darkness at depth, to being transported across the full gradient and back within a few minutes in a vigorously mixed surface layer. The steeper the light gradient in relation to the fluctuations in mixed depth, the more pronounced is the contrast between the extremes. The greater is the contrast, then the more likely it is that the dominant planktonic species will be one of the larger, strong-swimming or fast-migrating species. While the water remains fairly clear (range of mixed depth within euphotic depth), green algae (Volvocales) may dominate but, in relatively more turbid habitats (maximum mixed depth exceeding the euphotic depth), such conspicuously motile genera as *Ceratium*, *Peridinium* and *Microcystis* appear to be selectively favoured (Fig. 2c). Their ability to make rapid positional adjustments in unmixed water columns theoretically ensures quick recovery of optimal station when turbulence subsides; populations of species unable to make rapid movements are, in contrast, liable to be 'stranded' at depths where the light is either inadequate or (soon) inhibiting to the support of net photosynthesis (see Reynolds, 1989).

Effects of continuous disturbance

The logical continuation of this series of increasing frequency of disturbance culminates in the mergence of episodes and the exclusion of 'intermediate quiescence'. However, this step in logic is not helpful to the ecological theory. This is not because biological communities do not relate to the overriding effects imposed by physical conditions in continuously wind-mixed, turbid lakes, like Lough Neagh, Northern Ireland (Gibson *et al.*, 1971), the prerestoration Veluwemeer, Netherlands (Berger, 1975), the clay-rich Waco Reservoir, Texas (Kimmel & Lind,1972) and P.K. le Roux Reservoir, RSA (Allanson & Hart, 1979), or in high-flushing rivers (Reynolds, 1988c) and tidal estuaries (e.g. Joint & Pomroy, 1981); clearly, they do but the successful species which comprise them must be adapted, physiologically and/or behaviourally, to survive the exacting conditions. Rather, continuous disturbance, at the scale of generation times, represents a form of environmental constancy: successive generations build up populations in the analogous manner to the colonists of habitats. Indeed, the ecological disturbance to this *status quo* might be precipitated by an interruption to the otherwise continuous external physical forcing: the growth of *Anabaena* in Lough Neagh during episodes of clearer, warmer water conditions (Gibson & Fitzsimmons, 1982) or the development of large phytoplankton populations in the lower reaches of the River Thames, UK, during the 1976 drought (see Whitehead & Hornberger, 1984) provide good examples of the selection and organisation of communities in otherwise continuously dispersive environments, intermediately disturbed by the imposition of new conditions.

It is often less easy to discern in these continuously dispersive environments whether an analogous successional template governs the development of species dominance. There is evidence, discussed in Lund & Reynolds (1982) and developed in Reynolds (1988a), that uninterrupted mixing of lakes (say through temperature winter-spring periods) is likely to select progressively in favour of *Oscillatoria* spp. and against diatoms (Fig. 2(d)). That this transition is not often observed may be related to several local factors, including water temperature and wind variability, and to the biassing influence of comparative growth rates, flushing times and silicon concentrations.

Where the successional sequence may be observed however, progress is generally slow; its conclusion may be rarely achieved and it is not certain that the supposed equilibrium condition – that of overwhelming dominance by (e.g.) *Oscillatoria agardhii* Gomont represents an ecological climax. It might be rather more useful to take an opposite view which regards the ecological state of continuously dispersive environments as being likely to maintain sub-climactic communities for long periods, or perhaps indefinitely. In short, they might be understood to be the aquatic analogues of plagioclimax.

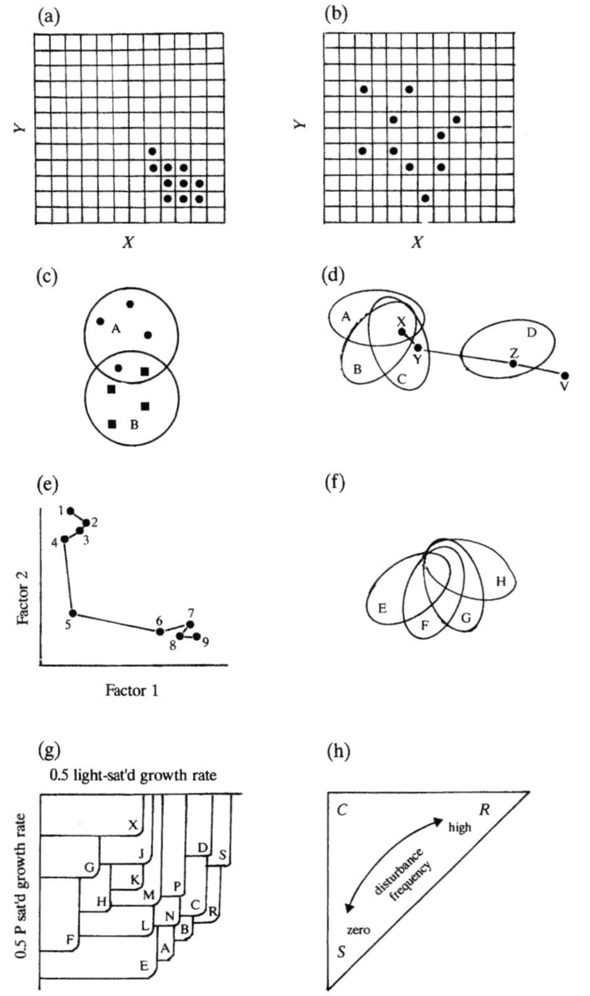

Fig. 3. Sketches to represent: (a,b) Booleian matrices with (a) low-and (b) high-diversity of information; (c) Venn symbol sets with subset of common behaviour; (d) time-tracks (X→Y→Z) in lead variables selecting for different species, or none at V; (e) time tracks of co-ordinates of two key variables; (f) skewed Venn complex through which (e) might select among species; (g) analogous plot of species-specific physiological ranges against axes defining phosphorus availability and hydraulic disturbance; (h) the same, reduced to a triangular environmental matrix, showing the apices preferred by *C*-, *S*- and *R*-species. for full explanation, see text.

The relationship between diversity and disturbance

If, as is implied by Hardin's (1960) competitive exclusion principle, species diversity is suppressed as ecological equilibrium is approached, then it follows that intermediate disturbances with the appropriate intensity, frequency and duration should prevent against competitive exclusion and so maintain species diversity.

The measurement of diversity using, for instance, Shannon's functions, derives from Information Theory (see, e.g., Juhász-Nagy, this volume): diversity is a spatial measure of the extent of a multi-dimensional envelope embracing the co-ordinates of each point of information relating to the attributes represented. Distributions having low and high diversities are sketched in Figs. 3(a) and 3(b), respectively.

Whether applied to individuals of a specific population or to a population of a species within a *de facto* assemblage of several such species, the distributions may be setted or sub-setted into associations sharing defined properties (Fig. 3c). The extents of these Venn envelopes might enclose the known ranges of preferences or tolerances of individual species with reference to the plane of representation and its dimensions. In this way, separate species having envelopes which enclose the co-ordinates of the contemporaneous variables, constitute a subset of common behaviours (selection); those which do not are now excluded. In Fig. 3d, a change in conditions from point X to Y, where A and B are no longer supported, nevertheless continues to favour C; species D, however, requires a more significant environmental shift (to Z) which, incidentally, now selects against A, B and C. Note that at V, the conditions are too hostile for any species and none is selected.

Sequential plots recording conditions obtaining at regular time intervals show the rate and intensity of change (see Fig. 3e). In the case shown, the changes are quite gradual and modest through steps 1–4, revealing a high level of constancy with respect to the factors maintained and, hence, continuously selecting species having envelopes that embrace those co-ordinates. A more extensive and rapid transition occurs between steps 4 and 6, before settling down again, perhaps within another subset of quite different species. Holzmann (1993) gives some excellent illustrations of this principle.

It is a simple step in logic to suppose that the Venn complexes of species subsets need be neither symmetrical nor regular. 'Good' conditions will be simultaneously optimal for relatively more species, so more might be expected to coexist simultaneously; more extreme conditions, however, will be more rigourously selective (Fig. 3(f)).

In an earlier attempt to discern pattern in phytoplankton community behaviour, Reynolds (1980) claimed that the physical structure of the environment and the availability of nutrients therewithin provided the most important variables influencing species selection. The original matrix was further refined with axes quantified in terms of the availability of phosphorus and by the daily integral of light received by organisms in mixed layers, the latter being a partial correlative of relative hydraulic disturbance (Reynolds, 1987c). The superimposition of rectangles defining the half-saturation constants of P-limited growth-rate and of light-limited photosynthesis for each species (Fig. 3(g)) is conceptually analogous to the skewed Venn complex. It may also be noted that its bottom right-hand corner, representing the contingency of low nutrient resources with frequent disturbance is excessively hostile and untenable to any species (Grime, 1979); the complex now becomes approximately triangular (Fig. 3(h)), with most species able to thrive in the upper left corner of high resources and low disturbance but with fewer and fewer specialist species able to tolerate either the conditions of severe nutrient depletion (stress tolerant S-species) or of relatively more frequent or more persistent hydrographic disturbance (ruderal R species). Whether driven internally, by chronic exhaustion of resources, towards the S-apex, or externally, by more frequent disturbance, towards the R-apex, the more persistent extremes, the number of potentially surviving species is reduced: The diversity is 'squeezed out' of the system.

Conversely, with frequent alterations of directions, environmental selectivity will track predominantly in the body of the triangle, periodically also alternating the selective advantage among a wider number of species for shorter periods of time (Reynolds, 1991). In this derivation, assemblage diversity depends less on whether the stimulus is autogenic or allogenic so much as that a stimulus of sufficient intensity is applied with sufficient frequency. From the evidence of Reynolds' (1980, 1984) investigation of the rates of compositional change and the simultaneous changes in diversity (Reynolds, 1988a), it is apparent that following both abrupt allogenic shifts and gradual autogenic transitions in dominance, both the index of change and the diversity index increase as the new species grow in number and the replaced species decline. The measurements are broadly correlated with each other but they are not equivalent. There is a point of maximum diversity and one of maximal rate of change, each of which will occur at some point in time between the first reproductions of the new species and, if equivalence between growth- and attrition-rates is assumed, the sixth to eighth generation. There is some evidence – from the studies of Trimbee & Harris (1983), Gaedeke & Sommer (1986) and Reynolds (1988a) – that the diversity peaks at approximately 3 generations or (depending on temperature) some 5–15 days after the stimulus is applied (see also the discussion of evidence in Sommer, 1993).

Conclusions

The preceding consideration establishes the applicability of the intermediate disturbance hypothesis to the understanding of community organisation in the plankton, to the maintenance of species diversity and to the productivity of pelagic ecosystems. In real time, community processes operate at very much shorter temporal scales than in the forest and reef systems to which Connell's (1978) explanation was originally directed. Both the sizes and the life-spans of the dominant vegetation type in the pelagic are appropriate to life in a relatively viscous, dense fluid with exacting optical properties, yet in which medium motion generated through the interaction with the atmosphere, whether the gravitation of precipitation or the forcing by wind, is almost always turbulent. The evolutionary ecology of phytoplankton is thus intimately related to the dominating feature of the water, that of its three-dimensional motion, and the frequency with which the scale of motion is altered. In this way, the plants of the plankton, as on the land, will largely complete their life cycles within the time spans normally available (see Fig. 4). If the time span lengthens and successive generations are permitted to develop a hierarchical internal organisation (succession), a generally clear understanding of the direction of its structural development (towards a climactic equilibrium) is now available. If, before equilibrium is achieved, an imposed change brings about a breakdown or weakening of structure and organisation, it may be construed to have been disturbed. Thus, depending upon the nature of the system, the severity and frequency with which it is disturbed and, indeed, the time that has elapsed since its last major disturbance, together play the primary role in establishing, at any given moment, the species composition, successional state, productivity, competitiveness and diversity of a given ecological system.

The greater the frequency of disturbance, then the weaker must be the organisation and the more important are the unfulfilled time lags in species replacement to the maintenance of diversity.

If the time span between significant environmental alterations is shortened, however, to substantially less than the generation time, the disturbances are perceived as a fluctuating component of the isotropic and, thus, constant environment. If it can be accommodated by the physiological flexibility of the organisms concerned they may well survive, reproduce and come to dominate the community. If it cannot, they will die out. This sort of community may achieve equilibrium with external forcing at suitable intensities and frequencies, yet is not a successional climax. Like a pasture in previously forested landscape, the plagioclimax is maintained by the application of external disturbance (management) too frequently to permit the growth of trees but which allows certain grasses the opportunity to increase

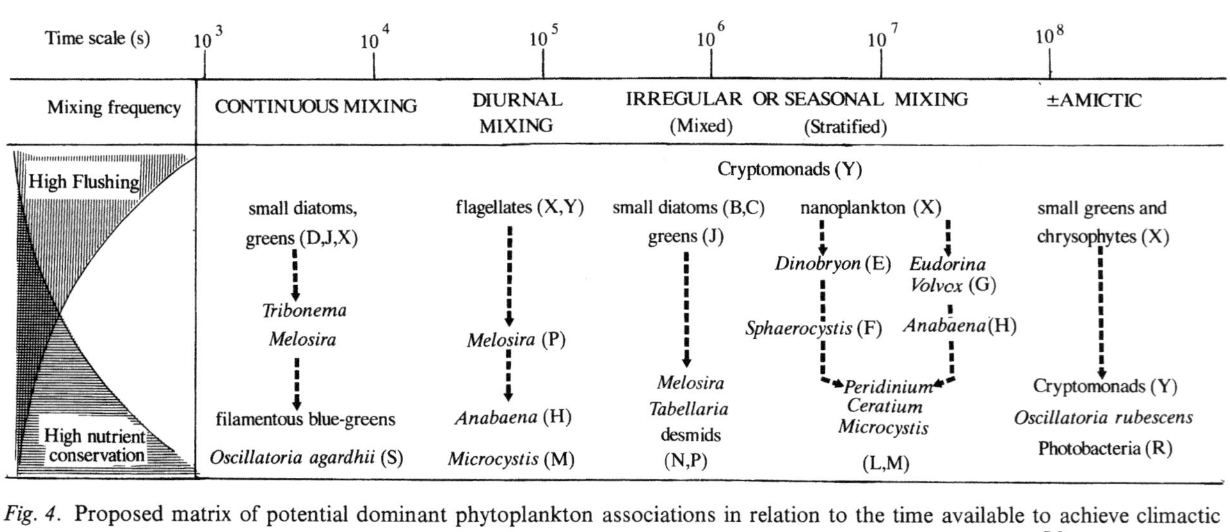

Fig. 4. Proposed matrix of potential dominant phytoplankton associations in relation to the time available to achieve climactic or plagioclimactic equilibrium (vertical axis) against the frequency of mixing events (horizontal axis, in s; $10^{7.5}$s $\sim$1 year). The assemblage labels (B, C, D....X, Y) are referred to in the text but are more fully elaborated in Reynolds (1988b).

and dominate. The planktonic analogue of continuous dominance by (e.g.) *Oscillatoria agardhii* in persistently wind-mixed shallow lakes completes the argument for the tenability of Connell's hypothesis. Not for the first time, the key time unit in plankton ecology is recognised to be the generation time (10^5–10^6 s).

Acknowledgements

I am most indebted to my editorial colleagues for improvements to the original manuscript and to Professor P. Juhász-Nagy for the benefit of his perceptive critique.

I am extremely grateful to Mrs Y. Dickens for her capable help in the preparation of the manuscript of this paper.

References

Allanson, B. R. & R. C. Hart, 1979. Limnology of P. K. le Roux Dam. Reports, Rhodes University Institute for Freshwater Studies 11: 1–3.

Berger, C., 1975. Occurrence of *Oscillatoria agardhii* Gomont in some shallow eutrophic lakes. Verh. int. Ver. Limnol. 19: 2687–2697.

Connell, J. H., 1978. Diversity in tropical rain forests and coral reefs. Science 199: 1302–1310.

Ferguson, A. J. D. & D. M. Harper, 1982. Rutland Water phytoplankton: the development of an asset or a nuisance? In D. M. Harper & J. A. Bullock (eds), Rutland Water – Decade of change. Developments in Hydrobiology 8. Dr W. Junk Publishers, The Hague: 117–133. Reprinted from Hydrobiologia 88.

Ganf, G. G., 1974. Diurnal mixing and the vertical distribution of phytoplankton in a shallow equatorial lake (Lake George, Uganda). J. Ecol. 62: 611–629.

Gaedeke, A & U. Sommer, 1986. The influence of the frequency of periodic disturbances on the maintenance of phytoplankton diversity. Oecologia 71: 98–102.

Gibson, C. E. & A. G. Fitzsimmons, 1982. Periodicity and morphology of planktonic blue-green algae in an unstratified lake (Lough Neagh, Northern Ireland). Int. Revue ges. Hydrobiol. 62: 459–471.

Gibson, C. E., R. B. Wood, E. L. Dickson & D. M. Jensen, 1971. The succession of phytoplankton in L. Neagh, 1968–1970. Mitt. int. Ver. Limnol. 19: 146–160.

Grime, P., 1979. Plant strategies and vegetation processes. Wiley-Intescience, Chichester.

Grygierek, E. & B. Wasilewska, 1979. Regulation of fishpond biocoenoses. Special Publications of the European Mariculture Society 4: 317–333.

Haffner, G. D., G. P. Harris & M. K. Jarai, 1980. Physical variability and phytoplankton communities. III. Vertical structure in phytoplankton populations. Arch. Hydrobiol. 89: 363–381.

Hamilton, S. K. & W. M. Lewis, 1990. Basin morphology in relation to chemical and ecological characteristics of lakes on the Orinoco River floodplain, Venezuela. Arch. Hydrobiol. 119: 393–425.

Hardin, G., 1960. The competitive exclusion hypothesis. Science. 131: 1292–1297.

Holzmann, R., 1993. Seasonal fluctuations in the diversity and compositional stability of phytoplankton communities in small lakes in upper Bavaria. In J. Padisák, C. S. Reynolds & U. Sommer (eds), Intermediate Disturbance Hypothesis in Phytoplankton Ecology. Developments in Hydrobiology 81. Kluwer Academic Publishers, Dordrecht: 101–109. Reprinted from Hydrobiologia 249.

Hutchinson, G. E., 1961. The paradox of the plankton. Am. Nat. 95: 137–147.

Ibelings, B. W., L. R. Mur & A. E. Walsby, 1991. Diurnal changes in buoyancy and vertical distribution in populations of *Microcystis* in two shallow lakes. J. Plankton Res. 13: 419–436.

Joint, I. R. & A. J. Pomroy, 1981. Primary production in a turbid estuary. Estuar. coast. mar. Sci. 13: 303–316.

Juhász-Nagy, P., 1993. Notes on compositional diversity. In J. Padisák, C. S. Reynolds & U. Sommer (eds), Intermediate Disturbance Hypothesis in Phytoplankton Ecology. Developments in Hydrobiology 81. Kluwer Academic Publishers, Dordrecht: 173–182. Reprinted from Hydrobiologia 249.

Kilham, P. & S. S. Kilham, 1980. The evolutionary ecology of phytoplankton. In I. Morris (ed.), The physiological ecology of phytoplankton. Blackwell Scientific Publications, Oxford: 571–597.

Kimmel, B. L. & O. T. Lind, 1972. Factors affecting phytoplankton production in a eutrophic reservoir. Arch. Hydrobiol. 71: 124–141.

Kořínek, V., J. Fott, J. Fuksa, J. Lellák & M. Pražáková, 1987. Carp ponds of Central Europe. In R. G. Michael (ed.), Managed Aquatic Ecosystems. Elsevier, Amsterdam: 29–62.

Lewis, W. M., 1978. Dynamics and succession of the phytoplankton in a tropical lake: Lake Lanao, Philippines. J. Ecol. 66: 849–880.

Lindholm, T., 1982. Dynamics of hydrography and primary production in three stratified coastal lakes on Åland (S. W. Finland). Acta Academia Aboensis B 42: 1–75.

Lund, J. W. G., 1949. Studies on *Asterionella*. I. The origin and nature of the cells producing seasonal maxima. J. Ecol. 37: 389–419.

Lund, J. W. G. & C. S. Reynolds, 1982. The development and operation of large limnetic enclosures in Blelham Tarn,

English Lake District and their contribution to phytoplankton ecology. In F. E. Round & D. J. Chapman (eds), Progress in Phycological Research Vol. I. Elsevier, Amsterdam: 1–65.

Margalef, R., 1961. Communication of structure in planktonic populations. Limnol. Oceanogr. 6: 124–128.

Odum, E. P., 1969. The strategy of ecosystem development. Science 164: 262–270.

Pennington, W., 1969. The history of the British vegetation. English Universities Press, London.

Petersen, R., 1975. The paradox of the plankton: an equilibrium hypothesis. Am. Nat. 109: 35–49.

Pickett, S. T. A., J. Kolasa, J. J. Armesto & S. L. Collins, 1989. The ecological concept of disturbance and its expression at various hierarchical levels. Oikos 54: 129–136.

Pollingher, U., 1988. Freshwater armoured dinoflagellates: growth, reproduction strategies and population dynamics. In C. D. Sandgren (ed.), Growth and reproductive strategies of freshwater phytoplankton. Cambridge University Press, New York: 134–174.

Price, P. W., 1984. Alternative paradigms in community ecology. In P. W. Price, C. N. Slobodchikoff & W. S. Gaud (eds), A new ecology: novel approaches to interactive systems. Wiley-Interscience, New York: 353–383.

Priscu, J. C., W. F. Vincent & C. Howard-Williams, 1989. Inorganic nitrogen uptake and regeneration in perennially ice-covered Lakes Fryxell and Vanda, Antarctica. J. Plankton Res. 11: 335–351.

Reynolds, C. S., 1976. The ecology of phytoplankton in Shropshire and Cheshire meres. Report Freshwater Biological Association 44: 36–45.

Reynolds, C. S., 1980. Phytoplankton assemblages and their periodicity in stratifying lake systems. Holarct. Ecol. 3: 141–159.

Reynolds, C. S., 1984. Phytoplankton periodicity: the interaction of form, function and environmental variability. Freshwater. Biol. 14: 111–142.

Reynolds, C. S., 1987a. Community organization in the freshwater phytoplankton. In J. H. R. Gee & P. S. Giller (eds), The organization of communities, past and present. Blackwell Scientific Publications, Oxford: 297–325.

Reynolds, C. S., 1987b. Cyanobacterial Water Blooms. In J. Calow (ed.), Advances in botanical research, Vol. 13. Academic Press, London: 67–143.

Reynolds, C. S., 1987c. The response of phytoplankton communities to changing lake environments. Schweiz. Z. Hydrol. 49: 220–236.

Reynolds, C. S., 1988a. The theory of ecological succession applied to the freshwater phytoplankton. Verh. int. Ver. Limnol. 23: 683–691.

Reynolds, C. S., 1988b. Functional morphology and the adaptive strategies of freshwater phytoplankton. In C. D. Sandgren (ed.), Growth and reproductive strategies of freshwater phytoplankton. Cambridge University Press, New York: 388–433.

Reynolds, C. S., 1988c. Potamoplankton: paradigms, paradoxes, prognoses. In F. E. Round (ed.), Algae and the aquatic environment. Biopress, Bristol: 285–311.

Reynolds, C. S., 1989. Relationships among the biological properties, distribution and regulation of production by planktonic Cyanobacteria. Toxicity Assessment 4: 229–255.

Reynolds, C. S., 1991. Lake Communities: an approach to their management for conservation. In I. F. Spellerberg, M. G. Morris & F. B. Goldsmith (eds), The scientific management of temperature communities for conservation. Blackwell Scientific Publications, Oxford: 199–225.

Reynolds, C. S. & J. W. G. Lund, 1988. The phytoplankton of an enriched, soft-water lake subject to intermittent hydraulic flushing (Grasmere, English Lake District). Freshwat. Biol. 19: 379–404.

Reynolds, C. S., V. Montecino, M.-E. Graf & S. Cabrera, 1986. Short-term dynamics of a *Melosira* population in the plankton of an impoundment in central Chile. J. Plankton Res. 8: 715–740.

Reynolds, C. S., J. G. Tundisi & K. Hino, 1983. Observations on a metalimnetic *Lyngbya* population in a stably stratified tropical lake (Lagoa Carioca, Eastern Brasil). Arch. Hydrobiol. 97: 7–17.

Reynolds, C. S., S. W. Wiseman & M. J. O. Clarke, 1984. Growth- and loss-rate responses of phytoplankton to intermittent artificial mixing and their potential application to the control of planktonic algal biomass. J. appl. Ecol. 21: 11–39.

Richerson, P., R. Armstrong & C. R. Goldman, 1970. Contemporaneous disequilibrium, a new hypothesis to explain the paradox of the plankton. Proc natn. Acad. Sci. USA 67: 1710–1714.

Round, F. E., 1971. The growth and succession of algal populations in freshwaters. Mitt. int. Ver. Limnol. 19: 70–99.

Smayda, T. J. 1980. Phytoplankton species succession. In I. Morris (ed.), The physiological ecology of phytoplankton. Blackwell Scientific Publications, Oxford: 493–570.

Sommer, U., 1985. Comparisons between steady state and non-steady state competition: experiments with natural phytoplankton. Limnol. Oceanogr. 30: 335–346.

Sommer, U., 1986. The periodicity of phytoplankton in Lake Constance (Bodensee) in comparison to other deep lakes of Central Europe. In M. Munawar & J. F. Talling (eds), Seasonality of Freshwater Phytoplankton – a Global Perspective. Developments in Hydrobiology 33. Dr W. Junk Publishers, Dordrecht: 1–7. Reprinted from Hydrobiologia 138.

Sommer, U., 1989. The role of competition for resources in phytoplankton succession. In U. Sommer (ed.), Plankton ecology. Springer-Verlag, Berlin: 57–106.

Sommer, U., 1993. Disturbance-diversity relationships in two lakes of similar nutrient chemistry but contrasting disturbance regimes. Hydrobiologia 249: 59–65.

Sommer, U., Z. M. Gliwicz, W. Lampert & A. Duncan, 1986. The PEG-model of seasonal succession of planktonic events in freshwaters. Arch. Hydrobiol. 106: 433–471.

Tallis, J. H., 1973. The terrestrialization of lake basins in north Cheshire, with special reference to the development of a 'Schwingmoor' structure. J. Ecol. 61: 537–567.

Tansley, A. G., 1939. The British Isles and their vegetation. Cambridge University Press, Cambridge.

Tilman, D., 1982. Resource competition and community structure. Princeton University Press, Princeton.

Tilman, D., S. S. Kilham & P. Kilham, 1982. Phytoplankton community ecology: the role of limiting nutrients. Ann. Rev. Ecol. Syst. 13: 349–372.

Trimbee, A. M. & G. P. Harris, 1983. Use of time-series analysis to demonstrate advection rates of different variables in a small lake. J. Plankton Res. 5: 819–833.

Tundisi, J. G., 1980. Ecológia aquática no Brasil: problemas e perspectivas. Interciencia 5: 373–379.

Uhlmann, D., 1971. Influence of dilution, sinking and grazing rate on phytoplankton populations of hyperfertilized ponds and microecosystems. Mitt. int. Ver. Limnol. 19: 100–124.

Vicente, E. & M. R. Miracle, 1988. Physicochemical and microbial stratification in a meromictic karstic lake of Spain. Verh. int. Ver. Limnol. 23: 522–529.

Vincent, W. F., 1981. Production strategies in Antarctic inland waters: phytoplankton eco-physiology in a permanently ice-covered lake. Ecology 62: 1215–1224.

Vincent, W. F., P. J. Neale & P. J. Richerson, 1984. Photoinhibition: algal responses to bright sunlight during diel stratification and mixing in a tropical alpine lake. J. Phycol. 20: 201–211.

Vincent, W. F. & C. L. Vincent, 1982. Factors controlling phytoplankton production in Lake Vanda (77° S). Can J. Fish. aquat. Sci. 39: 1602–1609.

Vollenweider, R. A., 1976. Advances in defining critical loading levels for phosphorus in lake eutrophication. Mem. Ist. ital. Idrobiol. 33: 53–83.

Whitehead, P. G. & G. M. Hornberger, 1984. Modelling algal behaviour in the River Thames. Wat. Res. 18: 945–953.

Zohary, T. & R. D. Robarts, 1989. Diurnal mixed layers and the long-term dominance of *Microcystis aeruginosa*. J. Plankton Res. 11: 25–48.

Hydrobiologia **249**: 173–182, 1993.
J. Padisák, C.S. Reynolds & U. Sommer (eds), Intermediate Disturbance Hypothesis in Phytoplankton Ecology.

Notes on compositional diversity

Pál Juhász-Nagy
Department of Plant Taxonomy and Ecology, Eötvös Lorand University, H-1083 Budapest, Ludovika tér 2, Hungary

Key words: binary modelling, biotic diversity, local diversity, associatum, characteristic scaling, Venn-complex

Abstract

This paper tries to show some ways and means of simple (binary) modelling, whereby diversity can be interconnected with other attributes of a community. A new type of scaling (called characteristic scaling) is introduced for further use.

Introduction; motivation

I remember quite well of the excitement, when I first read the interesting paper of Joe Connell (1978). *Ecce, voilà*, I felt; Intermediate Disturbance Hypothesis (IDH) maybe a good guide-line for a better orientation; or, it may serve as a satisfactory '*corpus theoreticum*' around which many observations, facts, evaluations, interpretations might be arranged more properly than beforehand.

But, alas, just afterwards, the Cartesian '*omnia dubitanda*' had started to work; the clouds of a sober scientific skepticism have accumulated further during the past decade. What does the shrewd adjective 'intermediate' really mean? How can one make an operative distinction between 'intermediate disturbances' and, say, 'large-scale perturbations' (Matson & Carpenter, 1990)? Using the titleforming question of Carpenter (1989), 'do we know what we are talking about?' Is it due only to the rather idiotic fluctuations of pseudo-scientific fashions and counter-fashions that Connell (1978) is even not mentioned in any paper edited by Matson and Carpenter?

One can develop even more pessimistic views, if one starts reading the huge body of literature on 'non-linear dynamics', including recent books on systems theory (e.g. Vincent *et al.*, 1990), or on the delightful (and frightening) topic of chaos (e.g. Cvitanovič, 1979; Steward, 1979). One of the most pertinent questions is: what is the true difference between disturbance and perturbation? If both processes contain 'harmful' and 'beneficiary' components as well, how can one make a successful distinction between these types of responses?

Cutting the story of a long and tedious speculation short, the author is motivated to write this paper by the following main aspects of our actual ignorance:

(a) Theoretical weakness
IDH is still too weak for a proof or a disproof. (Using a Popperian phrase, IDH, in its present form, is not suitable either for a verification or for a falsification. Unfortunately enough, the same is true for some modification of IDH; the author has tried several versions and failed).

(b) Methodological weakness
(a) is partly due to a number of methodological defects. IDH, for instance, is a 'diversity-centered' hypothesis. But diversity, however it is measured, is usually a single property or attribute of a particular community. There is a growing need to develop such simple models, where diversity can be interconnected properly with other attributes as well.

(c) Conceptual weakness
One of the obstacles of such a modelling is that our language is still too weak in both ways. One way is mathematical where, for instance, entropy-information relations are frequently obscured with each other; the other way is biological where, even the primitive scheme, flora→vegetation, has no proper equivalent in zoology, hydrobiology.

One feels that the appropriate order is (c)→(a); that is, theoretical weakness cannot be compensated in the future otherwise than by some primary reconsideration of our methods and concepts.

Notes on composition; an iteration

Composition, even in the simplest etymological sense of the word (*con-positio* – 'collective position'), refers always to some 'mutual positional relations' of some sets of components. Even this simplistic pseudo-definition is contrary to the pragmatic use of the term is ecology, where 'composition' refers usually to some sets or weighted sets (like frequency distributions). But, we must realize, that any composition (let it be either a piece of art or a master-piece of Nature) is much more than silly '%-spectra' or similar representations. Clearly, some kind of iteration is needed.

At a start, we can think of the well-known scheme,

$$\text{flora} \longrightarrow \text{vegetation} \tag{1}$$

or, in a much more general context, of (2),

$$\begin{matrix}\text{Basic sets} \\ \text{(alphabets)}\end{matrix} \longrightarrow \begin{matrix}\text{Compositional} \\ \text{structures}\end{matrix} \tag{2}$$

where (2) may refer to a hydrobiological situation of some kind, and where 'large arrows' try to indicate some 'epistemiological way' of our understanding.

In order to make 'large arrows' more specific, we may decompose (1)–(2) into several and more articulate states or steps. One way of such a decomposition is shown by Fig. 1, where certain object and operations are arranged in a proper order.

A point set (like: flora, fauna, biota) is just a list of the components involved, without any further specification. A simplex includes abundance estimates (usually as discrete frequency distributions, called 'species-abundance' relations) but does not include representation of 'interactions' among components. The last requirement is the job of a Venn-complex (used frequently in Set Theory texts). Some properties of V-complexes are used in the construction of an S-complex where a sorted complex may be either a topological tree, a dendrite, or, it may be some ordination diagram. If an S-complex is somehow allocated into the topographical ('real') space, then an A-complex, an allocated complex is gained (where classification can be regarded as a special, 'fortunate' case).

The simplest possible primary representations of some composition are binary tables (like the primitive ones in Fig. 2.). It is to be noted at once that if we use only the marginals of such tables, then some simplex is represented, only; in order to have a better insight, the 'inner configuration' of such a table is to be taken into account. In other words, there exists a contrast between 'simplex' and 'complex' representations of diversity with many further implications.

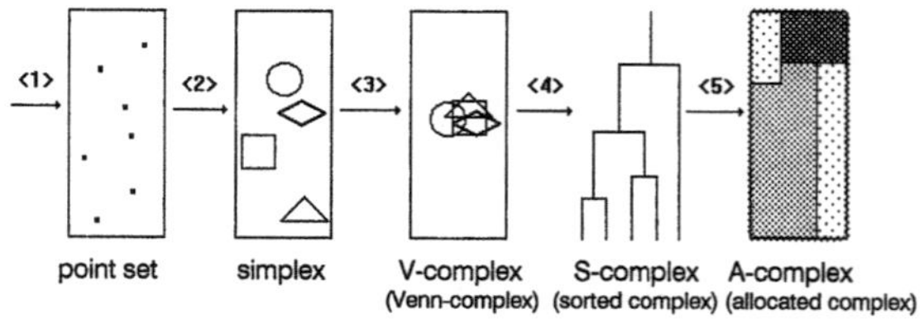

Fig. 1. Certain objects and operations in a proper order for a better specification of (1) and (2).

X_1

populations	sampling units 1	2	3	4	5	6	7	8	Σ
a	0	1	0	0	1	1	0	1	4
b	0	0	1	0	1	0	1	1	4
c	0	0	0	1	0	1	1	1	4
Σ	0	1	1	1	2	2	2	3	12

X_2

populations	sampling units 1	2	3	4	5	6	7	8	Σ
a	1	1	1	1	0	0	0	0	4
b	0	0	0	0	1	1	1	1	4
c	0	0	0	0	1	1	1	1	4
Σ	1	1	1	1	2	2	2	2	12

Fig. 2. Two binary tables of the $s \times m$ type, where s is number of populations and m is number of sampling units.

If one tries to follow carefully the most relevant publications in this field of study (to mention but a few: e.g. Margalef, 1958; MacArthur, 1965; Woodwell & Smith, 1969; Peet, 1974; Pielou, 1975; Grassle *et al.*, 1979; Mayr, 1984; Washington, 1984; Wilson, 1988; Magurran, 1988), then the critical reader can trace easily the funny oscillations – appearance and disappearance – of good ideals in a strange sequence. For instance, since the early guesses of Margalef and the activity of the MacArthur's group, it is quite clear (say, a scientific common-place) that without 'patterning' almost all diversity data are meaningless. But, surprisingly enough, this simple truth does not appear in the recent book edited by E. O. Wilson; as far as I can see, very few scientists are aware of it, even in the programme of IUBS, called 'Biodiversity Crisis'. One may ask: why is it so? Clearly, again, some reconsideration is needed.

Notes on diversity; another iteration

Such a reconsideration can be made concise enough by introducing a few conceptual and methodological oppositions.

§1. Scalars versus vectors

For too many people diversity means still 'number of species' (or other taxa); the 'only' tangible danger is being extinction. But, using the *ex verbis* terms of my late professor, Alfréd Rényi, 'number of components' (alias: 'richness') refers always to zero-ordered diversity, and extinction of one or more elements must considered to be a tragic end-result ('finality') of a more or less long stochastic process. If we want to study the process itself, then, to say the least, we need always vectorial representations, instead of scalars.

§2. Static versus dynamic representations

If the need for a dynamic approach is accepted, then a number of very difficult problems should be faced. As it is most commonly understood, the adjective 'dynamic' is used most frequently for temporal processes, as in the vast majority of differential or difference equations in theoretical physics and theoretical biology. In such representations, the space is still 'static' (by considering space as 'homogeneous', spatial heterogeneity as 'negligible' etc.). Contrary to this common and deceivingly convenient view, the most important biological processes (like evolution, succession etc.) are clearly spatio-temporal processes, where both space and time should be taken into account, even if the methodology involved is frequently troublesome. If we want to study diversity changes (say, during the process of a degradative succession), then we must find ways and means of comparing spatio-temporal coordinates (with special reference to the very neglected spatial processes).

§3. Single vectors versus classes of vectors

By reiterating argument § 1, it is fairly important to consider at least the contours of a 'singularistic' versus 'pluralistic' dilemma. Figure 2 shows how deceiving it can be to say that X_1 and X_2, two binary tables (where '1' and '0' mean presence and absence, resp.) are practically the same, because their raw marginals as vectors are identical and column marginals have only a small degree of difference. (De facto, using weighted Shannon's estimates, single raw marginals have 19.02

bits in both cases; single column marginals of X_1 and X_2 can be characterized by 32.27 and 35.02 bits, resp.) In order to have a better insight, it is relevant to see the difference between 'inner configurations' (composed of raw and column vectors, resp.); in some way, X_1 can be regarded as a 'random' composition, whereas X_2 as a 'regular' composition. In a more explicit way, let us consider first set Q, $Q = \{a,b,c\}$; secondly, Q', the power set of Q (set of all possible subsets), $Q' = \{0,a,b,c,ab,ac,bc,abc\}$; thirdly, the fact that X_1 uses all elements of Q' once, but X_2 uses only two elements of Q' (a,bc) four times. In consequence, $m\hat{H}_1 = 24$ and $m\hat{H}_2 = 8$ bits, where both quantities are weighted Shannon's entropy estimates for X_1 and X_2, resp.; and where $m = 8$ (number of sampling units). In a more detailed way,

$$m\hat{H}_1 = 8 \log_2 8 - 8\,(1 \log_2 1) = 24,$$

$$m\hat{H}_2 = 8 \log_2 8 - 2\,(4 \log_2 4) = 8.$$

§4. Diversity versus dependence

Even this over-simplistic case and the main properties of this type, say, 'combinatorial type' of diversity – which may be called biotic diversity – shows properly that diversity in most cases should not be used *per se* but it is to be interconnected (coupled) with other types of phenomena *viz*. processes. Let us consider now the simple (pairwise) comparison of raw vectors in Fig. 2 (say, as an interlocal comparison), using the well-known 2 × 2 contingency tables as they are shown by Fig. 3. It is easy to see that all contingency tables for all pairs of X_1 represent the clear-cut case of stochastic independence, whereas all tables for X_2 represent maximum degree of dependence (i.e. 'non-independence'). In a more explicit way, and concentrating on the (A,B)-pair, the values of joint entropy functions are:

$$m\hat{H}_1([A,B]) = 16 \text{ bits}; \; m\hat{H}_2([A,B]) = 8 \text{ bits};$$

the values of simple (marginal) entropy functions are clearly identical:

X_1

	1	0	Σ
1	2	2	4
0	2	2	4
Σ	4	4	8

X_2

	1	0	Σ
1	0	4	4
0	4	1	4
Σ	4	4	8

or

	1	0	Σ
1	4	0	4
0	0	4	4
Σ	4	4	8

Fig. 3. 2 × 2 contingency tables for X_1; and X_2 of Fig. 2.

$$m\hat{H}_1(A) = m\hat{H}_1(B) = m\hat{H}_2(A) = m\hat{H}_2(B) = 8 \text{ bits};$$

in consequence, the values of contingency information (say, association) functions are:

$$m\hat{I}_1(A,B) = m\hat{H}_1(A) + m\hat{H}_1(B) - m\hat{H}_1([A,B]) = 8 + 8 - 16 = 0;$$

$$m\hat{I}_2(A,B) = m\hat{H}_2(A) + m\hat{H}_2(B) - m\hat{H}_2([A,B]) = 8 + 8 - 8 = 8 \text{ bits}.$$

This relevant difference can be described by the usual notation of set theory as well [i.e. $m\hat{H}_1(A) \cap m\hat{H}_1(B) = \varnothing$; $m\hat{H}_2 \cap (A) \cap m\hat{H}_2(B) = m\hat{I}_2(A,B)$]; or, it can be depicted my means of the usual Venn-diagrams (see Fig. 4). In addition, Fig. 4 shows a third case, X_3, where the intersection of $m\hat{H}_3(A)$ and $m\hat{H}_3(B)$ is neither empty, nor 'full' than beforehand. Note that in all Venn-diagrams of this type diversity appears always as some 'envelope' and dependence does as some non-empty intersection.

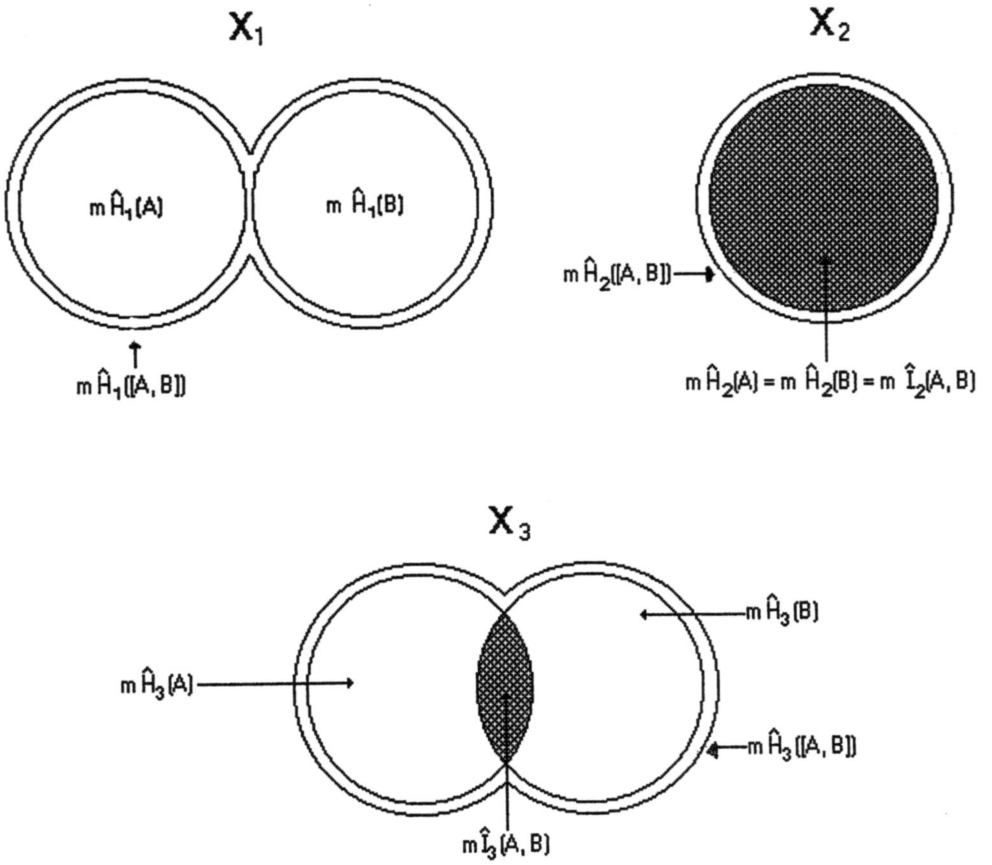

Fig. 4. Venn-diagrams for a pair of populations (a, b).

§5. Entropy versus information

The last statement remains true even if we do not consider elementary (pairwise) comparison. In the case of our primitive example, it is straightforward to introduce triple comparisons, for all elements of Q (by means of 2 × 2 × 2 tables). Such a comparison is depicted in Fig. 5, quite analogous to Fig. 4. It is easy now to identify the quantities at the end of §3; namely, $m\hat{H}_1 = m\hat{H}_1$ ([A,B,C]) = 24, $m\hat{H}_2 = m\hat{H}_1$ ([A,B,C]) = 8 bits, where the actual valuation is due to the fact that for X_1 all the possible intersections are empty (and, therefore, the value of joint-triple entropy is of *maximum*), while for X_2, since all intersections are being 'full' (with maximum values), therefore, the value of joint entropy is of a *minimum*. Even this primitive contrast shows why it can be very dangerous to interpret entropy-information relations in a vulgar way ('the less the uncertainly, the greater our information'). Instead, it is much better to think always in terms of *posse-esse* relations, as even Duns Scotus guessed it rightly, in the dawning period of our age (see e.g. de Muralt, 1991). In a more explicit way, we may introduce here the function ml_j (A,B,C), or, in brief, $m\hat{I}_j(\lambda)$,

$$m\hat{I}_j(\lambda) = sm\hat{H}_j(\mathrm{L}) - m\hat{H}_j \tag{3}$$

where $sm\hat{H}_j$(L), local diversity, is entropy estimate for all raw marginal values, where $m\hat{H}_j$ is again biotic diversity, and $m\hat{I}_j(\lambda)$, called associatum, is contingency information for all elements of a set Q. Since in the tables of Fig. 2. raw marginals are identical, $sm\hat{H}_1$(L) = $sm\hat{H}_2$(L) = 24 bits, therefore, $m\hat{I}_1$(A,B,C) = 0, and $m\hat{I}_2$(A,B,C) = 16 bits (as a special kind of *maximum* value). Again,

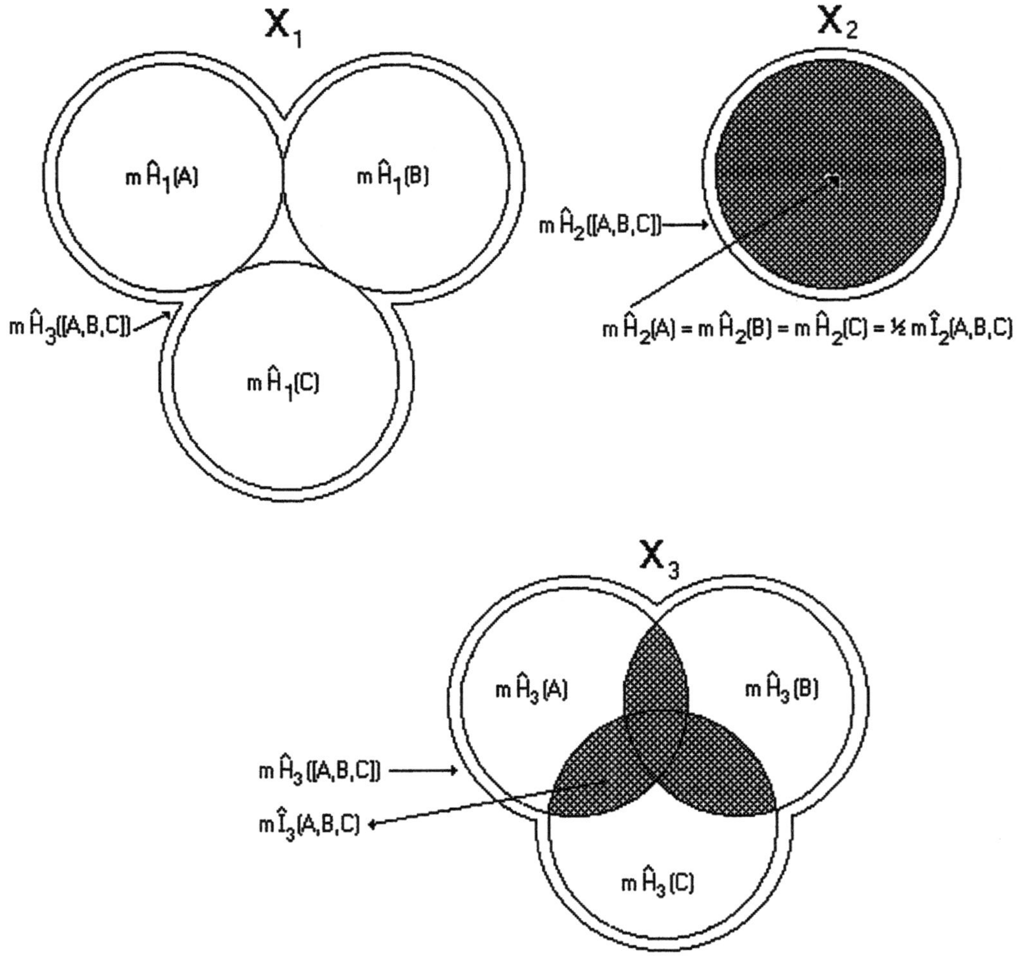

Fig. 5. Venn-diagrams for a triplet of populations (a, b, c).

if some intermediate third case, X_3, is introduced, then it is expected to have 'intermediate values' between the *extrema* indicated above. This shows clearly why it is so important to know always the proper *extrema*; how to manipulate with relevant relations (e.g. $\leq$, $\geq$); how to use the so-called 'sigma-conditions', proposed by the best authors in the field of information theory (e.g. Khinchin, 1957; Kullback, 1959; Rényi, 1962 etc.). Note that 'Σ-conditions' means roughly 'perfect' additivity and subadditivity relations within an envelope. Note also that these conditions and the introduction of proper inequalities make superfluous in most cases the majority of 'diversity versus evenness' arguments in ecology.

§6. Diversity versus complexity

We may call the configurations of Fig. 4 elementary Venn-complexes (where the adjective 'elementary' refers here to a pairwise comparison of some kind). If we have *s* kinds of populations (the cardinality of a set Q), and if $s > 2$, then we may get some non-elementary Venn-complex (see Fig. 6). Although it is much beyond the scope of this short paper to show the proper methodology (e.g. how to manipulate with $2 \times 2 \times \ldots{}^{s} \times 2$ tables), we may guess at least some simple properties of such a complex. First, it has like beforehand, an overall diversity envelop, represented by some joint entropy functions. (*Nota bene*: since such a complex has or may have several 'inner subenvelopes' of different orders as well, there-

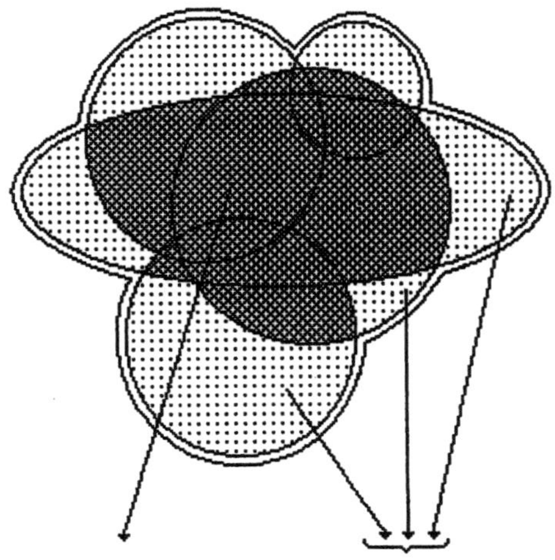

Fig. 6. A general Venn-diagram, where biotic diversity (as an envelope) can be subdivided into associatum and dissociatum.

fore, a relevant but still missing concept, subdiversity, should be taken into account.) Secondly, such a complex can be subdivided roughly into an 'overall intersectional region' and an 'overall outer region'; the former can be represented as a multivariate association (called associatum), the latter as a multivariate dissociation (called dissociatum), referring to the 'common' and 'uncommon' (particular) behavior of the populations, respectively. (*Nota bene*: due to the subdiversity relations of such a complex, a number of subassociata or subdissociata relations can be distinguished, representing coalitional relations of groups of populations.) Thirdly, the ordering relations mentioned above make possible to study at least some main features of complexity of our objects (including such seemingly innocent questions: how to relate 2×2 or $2 \times 2 \times 2$ etc. structures to $2 \times 2 \times \ldots^{s} \times 2$ structures.) Fourthly, it is very relevant to study a nonelementary Venn-complex 'in motion', when a number of its properties (like the sized of intersections) may change in a spatio-temporal referential system.

Without listing further possible oppositions, some tentative conclusions derived from § 1 § 6 can suggest

- that there are many types of diversity (i.e. diversity includes a highly 'diverse body' of phenomena);
- that a particularly relevant and fairly inclusive type of diversity may be termed compositional diversity (and biotic diversity is one example of such a category);
- that compositional diversity makes good sense only if it is related to other attributes (like dependence) of our object.

Surprisingly enough, the simplest binary representations can generate more consistent and much richer families of models than the non-binary (miscalled 'numerical') representations (of e.g. Juhász-Nagy,1976, 1984; Juhász Nagy & Podani, 1983).

Fragments of a case-study; some illustration

Some reasoning of the previous section can be illustrated, at least *pro parte*, by some results of an old case-study (cf. Dévai *et al.*, 1971; Juhász-Nagy *et al.*, 1973).

The basic data have been obtained from a survey of Cladocera fauna during ten days (summer, 1968), by counting 66 565 individuals collected from a shallow fish-pond at Bánhalma (in the Great Hungarian Plain, near the village Kenderes). The number of taxa has been ten ($s = 10$); e.g. *Bosmina longirostris* (O. F. Müller) (with four well-distinguished varieties), *Daphnia longispina* (O. F. Müller), *Chydorus sphaericus* (O. F. Müller), *Ceriodaphnia pulchella* G. O. Sars, etc.

The 3-dimensional shape of a sampling unit was approximately a cube. The sizes of sampling units (V_1, V_2, ...V_j...) were selected according to the increasing power of 2, from 1 to 8, that is, we took *ca.* 2, 4, 8, 16, 32, 64, 128, and 256 ml. water for the different sampling volumes. The number of sampling units was 64 ($m = 64$) for all sizes. The lay-out of sampling points was of the random type. The instructions were constructed by obtaining a table of pairs of random numbers, where the first column represented random numbers for a compass (0–360), the second one the number of oar-strokes to be made in a given direction. In such a way the sampling procedure was somewhat similar to a 'Brownian motion' in the surface of the lake; i.e. all points of the surface have been selected more or less equiprobably.

Omitting many details, it may be of some interest to see the values of the three functions (Table 1), labelled by § 3 in the previous chapter,

- where $\hat{H}$(L), local diversity, entropy estimates of raw marginals of 10 × 64 ($s \times m$) tables;
- where $m\hat{H}_j$ is biotic diversity;
- where $m\hat{I}_j(\lambda)$, associatum, overall association for 10 populations as a gain of information (divergence) measure, gained by the difference of the entropy estimates.

Note that all quantities are given here in *nits* ('*n*atural un*its*'); i.e. *log. nat.* was used.

The 'trends' of the three sequences of data are clear enough. First, it is to be noted that in V_8, where all components are present (and the table is completely full of '1'-*s*), all functions have zero value. This size (or, better to say, some size between V_7 and V_8) may be called min-volume (closely related to the concept 'minimal area', used from the beginning of this century in terrestrial plant coenology). Secondly, note that such a minimum cannot be expected at the 'starting points' of a V-scale; instead, even small sizes have considerable values. Thirdly, it is to noted that all sequences have some relative maxima ('peaks') in V_3 and V_5; these sizes may be termed max-volumes (realizing that min-max-volumes should have a mutual reference to each other; guessing that these maxima may be the optimal sizes for a number of analysis, like cluster analysis of some

Table 1. Associatum values gained as differences between values of local and biotic diversity.

	$m\,\hat{I}_j\,(\lambda)$	=	$sm\,\hat{H}_j$ (L)	–	$m\,\hat{H}_j$
j = 1	20.30		158.26		137.96
2	25.52		198.54		173.02
3	50.35		252.58		202.23
4	30.23		207.63		177.40
5	47.62		241.69		194.07
6	22.53		179.02		156.49
7	16.50		162.44		145.94
8	0.00		0.00		0.00

kind). The author, having an experience in land communities, has never get such a coincidence of maxima; usually, the maximum of associatum is due to a smaller point – or interval – of scale than the maximum of biotic diversity.

In order to have a still better insight of such sets of data, it is interesting to consider the '*posse-esse*' relations as they are shown by Table 2. Note that the vast majority of the $\binom{10}{k}$ possibilities (1,10,45,120,210,252,...) are never 'realized'; in other words, Nature never uses all the possible combinations of elements in any given situation. (This trivial statement corresponds well with the experience of field biologists.)

Some outlook; vistas and perspectives

In order to have some outlook of our urgent agenda, a possible way is to see first some simple generalizations of the previous results, and then to speculate a bit on the perspectives of our research area.

First of all, leaving alone now some more advanced ways of inference (see Cliff & Ord, 1981; Ripley, 1981, 1988; Podani, 1984), some simple means of a spatial processing is to be considered. Such a processing is realized many times by *T*, a topographical vector whose points represent the geometrical sizes of sampling units in an ascending order. For the sake of a reliable evaluation, it is always supposed that $s<$m, $m \to \infty$, and the lay-out is random. If we put *T* versus *F* (where *F* is a function, or a family of function of chapter 3), then we get some graphicons (vulgo: 'curves'), like in Fig. 7. Although, for the sake of simplicity, Fig. 7. shows two graphicons only, we may have here quite a number of graphic representations) e.g. according to the reasoning of § 6 in chapter 3).

Secondly, let us introduce now the simple convention of elementary scaling, where only max-min-values of such graphicons are used. As it is shown by Fig. 7, this convention generates in the axis *T* (or, in a more advanced case, in both axes) some sets of characteristic points, intervals, orderings of topographic nature which sets can

Table 2. Relations between possible and realized events. Possible events are given here by the 10th row of the Pascal-triangle.

Number of possible events	$\binom{10}{0}$	$\binom{10}{1}$	$\binom{10}{2}$	$\binom{10}{3}$	$\binom{10}{4}$	$\binom{10}{5}$	$\binom{10}{6}$	$\binom{10}{7}$	$\binom{10}{8}$	$\binom{10}{9}$	$\binom{10}{10}$
Sampling volume	Number of realized events										
V_1	1	6	8	0	0	0	0	0	0	0	0
V_2	1	6	10	7	4	0	0	0	0	0	0
V_3	0	2	7	9	8	3	1	0	0	0	0
V_4	0	0	1	4	7	5	4	2	0	0	0
V_5	0	1	1	2	5	7	7	6	1	0	0
V_6	0	0	0	0	1	4	4	4	3	3	1
V_7	0	0	0	0	0	0	1	3	5	4	1
V_8	0	0	0	0	0	0	0	0	0	0	1

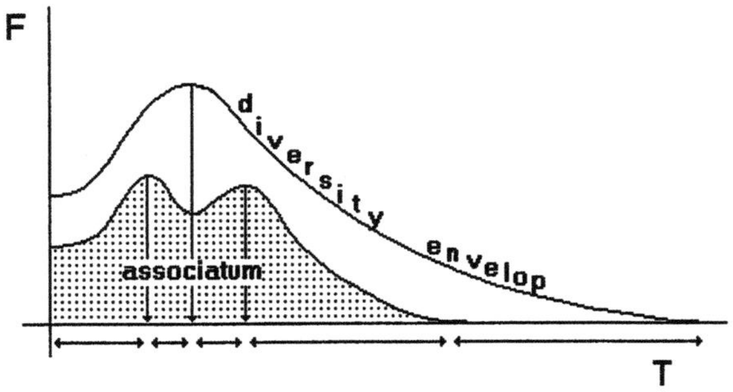

Fig. 7. Characteristic scaling in a plane of *T versus F*, where *T* is a topographical vector, *F* is a characteristic function. Only min-max values are used for scaling.

be regarded as relevant scaling attributes of a particular objective. (Needless to say, the more functions we have, the property of 'being characteristic' is more reliable.) In a somewhat metaphorical sense of the word, we can say that this type of characteristic scaling shows the ways how a community of some sort can 'sense for itself' different degrees of spatial heterogeneity. Note that in such a way, by means of some *T*-screening, can one detect scaling relations between maximal and minimal volumes, mentioned before.

Note also that some proper use of dynamical Venn-complexes makes possible to introduce diacretic modelling, where the dynamical behavior of particular populations or coalitions can be studied with relation to other (preferably: all) partner populations of a study.

Thirdly, a very relevant question is how characteristic scaling changes in time; how can one interconnect spatial attributes with results of temporal periodicity (see Reynolds, 1984). Such an interconnection would be imperative for a better understanding of 'mechanisms of succession' (see Reynolds, 1988); including such difficult problems as the activities of strategies and forced cycles (see Tilman, 1988; Sommer, 1991), or, the results of 'role analysis' of some kind (see Padisák, 1991). Clearly, some ways and means of a 'pattern and process' approach is indicated (see e.g. Juhász-Nagy, 1992), where spatial and temporal moments can be related to each other in an operative way.

Finally, a few remarks are needed in order to show why it is so important that these methodological efforts can reach a satisfactory theoretical framework as soon as possible. One of the most relevant problems of ecology, without doubt, is the acceptable explanation of stable coexistence of populations. But, because there are several types of coexistence and there are many types of stability (e.g. Lyapunov-types, assymptotic, interconnective, robust stability), we may guess that such a plurality needs a number of explanations (preferably: a minimal number). At the same time, one feels that the 'twelve explanations' of Wilson's (1990) fine paper represent somehow an 'over-plurality' but it is rather difficult to say how a proper reduction can be made; partly, because the verbal statements of Wilson's are far from being 'sharp enough'. It is somewhat depressing to see that since the period that Scudo and Ziegler (1978) have called the 'golden age', our theoretical tools (like Gause's hypothesis, IDH etc.)

were not able to reach the acceptable levels of testability. 'Towards more testable hypotheses'; I think that this way is indicated for theoretical ecology.

Acknowledgements

Many thanks are due to Dr. Judit Padisák for her kind help and for a number of fruitful discussion. I am very grateful to Dr. Colin Reynolds for his clever comments, corrections and criticism.

References

Carpenter, R. A., 1989. Do we know what we are talking about? Land Degradation and Rehabilitation 1: 1–3.

Cliff, A. D. & J. K. Ord, 1981. Spatial Processes: Models and Applications, Pion, London.

Connell, J. H., 1978. Diversity in tropical rain forests and coral reefs. Science 199: 1302–1310.

Cvitanovič, P., 1989. The Universality in Chaos. A. Hilger, Bristol.

Dévai, I., K. Horvath & P. Juhász-Nagy, 1971. Some problems of model-building in synbiology. Part 1. Spatial diversity process of the binary type in a simple situation. Annales Univ. Sci. Budapestiensis 13: 19–32.

Grassle, J. F., G. P. Patil, W. K. Smith & C. Taillie (eds), 1979. Ecological Diversity in Theory and Practice. Internat. Co-Operative Publ. House, Fairland (Maryland).

Juhász-Nagy, P., 1976. Spatial dependence of plant populations. Part 1. Equivalence analysis (an outline for a new model). Acta Bot. Hung. 22: 61–78.

Juhász-Nagy, P., 1984. Spatial dependence of plant populations. Part 2. A family of new models. Acta Bot. Hung. 30: 363–402.

Juhász-Nagy, P., 1992. Scaling problems; almost everywhere. An Introduction. Abstracta Botanica 16/1: 1–5.

Juhász-Nagy, P., I. Dévai & K. Horváth, 1973. Some problems of model-building in synbiology. Part 2. Associatum process in a simple situation. Annales Univ. Sci. Budapestiensis 15: 39–51.

Juhász-Nagy, P. & J. Podani, 1983. Information theory methods for the study of spatial processes and succession. Vegetatio 51: 129–40.

Khinchin, A. I., 1957. Mathematical Foundations of Information Theory. Dover, New York.

Kullback, S., 1959. Information Theory and Statistics. Wiley, New York.

MacArthur, R. H., 1965. Patterns of species diversity. Biol. Rev. 40: 510–553.

Magurran, A. E., 1988. Ecological Diversity and its Measurement. Princeton Univ. Press, Princeton.

Margalef, R., 1958. Information theory in ecology. General Systems 3: 36–71.

Matson, P. A. & S. R. Carpenter (eds), 1990. Statistical Analysis of Ecological Response to Large-Scale Perturbations. Ecology 71: 2037–2068.

Mayr, E., 1984. Die Entwicklung der biologischen Gedankenwelt (Vielfalt, Evolution und Vererbung). Springer, Berlin.

de Muralt, A., 1990. L'Enjeu de la Philosophie Médiévale (Études thomistes, scotistes, occamiennes et grégoriennes). Brill, Leiden.

Padisák, J., 1991. Relative frequency, seasonal pattern and possible role of species rare in phytoplankton in a large shallow lake (Lake Balaton, Hungary). Verh. int. Ver. Limnol. 24: 989–992.

Peet, R. K., 1974. The measurement of species diversity. Ann Rev. Ecol. Syst. 5: 285–307.

Pielou, E. C., 1975. Ecological Diversity. Wiley, New York.

Podani, J., 1984. Spatial processes in the analysis of vegetation: theory and review. Acta Bot. Hung. 30: 75–118.

Rényi, A., 1962. Wahrscheinlichkeitsrechnung, mit einem Anhang über Informationstheorie. VEB Deutscher Verlag der Wissenschaften, Berlin.

Reynolds, C. S., 1984. Phytoplankton periodicity: the interaction of form, function and environmental variability. Freshwat. Biol. 14: 111–142.

Reynolds, C. S., 1988. The theory of ecological succession applied to freshwater phytoplankton. Verh. int. Ver. Limnol. 23: 689–691.

Ripley, B. D., 1981. Spatial Statistics. Wiley, New York.

Ripley, B. D., 1988. Statistical Inference for Spatial Processes. Cambridge Univ. Press, Cambridge.

Scudo, F. M. & J. R. Ziegler, (eds) 1978. The Golden Age of Theoretical Ecology (1923–1940). Springer, Berlin.

Sommer, U., 1991. Phytoplankton: directional succession and forced cycles. In H. Remmert (ed), The Mosaic-Cycle Concept of Ecosystems. Springer Verlag, Berlin: 132–146.

Stewart, I., 1989. Does God Play Dice? – The Mathematics of Chaos. Blackwell, Oxford.

Tilman, D., 1988. Plant Strategies and the Dynamics and Structure of Plant Communities. Princeton Univ. Press, Princeton.

Vincent, T. L., A. I. Mees & L. S. Jennings, (eds) 1990. Dynamics of Complex Interconnected Biological Systems. Birkhäuser, Boston.

Washington, H. G., 1984. Diversity, biotic and similarity indeces. A review with special reference to aquatic ecosystems. Wat. Res. 18: 653–694.

Wilson, E. O., (ed.) 1988. Biodiversity. National Academy of Sciences, Washington (D.C.).

Wilson, J. B., 1990. Mechanisms of species coexistence: twelve explanations for Hutchinson's 'paradox of the plankton': evidence from New Zealand plant communities. New Zeal. J. Ecol. 13: 18–42.

Woodwell, G. M. & H. H. Smith, (eds) 1969. Diversity and Stability in Ecological Systems. Brookhaven Symp. Biol. No. 22.

Hydrobiologia **249**: 183–188, 1993.
J. Padisák, C.S. Reynolds & U. Sommer (eds), Intermediate Disturbance Hypothesis in Phytoplankton Ecology.

Intermediate disturbance in the ecology of phytoplankton and the maintenance of species diversity: a synthesis

C. S. Reynolds[1], J. Padisák[2] & U. Sommer[3]
[1] *Freshwater Biological Association, NERC Institute of Freshwater Ecology, Windermere Laboratory, Ambleside, LA22 0LP, UK;* [2] *Botanical Department of the Hungarian Natural History Museum, Budapest/Balaton Limnological Instiitute of the Hungarian Academy of Sciences, H-8237 Tihany, Hungary;* [3] *Institut für Biologie und Chemie des Meeres, Universität Oldenburg, Postf. 2503, D-2900 Oldenburg, Germany*

Key words: disturbance, quiescence, environmental variability, temporal scales, diversity, species richness

Abstract

This paper concludes a collection of contributions presented at the 8th Workshop of the International Association of Phytoplankton Taxonomy and Ecology. It derives a consensus as to the virtues and strengths of J. H. Connell's Intermediate Disturbance Hypothesis (IDH), its applicability to phytoplankton ecology and its theoretical and practical weaknesses. The view is expressed that the IDH is too useful a concept to reject and that, as a word model, it provides a powerful link between diversity and disturbance. The more robust investigations that are necessary to consolidate the tenancy of IDH need to concentrate upon the separation and quantification of the stimulus- and response-components of disturbance.

Introduction

The purpose of this short paper is to conclude the collection of articles contributed to the IAP Workshop on The Intermediate Disturbance Hypothesis, or IDH, (Padisák, Reynolds & Sommer, 1993) with particular reference to its applicability in plankton ecology. Here we attempt to summarize the general thrust and direction of the individual contributions. This article is not, however, intended to be an abstract of abstracts, or some convenient review of those papers. Rather, it is our joint statement of how the utility of the concept, founded in community ecology, is currently viewed by a group of contemporary plankton biologists. Moreover, our principal concern is not the concept itself, which is extensively discussed in the foregoing articles (Sommer *et al.*, 1993), by Connell (1978) himself and by others prior to that, albeit without the 'IDH' title (see e.g. Grime, 1973). We are concerned with its applicability to current problems in pelagic ecology. These include the extent and outcome of interspecific resource-based competition, the question of regulation by higher trophic levels, and the mechanisms contributing to the high biodiversity among planktonic organisms.

This discussion is arranged in four sections: first, why it is necessary to adopt an intermediate disturbance hypothesis (or something like it) into pelagic biology; second, to examine the relative merits of IDH; third, to consider its relative weaknesses; and, finally, we offer some deductions and recommendations.

The need for conceptual explanations

A part of modern biological science is directed towards the more subtle interrelationships of individual species with their environments and how these might respond to imposed change. Such questions assume increasing importance to the future of planetary ecosystems overexploited by our own species or damaged in consequence of changes wrought elsewhere. The answers generally involve, or imply, a level of understanding of ecosystem functioning which turns out to be mainly theoretical, unverified and (usually) hotly debated. One of the attractions of pelagic ecosystems for study is that, the vastness of the open-water habitats apart, they function at predominantly small spatial and temporal scales. Organisms are small (generally 10^{-2} to 10^{-7} m) and live short lives (generally 10^4 to 10^7 s) so that it is quite possible to follow, observe and experiment with the development of planktonic individuals, populations and communities. Curiously, this opportunity seems not to have attracted the attention of theoretical ecologists although several plankton biologists, for instance, G. Evelyn Hutchinson, Ramón Margalef & David Tilman became renowned theorists.

However, it is equally clear that the real world does not conform, either continuously or consistently, to the sum of experimentally-validated growth-, competition- and exclusion models. Yet, it is sufficiently structured and its processes are sufficiently reproducible for it not to be a fortuitous assembly of disorder and, in the common sense of the word, chaos. This apparent paradox reflects the tenous balance between biological organization, at every level from molecules to populations of organisms, and the counterposing entropic laws of dissipation (diffusion and thermodynamics). At the supraindividual level, the balance is poised between the organization of individuals into communities and variations in disruptive physico-chemical forcing.

Connell's (1978) Intermediate Disturbance Hypothesis is a fulcrum in this respect. It represents a mature and reasoned expression of the interactions among, on the one hand, the internally driven progress towards community organization and energetic equilibrium (*sensu* Odum, 1969) and, on the other, the externally-imposed stochasticity of environmental variability, operating at a variety of temporal (and, we infer, spatial) scales.

Strengths of the IDH concept

Two outstanding merits of the IDH concept come immediately to mind. One is the ease of its comprehension: as a word-model it is readily understood that, whereas natural communities under unchanged external conditions tend to become uniform, their progress towards the anticipated, equilibrated, competitively-excluded outcome can, at any time, be slowed, interrupted or overridden by forces emanating from outside and which are beyond the capability of the existing community to absorb. This first merit survives even the semantic differences of comprehension of Connell's view of intermediacy, *viz.* whether it refers to disturbances of intermediate strength or duration, as considered by Sommer *et al.* (1993), or of a disruptive event inserted firmly into the sequence of an internally-directed succession, prior to its conclusion, to the extent that it is halted, shifted or reversed. The second virtue of IDH is that it provides an explanation for the uncomfortable gulf that lies between what we expect to happen on the basis of controlled experimentation and what we generally observe. Simply, most recognizable ecosystems are far from being at steady state. Has not sufficient time elapsed for the majority of them to have come close to ecological equilibrium? Why have they not done so? Why do they support so many species? Can it be that, hitherto, their structures have been so frequently or extensively 'disturbed' in the recent past that they have failed to establish more than the elements of an equilibrated system? If equilibrium is, in fact, unusual, many of the contentious issues, like that of the 'top down vs. bottom up' control, become symptomatic dependents of the immaturity of the ecosystems in which they feature (Reynolds, 1993).

The problem may be viewed in the terms of the 'three patch' analogy. In this, a single hillside is envisaged, having a single aspect, macro-climate and geological foundation: a first vertical band is clothed by mature forest; a second was clear-filled and ploughed some years previously but was then abandoned to recolonization and has already developed the appearance of shrub and immature woodland; a third continues to be subject to regular livestock grazing. The appearances of the three areas are mutually contrasted, carrying communities corresponding approximately to Grimes (1979) S, C and R categories of plant species. Yet their differences lie essentially in the time that has elapsed since each was last 'disturbed' (in this case, by management). The first has progressed farthest towards its climactic condition. The second has begun to resemble the first, with conspecific juvenile trees developing within a richly diverse herb/shrub flora; the third looks quite open and its species composition is limited to species tolerant of the imposed management. In other words, each has acquired a species composition and an internal organization which is a function of the frequency of disturbance applied. If specific examples are preferred to this rather generalized consideration, then Grime's (1973) analysis of flowering-plant richness in herbaceous vegetation should be consulted. The idea works just as well for phytoplankton (Reynolds, 1993). Immaturity becomes a recognizable attribute of systems; equally, disturbance becomes a valid factor contributing to their productivity, species structure, biomass and organization.

Weakness of the IDH concept

Ranged against the deceptive simplicity of the Intermediate Disturbance Hypothesis are some quite fundamental questions of logic and/or practicality. Until they are adequately answered, the utility of the IDH remains in doubt. The difficulties are incisively exposed in Juhász-Nagy's (1993) paper, though certainly not in a manner which sought to discredit IDH in any way. On the contrary, any preconception that may be betrayed is supportive of IDH.

As discussed at the outset (Sommer *et al.*, 1993), the principal difficulty concerns the recognition and measurement of disturbance. This point is emphasized in most of the contributed papers (notably Rojo and Alvarez-Cobelas, 1993). Although it is becoming increasingly straightforward to measure a force imposed on an ecosystem, it is quite another thing to relate quantitatively the extent of the impact, if any, to that force. Whereas a sudden storm, or a flash flood attributable to sharply-enhanced summer rains, would be anticipated to have an immediate effect upon the biomass, species composition and productivity of a respondent system, it is in no way 'guaranteed'. Such a storm event might, for instance, be expected to interrupt abruptly the summer phytoplankton succession and stimulate the development of a diatom-desmid association (Reynolds, 1993). Yet the strength of stimulus to attain this is generally more effective in disturbing the development of (say) a midsuccessional *Sphaerocystis* population than of a well-established *Microcystis* bloom: its eradication may require more drastic and more persistent disturbance, although the same intensity of disturbance applied earlier in the development of the *Microcystis* could have delayed its establishment, perhaps indefinitely (Reynolds *et al.*, 1984). In other words, the imposition of the same external forcing may well invoke quite different community responses, depending upon the resilience shown by the development stage achieved and the strength of the internal linkages already established at the instant when it is applied (Padisák, 1993). For instance, Jacobsen & Simonsen (1993) showed that heavy rainfall and storms increased the loss from an *Aphanizomenon* population in Lake Godstrup but failed to break its dominance. The role of this acquired late-successional complexity in governing the responses to (e.g.) external forcing is highlighted in the papers contributed by Moustaka-Gouni (1993) and Eloranta (1993). In general, the sooner in the succession that a given stimulus is imposed, then the greater and more immediate is the compositional response. Late successional, senescent communities can become again more sensitive to environmental disturbances; however, senescence is difficult to observe in phytoplankton communities (cf. Sommer, 1991).

Similarly, the same forcing, even supposing it to be quantified in the appropriate vector units, can impact quite differently upon two adjacent systems. Several of the papers in this volume (Sommer, 1993; Trifonova, 1993; Holzmann, 1993; Chorus & Schlag, 1993), in fact, contrast the effects of what are assumed to be the same stimuli acting upon pairs of nearby lakes, in each case differing mutually either in area, depth or exposure. In the sites contrasted by Holzmann (1993) the responses in the oligotrophic Kautsee differ from those in the eutrophic Pelhamer See. In this instance, a more exacting or more persistent control (that of low nutrient concentration) acts on species structure and abundance throughout and which the external stimulus fails to disturb. Olrik & Nauwerck (1993) describe another case in which the reaction to external forcing was suppressed by internal conditions: severe carbon limitation brought to an end the dominance of a large *Limnothrix* population in Hjarbæk Fjord and, eventually, the initiation of a new successional sequence dominated by *Scenedesmus*. Olrik & Nauwerck (1993) and Padisák (1993) point out that stress and disturbance can produce similar outcomes. There were occasions, too, when the systems studied by Rojo & Alvarez-Cobelas (1993) and by Chorus & Schlag (1993) did not respond to forcing as expected.

These are truly instances when the pool of potential respondents may be viewed as being either 'squeezed into an apex' of a C.S.R. triangle (Reynolds, 1993) or the enforced matrix co-ordinates are encompassed by the Venn envelopes of a diminishing number of species (cf. Juhász-Nagy, 1993). The community becomes more specialist with fewer species of non-common behaviour (Kullback, 1959).

These concerns lead to a further, more fundamental difficulty to the tenancy of IDH. It is inescapable that disturbance is a phenomenon recognized and measured only as an effect: the stimulus may be generally external but its nature is judged exclusively by the reaction it engenders. This property is implicit in the definitions of Pickett *et al.* (1989): disturbance is a change in the minimal structure of an ecological object caused by a factor external to the level of interest (individual, population, community, etc.). It is equally clear that, since the 'minimal structure' is a system of lower entities permitting the persistence of the object, the application of the disturbance concept will always depend upon the perception of the structure of the ecological systems concerned.

The contributions on river communities remind us that it is less the intensity of the physical forcing that is critical to the development of ecological structure so much as the frequency with which it is applied. Intuitively, rivers may be regarded as highly kinetic and open systems in which structure will be difficult to elaborate. Not so, for the organisms tolerant of the conditions obtaining perceive them as being uniform and therefore undisturbed. Descy (1993), Carvajal-Chitty (1993) and Ács & Kiss (1993) all properly treat disturbance as an abrupt shift in the status quo. The intuitively disturbed state turns out to be structurally constant, pending flood events or, presumably, in the opposite direction, impoundment in a reservoir.

A similar cautionary remark has to be made for lakes:

Lake planktonologists tend to perceive sustained thermal stratification as the 'undisturbed state' of a lake. This preconception is for example apparent in Reynolds' (1987) adoption of Grime's plant strategy typology. Phytoplankton taxa characteristic for mixed water columns (e.g. diatoms) are termed 'ruderals' in analogy to terrestrial plants from disturbed sites. However, if vertical mixing is continuous the environment of the phytoplankton may be perceived as a more or less constant low light-high nutrient one. As long as at least some phytoplankton are still able to grow and to influence the transparency of the water there will be competitive exclusion of all species but those with minimal light requirements. Therefore, permanent circulation can be a low-disturbance status. Sudden stratification would then be a disturbance of this status. The results of Chorus and Schlag (1993) could best be interpreted this way.

Applications of the IDH concept

In spite of its unresolved weaknesses, the concept of intermediate disturbance remains too useful in its potential to reject. Its application to current problems in ecology and in management of ecosystems can advance as the understanding of the structural linkages is strengthened (Pickett *et al.*, 1989). The latter invokes the importance of selecting the correct spatial and temporal scales at which structure is established, information is assembled and entropy is diminished. As is made clear in several of the contributions, the scales are set by the organisms concerned. Within the context of seasonal change in phytoplankton species composition, our workshop agreed a working definition that '*disturbances are primarily non-biotic, stochastic events that result in distinct and abrupt changes in the composition and which interfere with internally-driven progress towards self-organization and ecological equilibrium; such events are understood to operate through the medium of (e.g.) weather and at the frequency scale of algal generation times*'.

In this sense, it is evident that undisturbed successions should eventually approach competitive exclusion and ecological equilibrium. This may well require the 12–16 generations suggested by Reynolds (1993) and hence occupy periods of 35–60 days (Sommer, 1985). On the other hand, external events interfering with that progress will promote a mix of contemporaneously occurring species, even though some may be decreasing while others are increasing. Several contributors (Padisák, 1993; Sommer, 1993) are agreed that the diversity – roughly the number of species over the numbers of individuals represented in a given space – is also highest early in successions, probably around the second-third generation (i.e. $\sim$5–15 days) before any achieves eventual dominance. Disturbances occurring at similar frequencies are adequate to explain both the failure of competitive exclusion and how the diversity might be maintained. At the same time, a plausible mechanism for driving contemporaneous disequilibrium (Richerson *et al.*, 1970) and for the 'paradoxical' structure of planktonic communities (Hutchinson, 1961) is projected.

However, despite the demonstrably close correlation between diversity and disturbance they are only indirectly linked, through the hierarchical structure of ecological systems. The bridging condition depends, on the one hand, upon the abilities and/or specialisms of individual species to process environmental signals and, on the other hand, the time taken by them to respond. While the external signals are perceived as being constant, succession proceeds towards competitive exclusion; when they alter frequently, early successional stages are supplanted by alternative early successional stages!

In this way, it is possible to substantiate the connectances in the general deductions presented in this volume:

– Diversity is high or increases when species replacement rates are rapid, as, for instance, in warm water; these are neither necessarily nor exclusively driven by disturbances, although communities repeatedly destabilized at frequencies in the order of three generation times are likely to support a high diversity of species. Diversity is promoted when fast-growing (i.e. usually small-sized) algae are abundant.

– Diversity is low or declines in advanced successions where a large biomass is dominated by a single, generally 'large' ($>200\ \mu$m) algal species. Diversity is rarely high in strongly selective environments, such as highly flushed systems or in lakes characterized by extremes of acidity, alkalinity, turbidity, oligotrophy or physical constancy (save by spatial niche differentiation). Without disturbance of the status quo, equilibrium dynamics predict an eventual total suppression of diversity.

Final remarks

The deductions and opinions advanced in this discussion are those of the authors and not necessarily shared by the contributors to this volume. Nevertheless, we are most grateful to them for the stimulus of their information and ideas. Besides, the topics of disturbance and diversity are scarcely exhausted; we hope that this article serves as a catalyst for further, more robust investigations that may consolidate the tenancy of the Intermediate Disturbance Hypothesis.

References

Acs, É. & K. T. Kiss, 1993. Effects of the water discharge on periphyton abundance and diversity in a large river (River Danube, Hungary). Hydrobiologia 249: 125–133.

Chorus, I. & G. Schlag, 1993. Importance of intermediate disturbances for species composition and diversity of phytoplankton in two very different Berlin lakes. Hydrobiologia 249: 67–92.

Carvajal-Chitty, H. I., 1993. Some notes about the Intermediate Disturbance Hypothesis and its effects on the phytoplankton of the middle Orinoco river. Hydrobiologia 249: 117–124.

Connell, J., 1978. Diversity in tropical rain forests and coral reefs. Science 199: 1304–1310.

Descy, J.-P., 1993. Ecology of the phytoplankton of the River Moselle: effects of disturbances on community structure and diversity. Hydrobiologia 249: 111–116.

Eloranta, P., 1993. Diversity and succession of the phytoplankton in a small lake over a two-year period. Hydrobiologia 249: 25–32.

Grime, J. P., 1973. Competitie exclusion in herbaceous vegetation. Nature 242: 344–347.

Grime, J. P., 1979. Plant stratiegies and vegetation processes. Wiley-Interscience, Chichester.

Holzmann, R., 1993. Seasonal fluctuations in the diversity and compositional stability of phytoplankton communities in small lakes in upper Bavaria. Hydrobiologia 249: 101–109.

Hutchinson, G. E., 1961. The paradox of plankton. Am. Nat. 95: 137-147.

Jacobsen, B. A. & P. Simonsen, 1993. Disturbance events affecting phytoplankton biomass, composition and species diversity in a shallow, eutrophic, temperate lake. Hydrobiologia 249: 9–14.

Juhász-Nagy, P., 1993. Notes on compositional diversity. Hydrobiologia 249: 173–182.

Kullback, S., 1959. Information Theory and Statistics. Wiley, New York.

Moustaka-Gouni, M., 1993. Phytoplankton succession and diversity in a warm monomictic, relatively shallow lake: Lake Volvi, Macedonia, Greece. Hydrobiologia 249: 33–42.

Odum, E. P., 1969. The strategy of ecosystem development. Science 164: 262–270.

Olrik, K. & A. Nauwerck, 1993. Stress and disturbance in the phytoplankton community of a shallow, hypertrophic lake. Hydrobiologia 249: 15–24.

Padisák, J., 1993. The influence of different disturbance frequencies on the species richness, diversity and equitability of phytoplankton in shallow lakes. Hydrobiologia 249: 135–156.

Padisák, J., C. S. Reynolds & U. Sommer (eds), 1993. Intermediate Disturbance Hypothesis in Phytoplankton Ecology. Developments in Hydrobiology 81. Kluwer Academic Publishers, Dordrecht. Reprinted from Hydrobiologia 249.

Pickett, S. T. A., I. Kolasa, I. I. Armesto & S. L. Collins, 1989. The ecological concept of disturbance and its expression at various hierarchical levels. Oikos 54: 129–136.

Reynolds, C. S., 1987. The response of phytoplankton communities to changing lake environments. Schweiz. Z. Hydrol. 49: 220–236.

Reynolds, C. S., 1993. Scales of disturbance and their role in plankton ecology. Hydrobiologia 249: 157–171.

Reynolds, C. S., S. W. Wiseman & M. J. O. Clarke, 1984. Growth- and loss-rate responses of phytoplankton to intermittent artificial mixing and their potential application to the control of planktonic algal biomass. J. appl. Ecol. 21: 11–39.

Richerson, P., R. Armstrong & C. R. Goldman, 1970. Contemporaneous disequilibrium, a new hypothesis to explain the 'paradox of plankton'. Proc. Natl. acad. Sci. 67: 1710–1714.

Rojo, C. & M. Alvarez Cobelas, 1993. Hypertrophic phytoplankton and the Intermediate Disturbance Hypothesis. Hydrobiologia 249: 43–57.

Sommer, U., 1985. Comparisons between steady state and non-steady state competitions: experiments with natural phytoplankton. Limnol. Oceanogr. 30: 335–346.

Sommer, U., 1991. Phytoplankton: directional succession and forced cycles. In H. Remmert (ed), The Mosaic-Cycle Concept of Ecosystems. Springer Verlag, Berlin: 132–146.

Sommer, U.,1993. Disturbance-diversity relationships in two lakes of similar nutrient chemisty but contrasting disturbance regimes. Hydrobiologia 249: 59–65.

Sommer, U., J. Padisák, C. S. Reynolds & P. Juhász-Nagy, 1993. Hutchinson's heritage: the diversity-disturbance relationship in phytoplankton. Hydrobiologia 249: 1–7.

Trifonova, I., 1993. Seasonal succession of phytoplankton and its diversity in two highly eutrophic lakes with different conditions of stratification. Hydrobiologia 249: 93–100.

General index

GPSR Compliance

The European Union's (EU) General Product Safety Regulation (GPSR) is a set of rules that requires consumer products to be safe and our obligations to ensure this.

If you have any concerns about our products, you can contact us on ProductSafety@springernature.com

In case Publisher is established outside the EU, the EU authorized representative is:

Springer Nature Customer Service Center GmbH
Europaplatz 3
69115 Heidelberg, Germany

Batch number: 10762211

Printed by Printforce, the Netherlands